Organic Molecules in Action

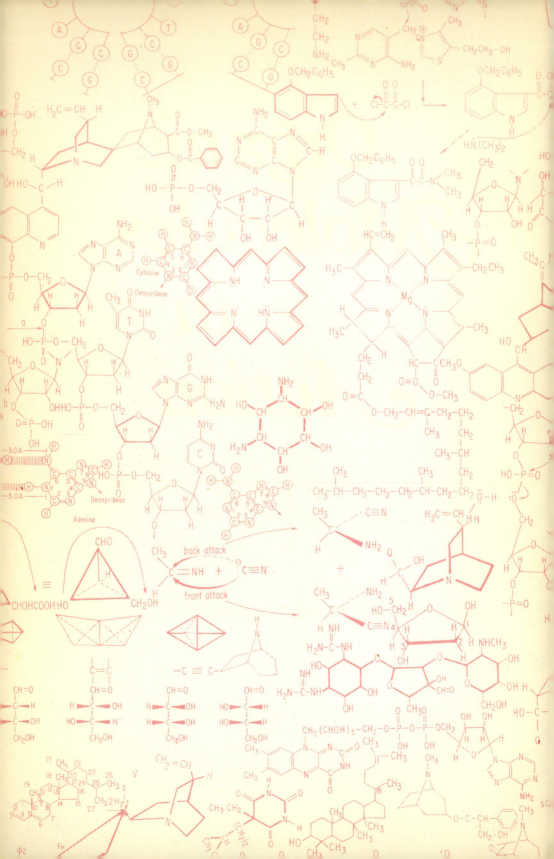

Organic Molecules in Action

MURRAY GOODMAN
University of California, San Diego

and

FRANK MOREHOUSE

GORDON AND BREACH SCIENCE PUBLISHERS
New York   London   Paris

First Printing— 1973
Second Printing with corrections— 1974

PREFACE

THIS BOOK grew out of our belief that all educated people should have an understanding of science. Much of our modern world is based on scientific discovery and the resulting technological applications. Scientists, social scientists and humanists must communicate with and understand each other since the survival of mankind requires the development of scientific programs whose social implications and human consequences have been taken into account.

Within the domain of science, chemistry plays a central role because it deals with the elements and molecules composing the universe. Organic molecules derive from compounds of carbon and other elements such as hydrogen, oxygen, nitrogen, sulfur, etc. Millions of combinations exist in nature, or have been synthesized, including those upon which life itself is based.

The book commences with a consideration of how life began and evolved. We present an overview of the code of life, emphasizing that studies of molecular size and shape have revolutionized our understanding of biological phenomena. These studies have relied heavily on the use of x-rays to show the details of the three-dimensional structure of organic molecules.

In our text we demonstrate the relationship of organic chemistry to the world about us. To accomplish this task, we discuss polymers in order to explain how so many of the synthetic materials used today are prepared. We place some scientific discoveries such as those of quinine, aspirin, hormones, vitamins and hallucinogens in an historical context. We point out that modes of biological action for most drugs are unknown. We give examples of accidental observations which lead to significant technological breakthroughs (e.g., the discovery of penicillin, the initial syntheses of guncotton and nylon). We stress that discovery in chemistry is a continuing process and that more scientific understanding is essential in the building of a better world.

On numerous occasions we have encountered undergraduates who recall absolutely nothing from their organic chemistry studies shortly after completing their courses. We found that in most cases the course content represented a watered-down exposure to a curriculum used to train chemists. We therefore undertook a completely different approach by choosing representative areas of chemistry that we believe are important to educated people. By definition, such a book can never be complete. However, each subject is presented in substantial detail, and therefore this book should be appropriate as a general introductory chemistry text and also as an informative reference for more advanced students.

Many people helped in the creation of this book. The text was perused at various stages by colleagues, Professors Martin Kamen, Paul Saltman, Stanley Miller and F. Thomas Bond. Through their constructive criticism, we were able to improve the organization and presentation of the material. Helpful comments were received on specific chapters by Drs. Michael Verlander and Ugo Lepore. We want to call particular attention to the discussions and suggestions of Dr. Jerrold Meinwald, who helped us formulate the chapter on Molecules of the Senses, and to Drs. F. R. Salemme and Jens Birktoft who helped us with the chapter on Molecular Structure by X-ray Vision. The guidance we received from Dr. Stephen Freer was indispensable in helping us make aspects of x-ray diffraction not only intelligible but readable. Dr. Eugene Cordes of the University of Indiana and Dr. Harold Moore of the University of California, Irvine, examined the entire text and made most useful suggestions for revisions and modifications. The chemical figures and many of the illustrations are the work of Mr. Theodore Velasquez to whom we are indeed grateful.

M. G. and F. M.

CONTENTS

FIGURE 1.1   The Crab Nebula represents the remains of a supernova explosion that was visible and recorded in the year 1054 in China and Japan. It went completely unnoticed in the western world until the eighteenth century. [Photograph courtesy of the Hale Observatories.]

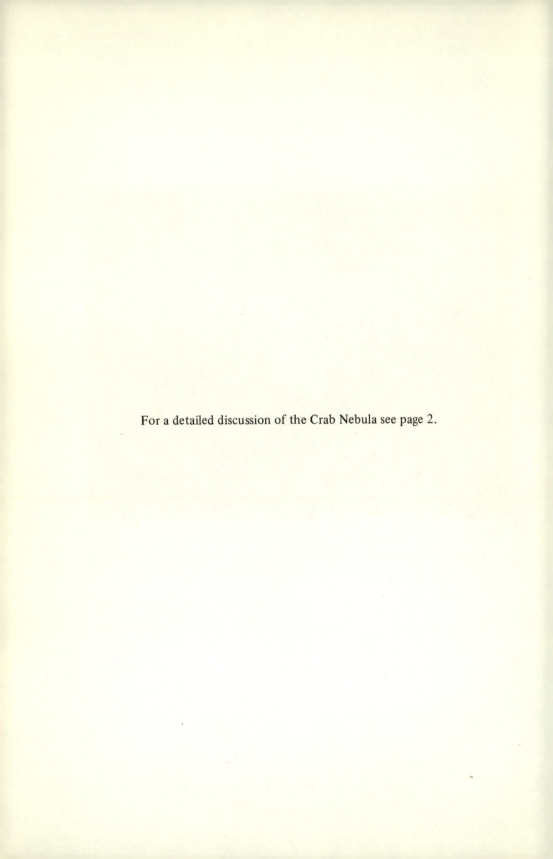

For a detailed discussion of the Crab Nebula see page 2.

The photograph on the following page shows several specimens of *Amanita muscaria*. For the story of the "divine mushroom" see page 153.

FIGURE 8.1 *Amanita muscaria*—"The Divine Mushroom". This species of mushroom, long known throughout Europe and Asia as a fly agaric is believed by some to be the source of Soma, the hallucinogen of Vedic culture.

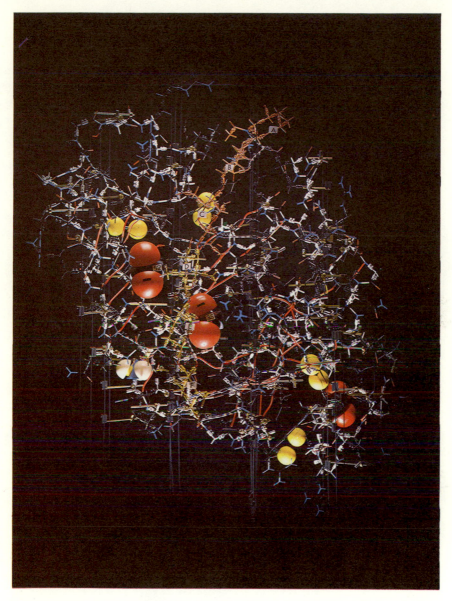

FIGURE 11.11   Photograph of a model of lysozyme. [Courtesy of D. C. Phillips, Oxford University.]

The structure of lysozyme shown in the preceding photograph is discussed in detail on page 256.

A full explanation of the α-helix model shown on the following page can be found on pages 261-263.

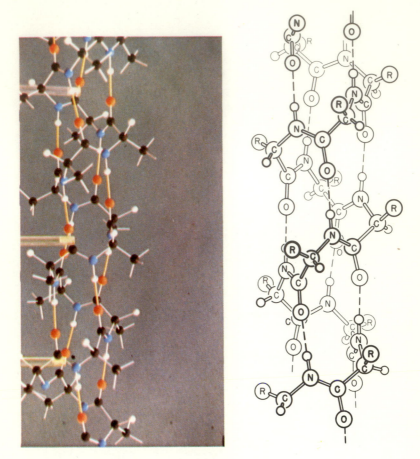

FIGURE 11.18   The α-helix model.

CHAPTER I

PREBIOTIC SYNTHESIS
Origins of Biological Molecules

THE STORY of organic molecules begins more than five billion years ago, before the sun and the earth and the other planets in our solar system formed. At that time, somewhere in the star-packed core of our galaxy, an ancient star was dying. The death of this old star—a blinding supernova explosion—signalled the coming of the solar system. It produced and threw out the atoms which now constitute the solar system, including the earth and all its inhabitants.

But between the supernova and the peopling of earth lies a fascinating story—the story of the formation and evolution of the organic molecules responsible for the appearance of life on earth. How did it all happen? What is the connection between the death of a star and the birth of man? The two events are separated by 5 billion years, but the atoms synthesized by the supernova are the very atoms that now make up our modern world. We can follow the paths of the atoms from the simplest organic molecule—methane—to the ultimate earth creature known today—man—through the molecular transformations we know as organic reactions.

If the whole story were written chronologically in ten volumes of 500 pages each, each page would represent 1 million years. Man would not appear until the last page of the last volume.

Let us look first at the beginning of the story, how the molecules of life came to exist on earth before the first living organism appeared in the maternal waters.

The Supernova

Our galaxy, the Milky Way, is between 10 and 15 billion years old. It is saucer-shaped and the central region is crowded with billions of large stars. When one of these stars dies, it goes out with a bang and not a whimper—to paraphrase the poet, T. S. Eliot. It explodes with an unearthly violence, releasing particles with enough energy to escape to the spiral fringes of the galaxy. These particles are also sufficiently energetic to undergo all the nuclear reactions needed to populate the entire periodic table.

But why should a star explode on its death bed? A star is mostly a cloud of hydrogen gas. As it ages, internal gravitational forces cause it to contract. As it contracts, its density increases. The increasing density accelerates the contraction until the core of the star approaches the density of nuclear matter. At this critical stage, tremendous repulsive forces build rapidly, and the core explodes as a consequence of the contraction.

The debris consists of cosmic gas and dust, probably containing all the elements of the periodic table, but still mostly hydrogen (about 90%). The non-hydrogen portion is mainly helium, the other inert gases, and methane. The other elements constitute the dust particles and amount to only a few thousandths of a percent of the total mass of gas and dust. But the quantity of material expelled by a supernova is so great that even this tiny portion contains enough material to furnish a sun with its planets.

The supernova which provided the material for our solar system occurred more than 5 billion years ago, but only 900 years ago, man had the privilege of actually observing a supernova. It was so bright at first that it could be seen with the naked eye in broad daylight. Most of the debris has since drifted away into space, but the remnants are still visible at night through a telescope. This luminous cloud of gas and dust is called the Crab Nebula (Figure 1.1). The event was viewable in the year 1054 and was duly reported by Japanese and Chinese astronomers but went unnoticed in Europe. It is now known that the Crab Nebula is 6000 light years from earth and therefore the explosion occurred 6000 years before the oriental scientists' eyes gazed upon it and noted that it was brighter than Venus. This supernova could be seen with the naked eye for well over a year. It then faded into heavenly obscurity.

In 1731 John Brevis rediscovered the event with the aid of a telescope and correctly deduced that it was the remnant of the 1054 supernova. Supernovae can be as bright as 100 million of our suns. The Crab Nebula which is not very bright today is important because it emits not only visible light but also x-rays and radiowaves. The source of the radiowaves lies in a pulsar which is a rapidly rotating, highly condensed star. The Crab Nebula contains the only pulsar so far clearly seen by optical telescope.

Some of the cosmic gas and dust from the ancient supernova drifted to the present location of our solar system on the periphery of the galaxy about 5 billion years ago and gathered itself together into a diffuse cloud. Again, internal gravitational forces caused the inner denser portion of the cloud to compact itself and form a new smaller star—our sun.

The remainder of the cosmic material was held by the gravitational pull of the sun and gradually coalesced to form the earth and the other planets. The material of the earth did not have sufficient mass to hold its cosmic gas and most was lost to space. At first, then, the earth was just an accumulation of cosmic dust, relatively cool, and without much of an atmosphere. This was about 4.8 billion years ago.

The ball of cosmic dust—the new earth—was made up of metals and their oxides, water as ice and as hydrates, ammonium compounds, silicon and sulfur compounds, and carbides. The ball became more and more compact because of its own gravity and the tremendous pressure of compaction, particularly at the center, raised the temperature of the mass. Chemical reactions, set off by the temperature increase, contributed to the heating effect, and the earth gradually differentiated into a dense molten core, a mantle, and a crust. The heat drove volatile materials and gases to the surface. These molecules formed the primitive atmosphere of the earth. This atmosphere was not a nitrogen and oxygen atmosphere such as we have today. It was a methane and ammonia atmosphere, and very steamy until the crust of the earth cooled enough to allow the primitive oceans to form.

The new earth was barren rock and water, basking in the brilliance of the new sun. Unfiltered as yet by our present layer of ozone, the sun provided the high-energy ultraviolet light needed to convert the simple molecules of the primitive atmosphere into new molecules capable of reacting further to build the complex molecules of life. Two of the most important of these new molecules which formed photolytically are formaldehyde and hydrogen cyanide. Formaldehyde forms from methane and water irradiated by ultraviolet light.

$$CH_4 + H_2O \xrightarrow{h\nu} H_2C{=}O + 4H\cdot$$

Stepwise sequence:

Hydrogen cyanide forms when methane and ammonia are irradiated, by the same kind of sequence.

$$CH_4 + NH_3 \xrightarrow{h\nu} H-C \equiv N + 6H\cdot$$

Formaldehyde and hydrogen cyanide provide a clue to the mystery of how the two universal constituents of life—proteins and nucleic acids—could have formed in the lifeless primitive oceans.

A protein is a long-chain molecule made of amino acids hooked together head to tail. The simplest amino acid is glycine.

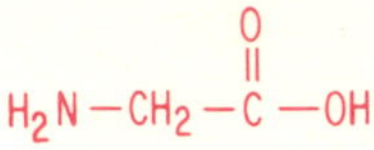

$$H_2N-CH_2-\overset{\overset{\textstyle O}{\|}}{C}-OH$$

A protein may contain up to twenty different kinds of amino acids. They are all like glycine except that each kind has its own special side chain. In the following general formula, R indicates one of twenty different side chains.

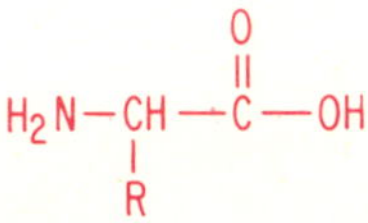

$$H_2N-\underset{\underset{\textstyle R}{|}}{CH}-\overset{\overset{\textstyle O}{\|}}{C}-OH$$

When $R = H$, the amino acid is glycine. If $R = CH_3$, the amino acid is alanine. When the head (carboxylic acid end) of one amino acid is bonded to the tail (amine end) of another amino acid, a molecule of water is released and an amide or peptide bond is formed.

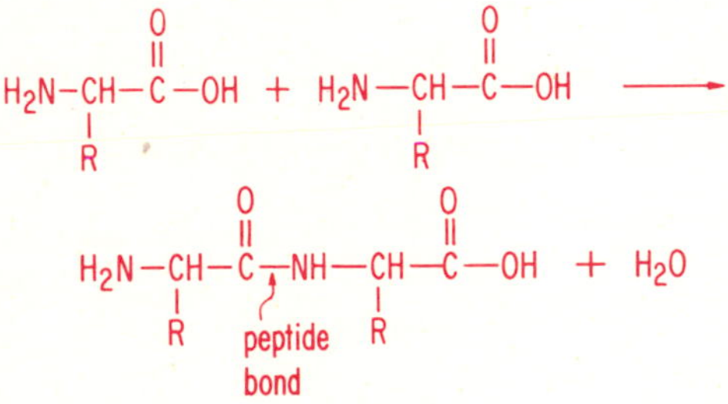

$$H_2N-\underset{\underset{\textstyle R}{|}}{CH}-\overset{\overset{\textstyle O}{\|}}{C}-OH \;+\; H_2N-\underset{\underset{\textstyle R}{|}}{CH}-\overset{\overset{\textstyle O}{\|}}{C}-OH \longrightarrow$$

$$H_2N-\underset{\underset{\textstyle R}{|}}{CH}-\overset{\overset{\textstyle O}{\|}}{C}-NH-\underset{\underset{\textstyle R}{|}}{CH}-\overset{\overset{\textstyle O}{\|}}{C}-OH \;+\; H_2O$$

peptide bond

Therefore, a protein is sometimes described as a polypeptide. A small portion of a protein molecule is shown below.

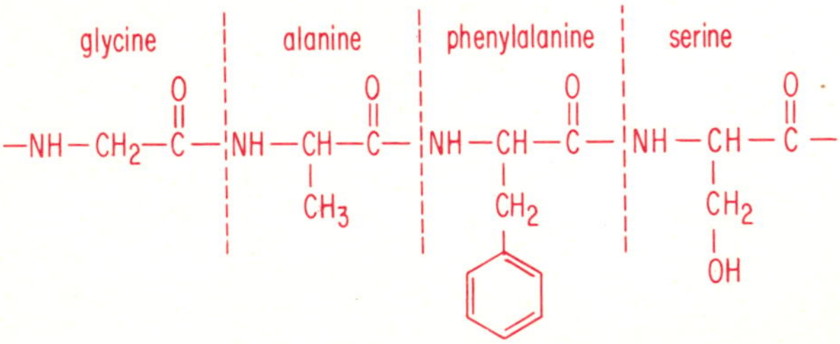

| glycine | alanine | phenylalanine | serine |

$$-NH-CH_2-\overset{\overset{\textstyle O}{\|}}{C}-NH-\underset{\underset{\textstyle CH_3}{|}}{CH}-\overset{\overset{\textstyle O}{\|}}{C}-NH-\underset{\underset{\textstyle CH_2}{|}}{CH}-\overset{\overset{\textstyle O}{\|}}{C}-NH-\underset{\underset{\textstyle OH}{\underset{\textstyle |}{CH_2}}}{CH}-\overset{\overset{\textstyle O}{\|}}{C}-$$

But how could an amino acid have formed in the primitive oceans from formaldehyde, hydrogen cyanide, ammonia, and water? A possible synthesis for glycine involves only three steps.

$$H_2C=O \ + \ NH_3 \longrightarrow H_2C=NH \ + \ H_2O$$
methylene
imine

$$H_2C=NH \ + \ H-C\equiv N \longrightarrow H_2N-CH_2-C\equiv N$$
amino   nitrile

$$H_2N-CH_2-C\equiv N \ + \ 2 \ H_2O \longrightarrow H_2N-CH_2-\overset{\displaystyle O}{\overset{\|}{C}}-OH \ + \ NH_3$$
glycine

The first step begins with a nucleophilic attack by ammonia on the carbonyl group of formaldehyde.

Proton transfer neutralizes the charges, possibly through a molecule of water.

Proton transfer then assists in releasing the imine from its unstable hydrate.

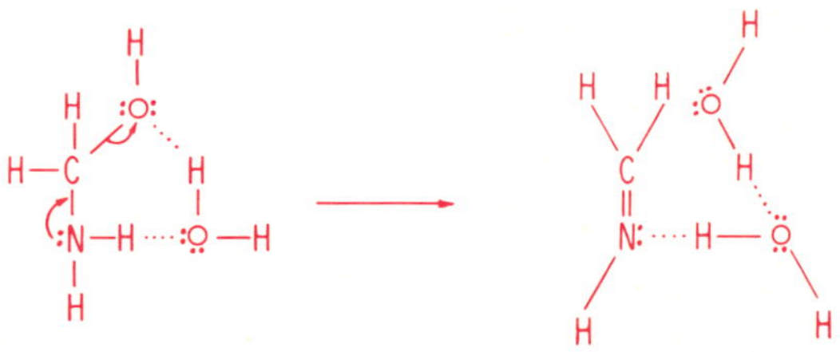

Step 2 requires nucleophilic attack by the cyanide ion on the imine carbon atom. The carbon end of the cyanide ion is the more negative end of the ion, and is therefore a more powerful nucleophile than the nitrogen end.

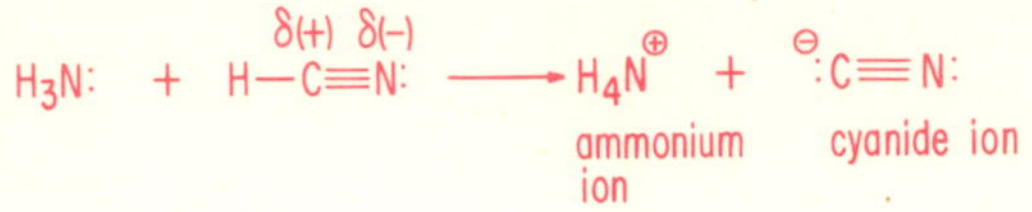

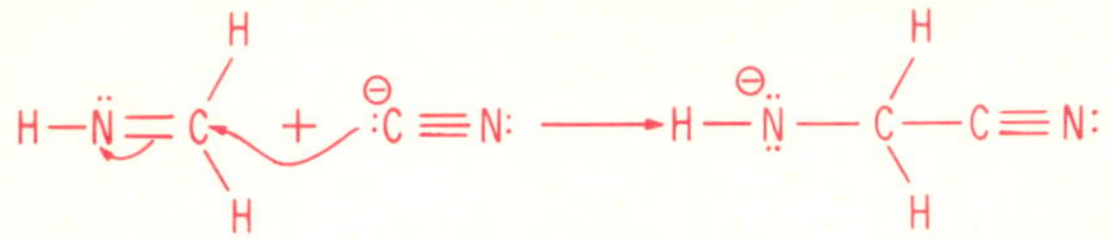

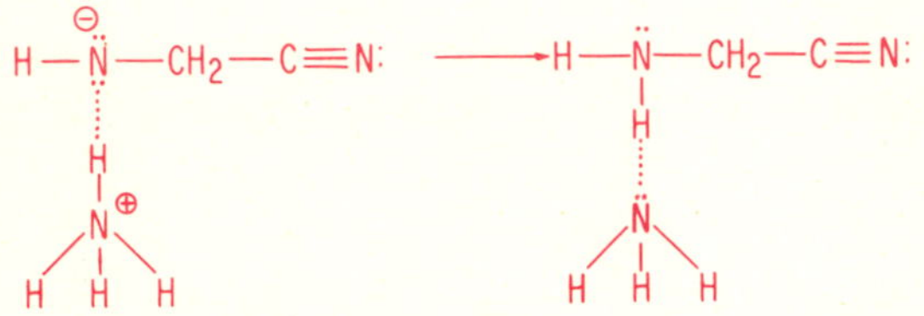

The third and final step in the synthesis of glycine is the hydrolysis of the amino nitrile to the amino acid. This also involves nucleophilic attack, but this time the nucleophile is water. In the reactions below, (a) is nucleophilic attack by water, (b) is proton transfer.

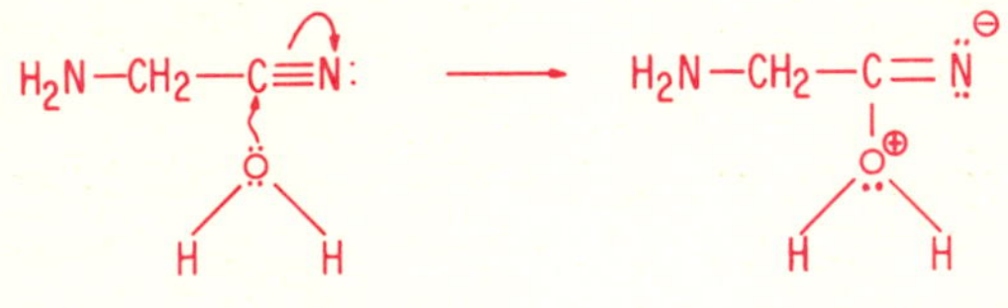

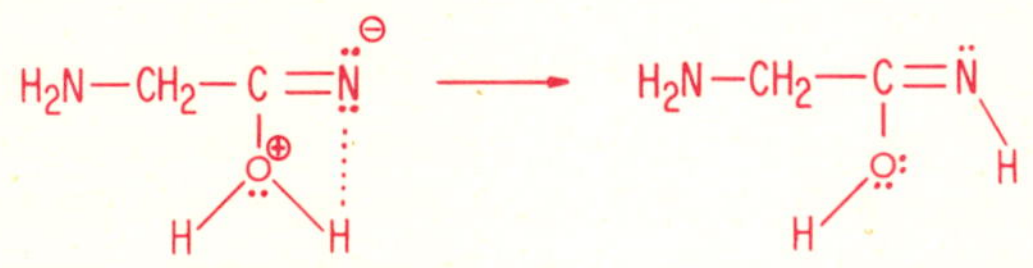

The carbon atom, now doubly bonded to the nitrogen atom, is still vulnerable to nucleophilic attack, and the sequence repeats.

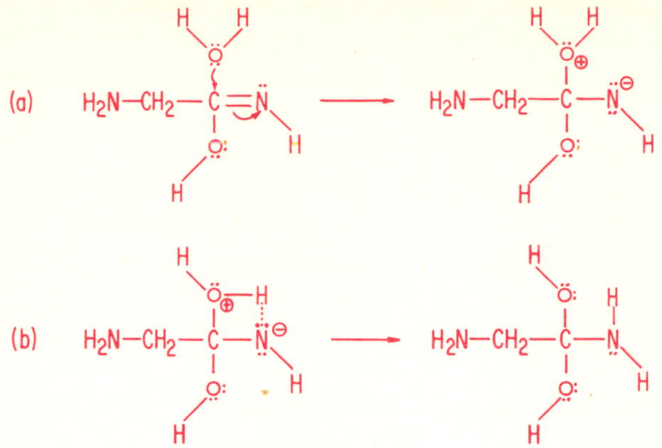

(a) $H_2N-CH_2-\overset{}{\underset{}{C}}=\overset{}{\underset{}{N}}$ $\longrightarrow$ $H_2N-CH_2-\overset{}{\underset{}{C}}-\overset{\oplus}{N}$

(b) $H_2N-CH_2-\overset{}{\underset{}{C}}-\overset{\ominus}{N}$ $\longrightarrow$ $H_2N-CH_2-\overset{}{\underset{}{C}}-N$

This amino dialcohol is unstable and may release either a molecule of water or a molecule of ammonia. If it releases water, it merely reverts to the imino alcohol it came from. This imino alcohol is a less stable form of glycin-amide. Both forms, however, are susceptible to nucleophilic attack.

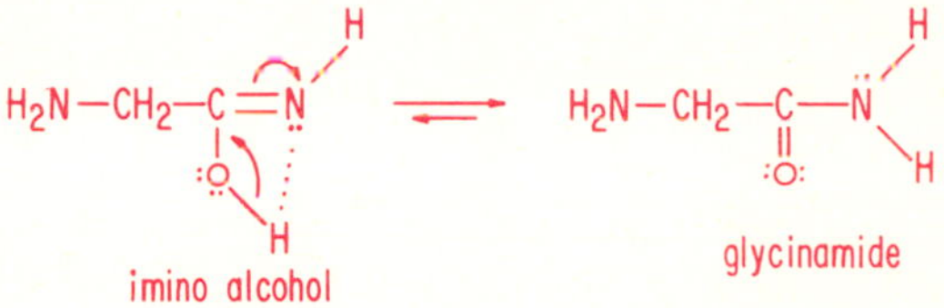

$H_2N-CH_2-C=N$ $\rightleftharpoons$ H_2N-CH_2-C-N

imino alcohol   glycinamide

If the amino dialcohol loses ammonia instead of water, the product is glycine.

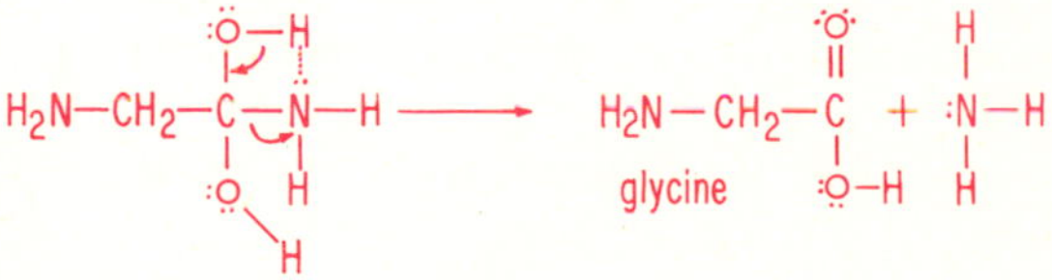

$H_2N-CH_2-C-N-H$ $\longrightarrow$ H_2N-CH_2-C $+$ $:N-H$

glycine

Coupling of these units to form polypeptide chains may have occurred in the primitive oceans at the amino nitrile stage, that is, before hydrolysis of the nitrile to the carboxylic acid. Indeed, this may have been the easiest route to primitive polypeptides, since nitrile hydrolysis and amino acid condensation require quite severe or special reaction conditions. Amino nitriles could have polymerized and hydrolyzed under the milder conditions presumably present at the time.

Possible Prebiotic Formation of Peptides

$$H_2N-CH_2-C\equiv N: \quad + \quad H-\overset{\displaystyle H}{\underset{\displaystyle |}{N}}-CH_2-C\equiv N:$$

$$H_2N-CH_2-\overset{\overset{\displaystyle :\overset{\ominus}{N}:}{\|}}{C}-\overset{\oplus}{\underset{}{N}}H-CH_2-C\equiv N:$$

$$H_2N-CH_2-\overset{\overset{\displaystyle :NH}{\|}}{C}-\ddot{N}H-CH_2-C\equiv N:$$

$$\Big\downarrow H_2O$$

$$H_2N-CH_2-\overset{\overset{\displaystyle O}{\|}}{C}-NH-CH_2-C\equiv N: \quad + \quad NH_3$$

peptide bond

The hydrolysis step is analogous to the imine-aldehyde equilibrium. Each end of the coupled molecule still contains a reactive group for extending the chain indefinitely.

Laboratory experiments designed to simulate the primitive earth atmosphere and oceans have been carried out to test the hypothesis that amino acids and other life molecules might have originated spontaneously under the conditions believed to have existed then. In 1953, Stanley Miller at the University of Chicago subjected methane, ammonia, water, and hydrogen to electric discharges for a week. He found at the end of this time that hydrolysis of the reaction mixture produced glycine and other amino acids. In fact, more than 2% of the methane in the original mixture had been converted to glycine. This remarkable result shows clearly that something exists in nature, at least under those particular reaction conditions, which favors the production of life molecules far beyond the degree which would be expected statistically. This experiment and others done since indicate that amino acids and other life molecules *could* have formed abiotically. But did they? Is there any purely *natural* evidence that this ever happened either on earth or anywhere else in the universe? The answer is yes.

In 1970, another scientist interested in the origins of life molecules, found amino acids on a meteorite which had fallen in Australia. Of these amino acids, six kinds are the same kinds found in living things on earth. This discovery points to one of two conclusions. Either amino acids have actually formed somewhere in the universe abiotically, or extraterrestrial life exists. Of course, the discovery of the extraterrestrial amino acids does not disprove either of these two possibilities. Both could be true. The discovery merely proves that at least one of them is true.

If the first is true, it answers the question posed earlier. Amino acids have actually arisen abiotically in nature without man's assistance. Scientists believe that the discovery points to this first conclusion, because the extraterrestrial amino acids are optically inactive. They do not rotate plane-polarized light. Amino acids of biological origin are optically active. This is always observed in terrestrial life forms and it is presumed to be true universally. The amino acids synthesized abiotically in the simulated primitive earth experiments are also optically inactive.

How can an amino acid be optically active in one instance and optically inactive in another? The answer lies in the structure of the molecule. The generalized structure for an amino acid contains a carbon atom bonded to four different atoms or groups.

$$H_2N-\overset{\overset{\displaystyle H}{|}}{\underset{\underset{\displaystyle R}{|}}{C}}-CO_2H$$

Because of the tetrahedral geometry of this carbon atom, two configurations (or spatial orientations) are possible for the molecule.

The relationship between these two molecules is the same as the relationship between a left hand and a right hand. They are very similar, but they are not identical. They cannot be superimposed, one upon the other, atom for atom. They are so similar that they have identical physical properties, except that one rotates plane-polarized light clockwise, and the other rotates plane-polarized light by the same amount counterclockwise. For this reason, they are called optical antipodes or enantiomers.* They are also called stereoisomers to indicate that they differ only in the three-dimensional orientation of their atoms in space. They do not differ in the order of attachment of their atoms as constitutional isomers do.

---

* The words optical antipodes or enantiomers refer to the specific optical isomers which are asymmetric and nonsuperimposable. See appendix.

The carbon atom attached to the four different groups is called an asymmetric carbon atom. The presence of this atom in the molecule makes the molecule as a whole dissymmetric, that is, capable of having a non-superimposable mirror-image.

An optically inactive amino acid contains both optical antipodes in equal amounts. This mixture is called a racemic mixture. It does not rotate plane-polarized light because the rotatory contribution of one antipode cancels the rotatory contribution of the other antipode. (Glycine is an exception. It is optically inactive because its mirror image is superimposable.)

In contrast, an amino acid obtained from a protein contains only one of the two possible antipodes. Therefore, it rotates plane-polarized light either counterclockwise or clockwise, that is, it is either levorotary or dextrorotary. But regardless of which way it rotates plane-polarized light, and regardless of which amino acid it is, the orientation of groups about the asymmetric carbon atom is always the same. This particular orientation or configuration is designated L, because if the amino acid were converted to glyceraldehyde without changing its configuration, the glyceraldehyde produced would be levorotary. The antipodes of the protein amino acids would be designated D if they would produce dextrorotary glyceraldehyde.

The antipodes of glyceraldehyde were assigned the above orientations arbitrarily, before techniques for three-dimensional structure determination were available. X-ray crystallography later verified the assignment (see Chapter XI).

All amino acids from proteins always have the L-configuration, regardless of whether they are levorotary or dextrorotary since they are stereochemically related to L-glyceraldehyde. Levorotary and dextrorotary are signified by lower-case letters l and d, or by the symbols $(-)$ and $(+)$.

The universality of the L-amino acids in proteins is difficult to explain. It is not so difficult to rationalize the existence of an all-L protein, or even an all-L organism, as opposed to a mixed L and D protein or organism. Natural selection can account for this on the basis of the energetic nonequivalence of all-L versus mixed L and D. An all-L protein might work much better than a mixed L and D protein. However, an all-D protein or an all-D organism would be completely equivalent to its looking-glass twin. Therefore, the spontaneous generation and development of L-life ought to have been paralleled by D-life. Evidently it was not, and science has no answer to the mystery.

Another observation which appears to hold true universally is that optically inactive materials always produce optically inactive products. For example, if the amino acid alanine were synthesized in the laboratory from acetaldehyde, ammonia, hydrogen cyanide, and water, it would come out as a racemic mixture of antipodes.

(a) $CH_3-CH=O$ + NH_3 $\longrightarrow$ $CH_3-CH=NH$ + H_2O
            acetaldehyde   imine

(b) CH_3-CH + $H-C\equiv N$ $\longrightarrow$ $CH_3-CH-C\equiv N$
                  $\parallel$ $\vert$
                  NH NH_2 amino nitrile
                  imine

(c) $CH_3-CH-C\equiv N$ + 2 H_2O $\longrightarrow$ $CH_3-CH-CO_2H$ + NH_3
                  $\vert$ $\vert$
                  NH_2 NH_2
                                                                          alanine
                                                                          (both antipodes)

The principal reason for the production of both stereoisomers of alanine is that cyanide ion can attack either face of the planar imine molecule with equal likelihood.

However, when alanine is produced within an organism, the reaction takes place on the surface of a dissymmetric catalyst called an enzyme. If the enzyme always attaches itself to the front face of the imine, cyanide ion can attack only from the back side. Thus, only one of the two antipodes of alanine would form. This kind of stereospecific enzyme model accounts for the observation that biosynthesis always produces only one antipode.

Prebiotic Origins of Other Biological Molecules

Amino acids are not the only essential precursors to life. All organisms contain not only protein, but also nucleic acids. With the discoveries that genetic information is transmitted by nucleic acids and that protein synthesis is directed by nucleic acids, the old notion that protein is the primary stuff of life was expanded to include nucleic acids.

Like proteins, nucleic acids are long-chain molecules. Instead of using amino acids as building blocks, however, nucleic acids use nucleotides. A nucleotide has three component parts—a purine or pyrimidine base, a ribose sugar, and phosphoric acid.

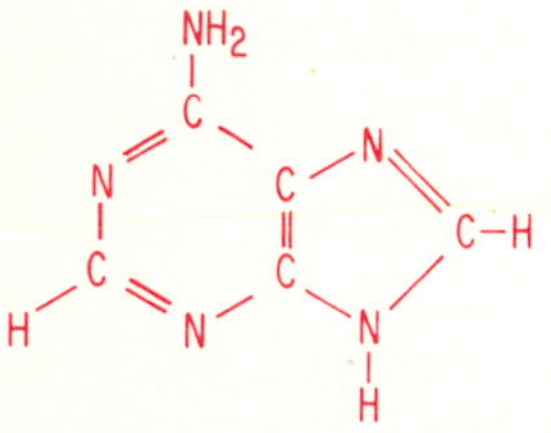

adenine   (a purine base) 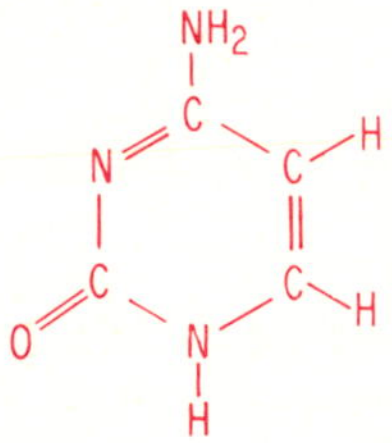   cytosine   (a pyrimidine base)

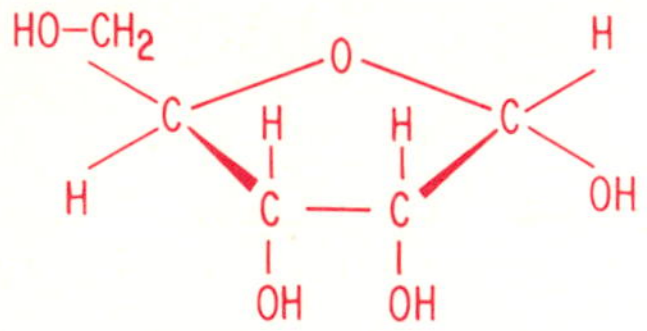

ribose   (cyclic form) 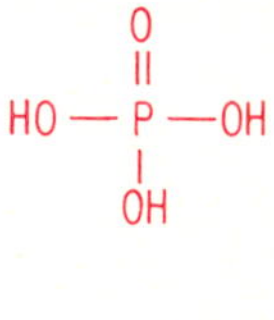   phosphoric acid

Representative Components of Nucleic Acids

The three components combine to form a nucleotide with the sugar in the middle.

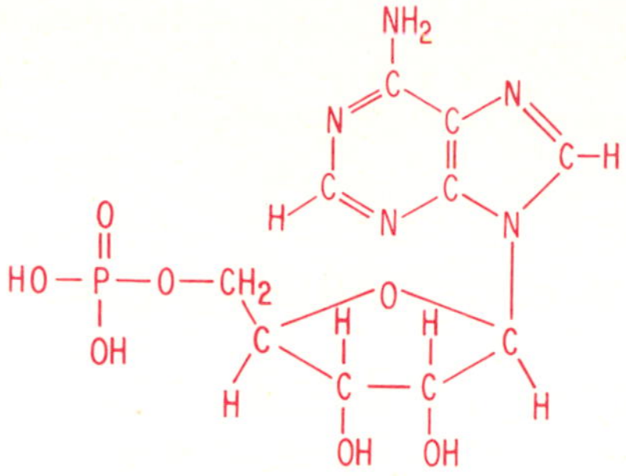

Adenylic Acid (a nucleotide)

When the nucleotides combine to build a nucleic acid, the giant molecule contains a sugar-phosphate backbone with purine and pyrimidine side chains.

Constitution of Nucleic Acids

The purine and pyrimidine bases and the sugars probably formed pre-biotically from the simple molecules available at the time, including hydrogen cyanide, formaldehyde, ammonia and water. Adenine has a molecular formula $C_5H_5N_5$, which corresponds to five molecules of hydrogen cyanide. The following sequence of reactions represents a reasonable route from hydrogen cyanide to adenine. Enolization steps have been omitted:

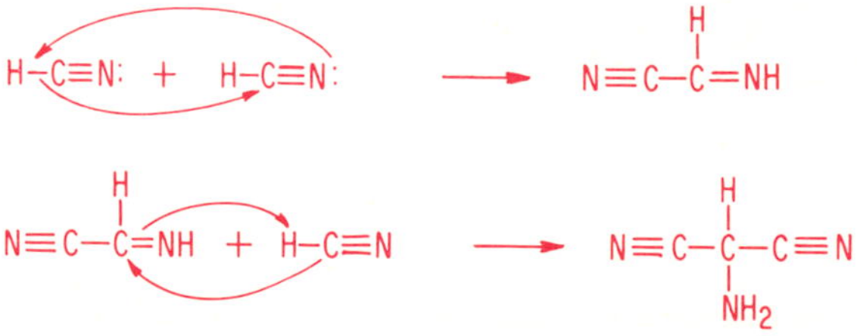

The above reactions depict self additions of hydrogen cyanide. Ammonia can also add to hydrogen cyanide or its products:

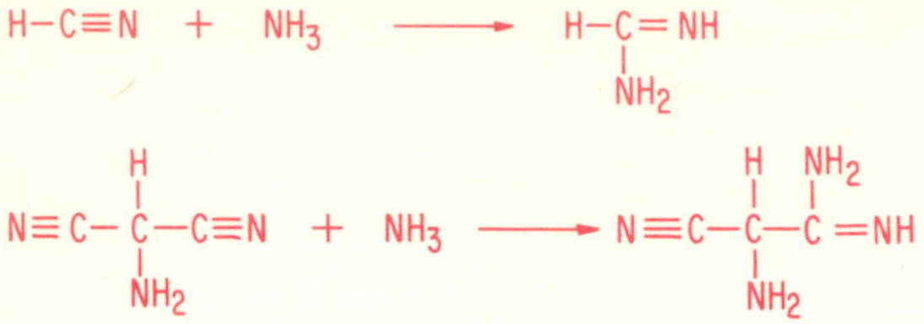

$$H-C\equiv N \; + \; NH_3 \longrightarrow H-C=NH \atop \quad\quad\; NH_2$$

$$N\equiv C-\underset{NH_2}{\overset{H}{C}}-C\equiv N \; + \; NH_3 \longrightarrow N\equiv C-\underset{NH_2}{\overset{H}{C}}-\overset{NH_2}{C}=NH$$

These two products can react with each other in a displacement fashion:

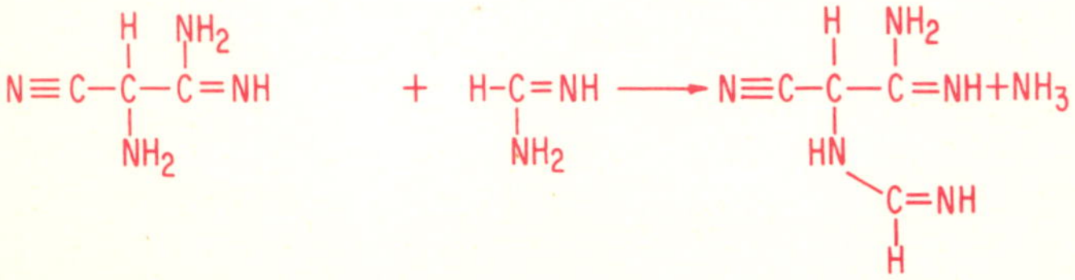

$$N\equiv C-\underset{NH_2}{\overset{H}{C}}-\overset{NH_2}{C}=NH \;\; + \;\; H-C=NH \atop \quad\quad\quad\quad NH_2 \longrightarrow N\equiv C-\overset{H}{C}-\overset{NH_2}{C}=NH + NH_3$$

A ring-closing reaction can be proposed:

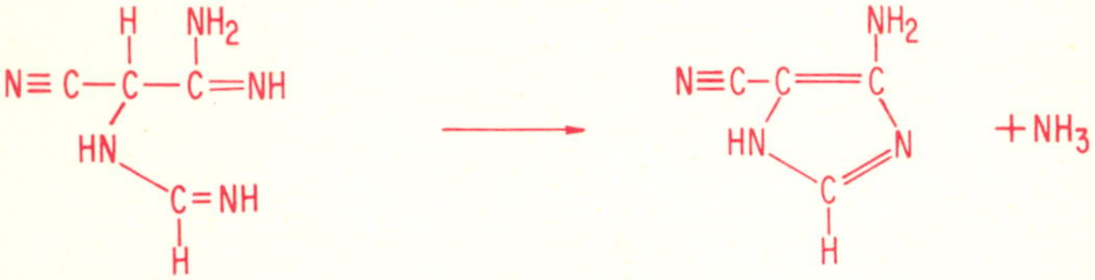

$$N\equiv C-\underset{HN}{\overset{H}{C}}-\overset{NH_2}{C}=NH \atop \quad\quad C=NH \atop \quad\quad H \longrightarrow N\equiv C-C=C-NH_2 \;\; (ring) \;\; + NH_3$$

A further displacement can occur:

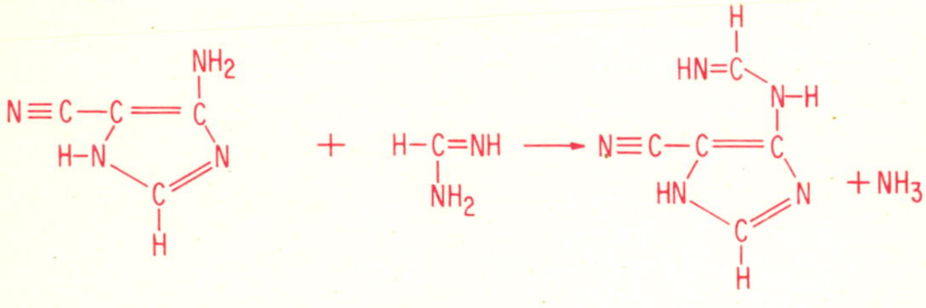

$$N\equiv C-C=C-NH_2 \;(ring)\; + \; H-C=NH \atop \quad\quad\quad\quad NH_2 \longrightarrow N\equiv C-C=C \;(ring) \;\; + NH_3$$

From this intermediate, a second ring closure can take place leading to adenine.

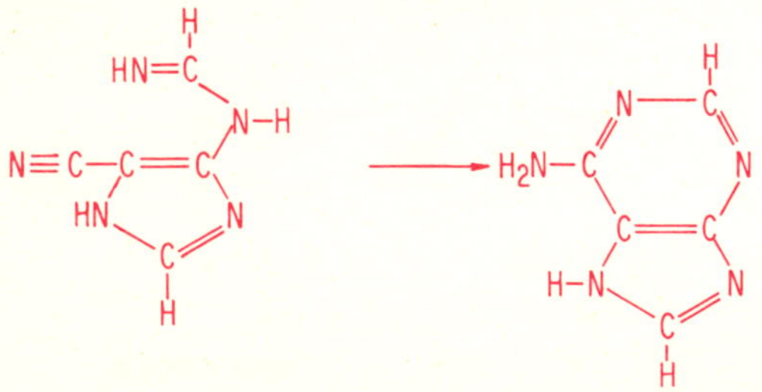

$$(intermediate) \longrightarrow H_2N-C \cdots (ring) \quad \text{adenine}$$

The ribose molecule is similar in that its molecular formula, $C_5H_{10}O_5$, corresponds to five molecules of formaldehyde, $(H_2C{=}O)_5$. The formation of ribose and other sugars in the primitive oceans may have begun with a formaldehyde anion attack on a formaldehyde molecule. This would form the two-carbon sugar glycoaldehyde.

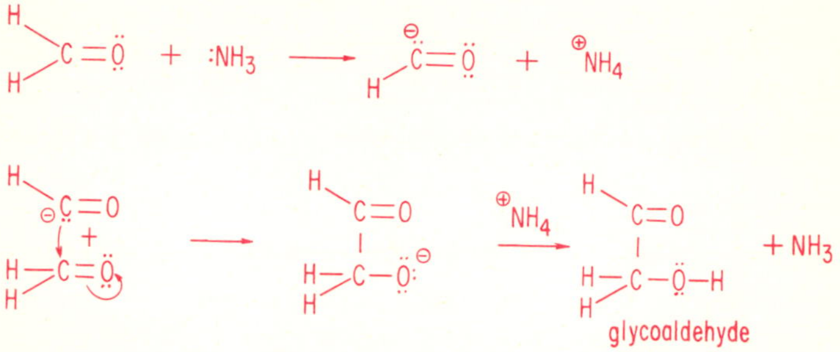

Glycoaldehyde forms an anion more easily than formaldehyde does, because of the acidifying influence of the adjacent carbonyl group.

The carbonyl group is acidifying because it allows resonance-stabilization of the anion, that is, delocalization of electrons.

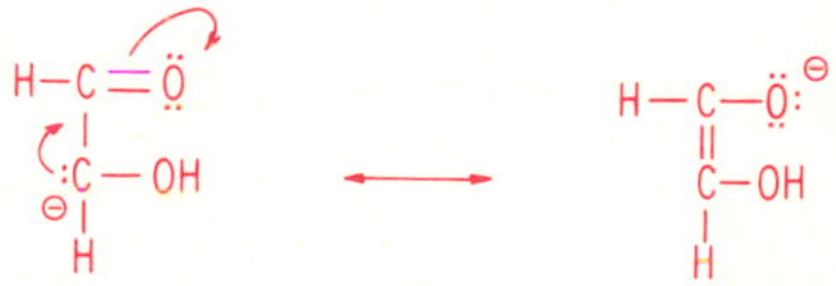

The glycoaldehyde anion can attack another formaldehyde molecule to form the three-carbon sugar glyceraldehyde.

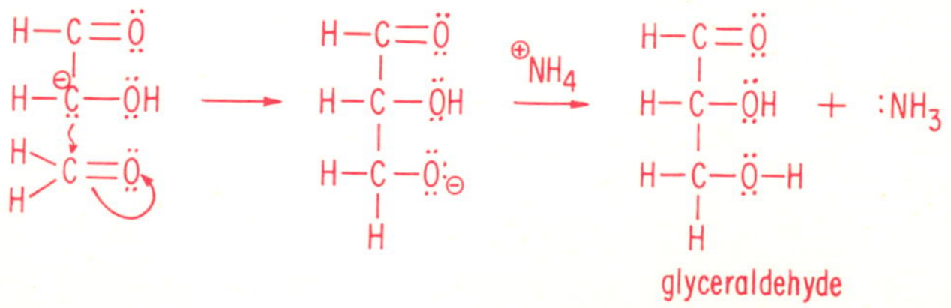

glyceraldehyde

Attack by a glycoaldehyde anion on glyceraldehyde would produce five-carbon sugars, including ribose.

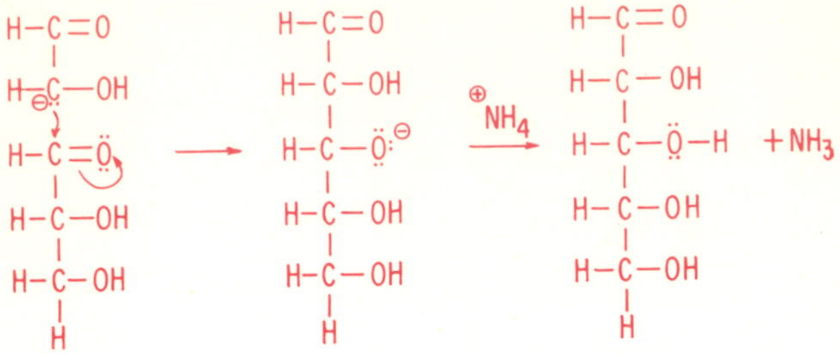

Five- and six-carbon sugars tend to cyclize. By so doing, they protect their carbonyl groups from further nucleophilic attack.

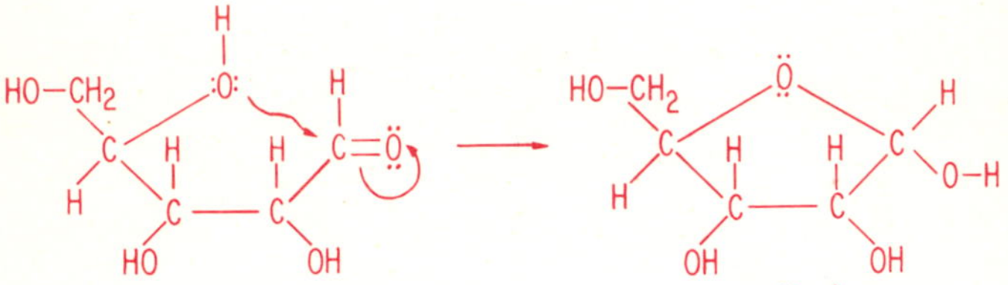

ribose   cyclic form

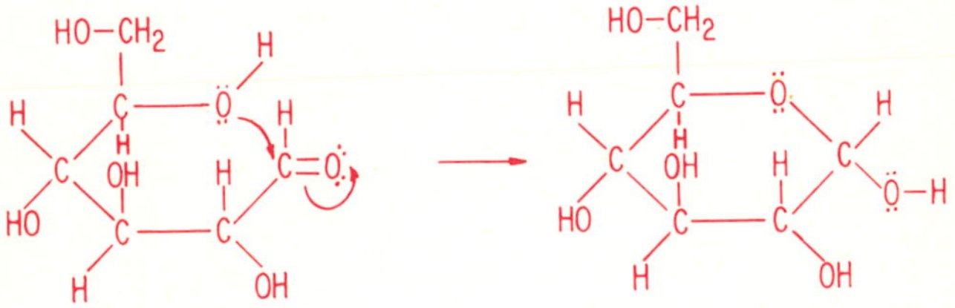

glucose   cyclic form

SUMMARY

We have seen that the primitive earth most probably contained such molecules as methane, formaldehyde, hydrogen cyanide and water. From these simple molecules, amino acids, simple sugars, purines and pyrimidines can be built using reasonable reaction pathways. It has been possible to produce some of these molecules in the laboratory under conditions resembling the prebiotic earth.

Nature's molecules go into living systems with the most refined stereo-selectivity. Although we cannot explain the origin of optically active molecules by simple reactions of the molecules noted above, we can speculate that asymmetric mineral surfaces existed at prebiotic times. Adsorption of the amino acids precursors could have led to optically active amino acids. Once formed, it is not unreasonable to expect preferential formation of other optically active molecules. In the next chapter we will consider the origin of life which took a remarkably short time from these tiny beginning steps.

SOURCE MATERIAL AND SUGGESTED READING

Margulis, L., ed., 1970. *Origins of Life* (proceedings of the First Conference), Gordon and Breach, New York.

Miller, S. L., and N. H. Horowitz, 1966. "The Origin of Life," Chapter 2, in *Biology and the Exploration of Mars*, Nat. Acad. Sci., Nat. Res. Council Pub. # 1296.

Ponnamperuma, C., and N. Gabel, 1968. "Current Status of Chemical Studies on the Origin of Life," *Space Life Science*, **1**, p. 64.

CHAPTER II

ORIGIN OF THE CELL
Life Appears On Earth

UNTIL A few years ago, scientists believed that the spontaneous generation of the first living things from individual molecules took a long time, perhaps one or two billion years. This extended period was considered necessary for the trial-and-error organizations of a reasonable number of combinations of molecules during the tremendous transition from individual molecules to highly organized living cells. However, recent evidence indicates that the first living organisms appeared as soon as the conditions necessary for their survival were established (see *Organic Chemistry: The Fossil Record in Origins of Life*, edited by L. Margulis).

The Timetable

The sun is about 5 billion years old. The earth formed and began to compact about 4.8 billion years ago. The crust of the hot earth differentiated and began to cool about 4 billion years ago. The oceans and the primitive reducing atmosphere began to accumulate about 3.5 billion years ago. The primary organic molecules such as formaldehyde and hydrogen cyanide began to form in this primitive reducing atmosphere. These molecules then evolved into the amino acids, sugars, and organic bases required for making proteins and nucleic acids, probably in the oceans. The build-up of complex organic molecules in the oceans, then, must have occurred less than 3.5 billion years ago.

The most remarkable thing about this history is the point where life appears. Bacteria-like particles have been isolated from Swaziland shale and dated as 3.2 billion years old. This means that the first life forms arose more than 3.2 billion years ago. Considering the age of the oceans, living things must have formed as soon as their constituent molecules became available.

Cyril Ponnamperuma, who identified the extraterrestrial amino acids on the meteorite (Chapter I, p. 8) believes that the problem of precellular organization is of paramount importance, because it bridges the gap between chemical evolution and biological evolution. "In essence, the problem is to envision how the now biologically important chemicals spontaneously organized themselves into a three-dimensional matrix, which would eventually acquire the characteristics of a readily agreed-upon life form."

Some of the most interesting work on this problem comes from the laboratories of Alexander I. Oparin, one of the originators of scientific studies on the molecular origins of life. Some 50 years ago, before the chemically reducing nature of the universe was established, Oparin postulated the existence of a reducing atmosphere as necessary for prebiotic organic synthesis on earth. This was 30 years before Stanley Miller's spark synthesis of amino acids.

Oparin has shown that when proteins and nucleic acids reach critical concentrations in water, they separate from solution as tiny droplets (called coacervate droplets). The concentrations of proteins and nucleic acids in the droplets are much higher than in the original solution, and the coacervate droplets have a much higher degree of organization. Furthermore, addition of an enzyme* results in selective absorption of the enzyme into the droplets, where the enzyme becomes more active than in solution. This is exciting because it is the kind of behavior observed in living cells, but in a much simpler system.

In *The Origins of Prebiological Systems*, edited by Sidney W. Fox and published by Academic Press in 1965, Oparin says

Life is characterized by the fact that it is not simply dispersed in space but is represented by individual systems—organisms separated from the external world. The appearance of these beings could have taken place only on the basis of a long-term evolution, of gradual perfection of some much simpler initial systems which separated from the primeval homogeneous soup.

In this soup, as in a simple aqueous solution of organic substances and mineral salts, chemical conversions are not specifically organized, but proceed independently in all directions, intercrossing chaotically. In living beings, however, individual chemical reactions are strictly coordinated and proceed in a certain sequence, which as a whole forms a network of biological metabolism directed toward the perpetual self-preservation, growth, and self-reproduction of the entire system under the given environmental conditions.

He goes on to say that the task is to obtain, experimentally, possible pathways for the development of such initial systems, which could interact with the environment as open systems do, and in which chemical conversions would become ever more organized, approaching in the process of their evolution the biological order of metabolism.

Oparin has shown that polymerization of a nucleotide in a pure solution produces only peculiar aggregations of polynucleotides. But if the polymerization occurs in the presence of another polymer, for example, a protein, the polynucleotides produce polymolecular complexes with the other polymer.

---

* An enzyme is a protein which acts as a catalyst, usually for a single, specific chemical reaction.

"These complexes, once separated from the surrounding solution, seem to be initial systems that interact with their environment in the process of evolution and give rise to primary living beings endowed with metabolism."

Oparin does not mean that his coacervate droplets become living beings in the laboratory. He means that the process of coacervation may have been involved in the giant step linking chemical evolution to biological evolution.

In one experiment, Oparin introduced potato phosphorylase into a coacervate droplet. (Potato phosphorylase is an enzyme that catalyzes the polymerization of glucose-1-phosphate, a six-carbon sugar, into starch.) Glucose-1-phosphate is then added to the liquid surrounding the coacervate droplet. It diffuses into the droplet, comes into contact with the potato phosphorylase, and polymerizes to starch. The starch accumulates within the droplet, causing the droplet to grow to three times its original volume in less than 2 hours.

Now a second enzyme, β-amylase, is introduced into the droplet. This enzyme splits the giant starch molecules down into "diglucose" units, called maltose, a twelve-carbon sugar. The maltose molecules diffuse out of the coacervate droplet*, and the droplet shrinks to its original size (Figure 2.1).

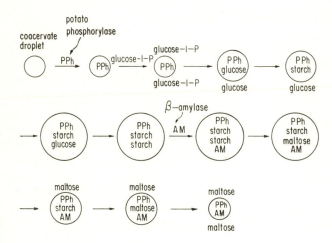

FIGURE 2.1 Diagrammatic representation of Oparin's experiment on enzyme activity in coacervate droplets. PPh represents the enzyme, potato phosphorylase; AM, the enzyme, β-amylase; and glucose-1-P is glucose-1-phosphate

---

* A coacervate droplet is defined as an ordered aggregate of colloidal droplets which is held together by electrostatic forces.

Thus, Oparin says,

We possess models of open multimolecular systems, which as a result of acceleration of the processes going on within them may grow on account of the surrounding solution, or on the contrary, may undergo disintegration. It is evident that systems similar to our coacervate models, besides existing for a long time, could also become larger, growing in solutions of substances whose presence and even abundance can be conceived for the primeval soup of the earth. Single drops would scarcely grow the whole time as entities. Under conditions of the primary hydrosphere of the earth, they would necessarily break into fragments under the effect of external mechanical forces, such as waves and tide, as emulsion drops break upon shaking. Such systems, interacting with the external medium, growing and increasing in number, would ever improve due to the action of natural selection, and so organize their metabolism.

The key to the explanation of the appearance of life on earth seems to lie in an understanding of the phenomenon known as "natural selection." Natural selection must be the basis for the organization of a "metabolism" from a collection of random chemical reactions. Possession of a metabolism is the earmark of a living thing. It is a series of molecular transformations from which an organism obtains the energy required to stay alive, to grow, and to reproduce.

The Primary Metabolism

What must the early metabolism of the first living organisms have been? These organisms arose "in the absence of air," as we know air today. Their atmosphere contained methane and ammonia, rather than oxygen and nitrogen. Therefore, they must have obtained energy by degrading food molecules anaerobically, that is, without using molecular oxygen. Oddly enough, the ability to do this has persisted in all living things down to the present day, even though newer, more efficient ways of obtaining energy have been added to the repertory of life-sustaining metabolic pathways.

This ancient metabolism involves a set of enzymes for breaking a molecule of glucose down into two molecules of pyruvic acid. Every living thing in the world—bacterium, plant, and animal—has this set of enzymes. This is a startling testimony to the unity of life on earth.

$$\underset{\text{glucose}}{CH_2-CH-CH-CH-CH-CH=O} \;\;\;\overset{\text{enzymes}}{\longrightarrow}\;\;\; \underset{\text{pyruvic acid}}{2\; CH_3-C-C-OH}$$

The anaerobic transformation of glucose to pyruvic acid requires several steps, each of which is catalyzed by a specific enzyme. In the first step, glucose becomes glucose-6-phosphate, having obtained the phosphate from the very important "energy rich" molecule, adenosine triphosphate (ATP) (see below).

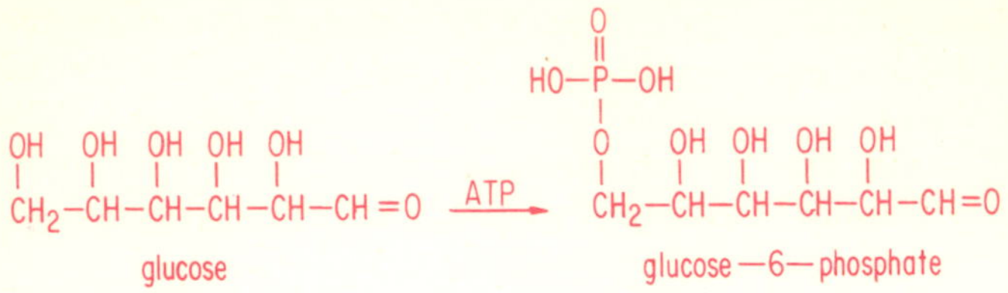

glucose   ATP →   glucose—6—phosphate

Then glucose-6-phosphate isomerizes to fructose-6-phosphate.

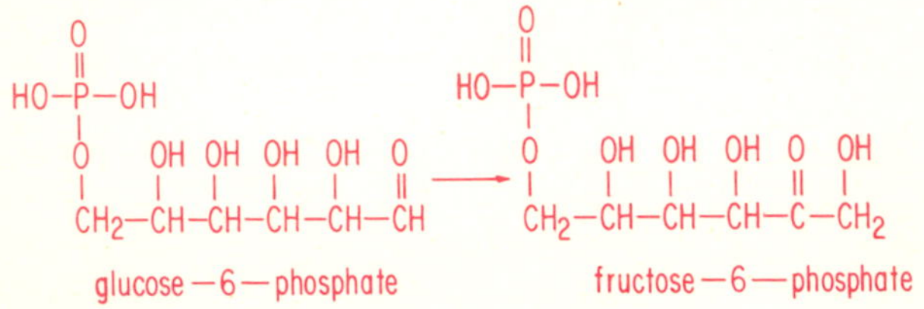

glucose—6—phosphate   →   fructose—6—phosphate

This is followed by a second phosphorylation, forming fructose-1,6-di-phosphate.

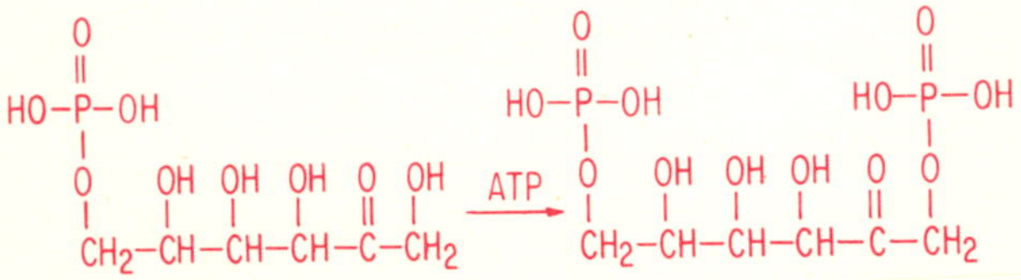

Neither of these phosphorylation reactions generates any energy for the organism. In fact, each phosphorylation uses up an "energy-rich" molecule of adenosine triphosphate (ATP).

Adenosine triphosphate (ATP)

(adenine)

(ribose)

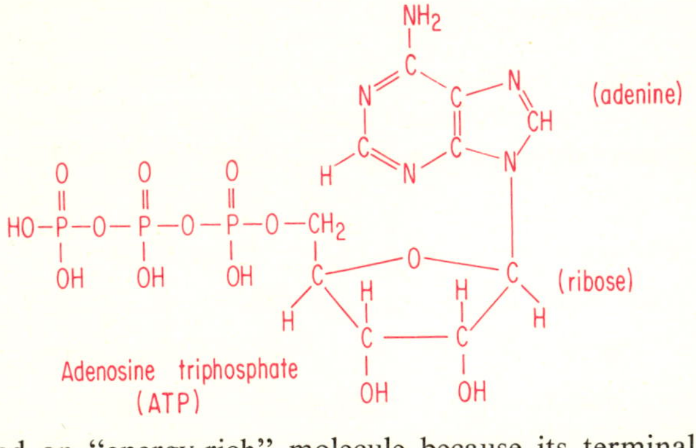

ATP is called an "energy-rich" molecule because its terminal phosphate–phosphate anhydride linkage is extremely reactive. It provides the energy or driving force for many metabolic reactions. This molecule is the universal storehouse of readily available energy for all living things.

ATP "spends" its energy-rich bond by hydrolyzing to adenosine diphosphate (ADP) and phosphoric acid. This reaction can be summarized by ATP $\longrightarrow$ ADP + P. However, in the formation of the sugar phosphates, this energy is not spent in hydrolysis but rather used to form phosphates by "phosphoryl transfer" reactions to the sugar hydroxyl groups and not to water.

The remainder of the ancient metabolic sequence—from fructose-1,6-diphosphate to pyruvate—converts four molecules of ADP into four molecules of ATP. Therefore, every time an organism breaks down a molecule of glucose anaerobically to pyruvate, it increases its energy storehouse by two molecules of ATP.

First, fructose-1,6-diphosphate splits, producing two three-carbon fragments which are interconvertible.

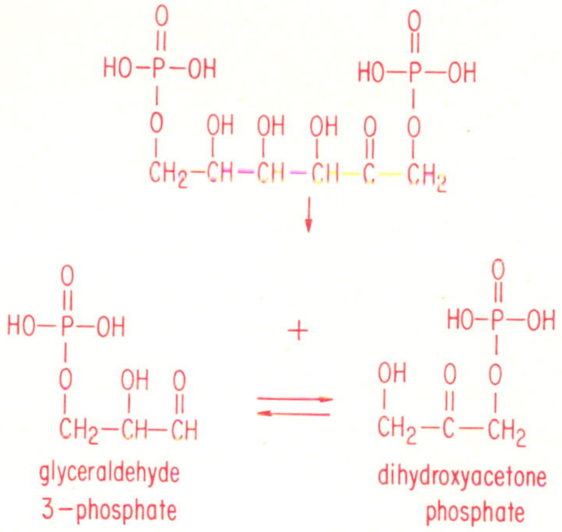

glyceraldehyde 3–phosphate

dihydroxyacetone phosphate

The two new molecules interconvert by the same kind of isomerization seen between glucose-6-phosphate and fructose-6-phosphate. The carbonyl group shifts over one atom through an unsaturated diol intermediate.

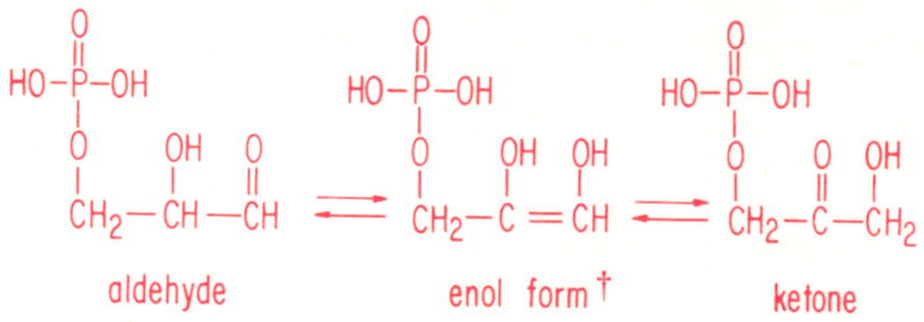

aldehyde   enol form†   ketone

---

† A discussion of the process of enolization is contained in the appendix.

The metabolic sequence continues through the aldehyde form, in this case, glyceraldehyde-3-phosphate. The next step is another phosphorylation, but a different type from the two seen earlier. This one is an oxidative phosphorylation and does not use ATP. Instead, it uses ordinary phosphoric acid and an extremely important oxidizing agent, seen universally in nature, known as NAD^+, or nicotinamide–adenine–dinucleotide. The molecule contains the following units in the following sequence:

adenine–ribose–phosphate–phosphate–ribose–nicotinamide$^+$

NAD^+

The new unit in this structure is the nicotinamide, the most important part, because it is the oxidizing site. We can represent the molecule as a nicotinamide connected to an adenosine diphosphate and ribose (shown as the group R in the following formulae):

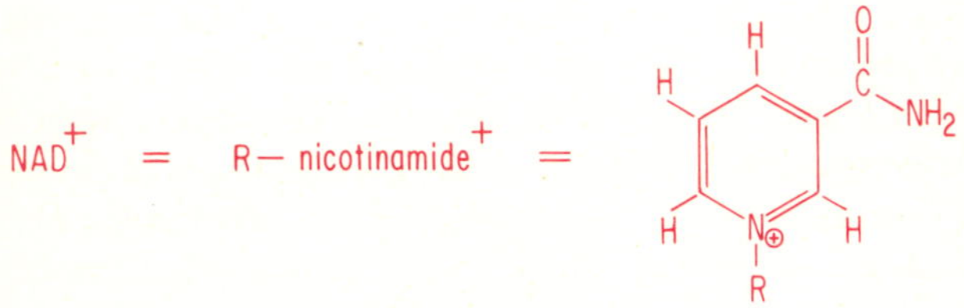

$$NAD^+ \;=\; R - nicotinamide^+ \;=\;$$

NAD^+ oxidizes a molecule by removing two hydrogen atoms from the molecule. NAD^+ thereby becomes $NADH + H^+$.

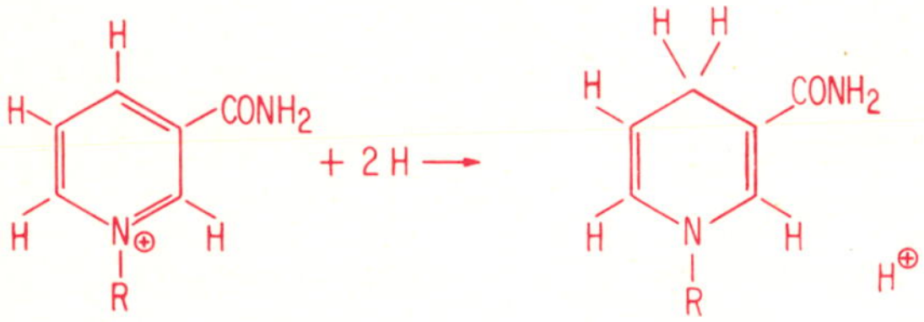

$$+\,2\,H \longrightarrow$$

$$NAD^+ \qquad\qquad NADH + H^+$$

When glyceraldehyde-3-phosphate is phosphorylated oxidatively by NAD^+ and phosphoric acid, it becomes the phosphoric anhydride of 3-phosphoglyceric acid. (The aldehyde group is oxidized to a carboxylic acid.)

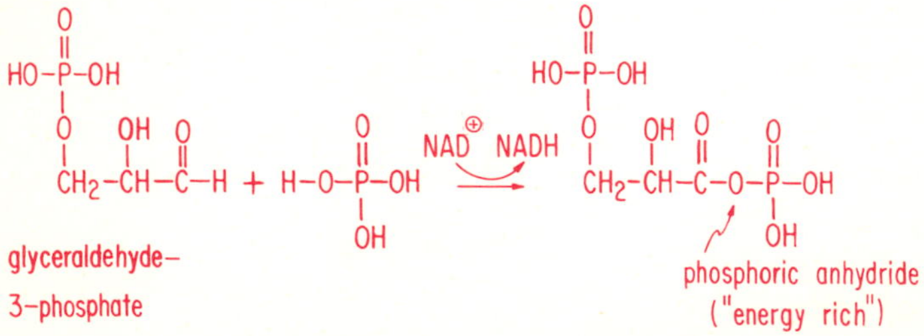

glyceraldehyde–
3–phosphate

phosphoric anhydride
("energy rich")

The significance of this step, as far as the organism is concerned, is that the phosphoric anhydride group, like that in ATP, is "energy rich," and can be used to convert ADP into ATP for energy storage. This happens next.

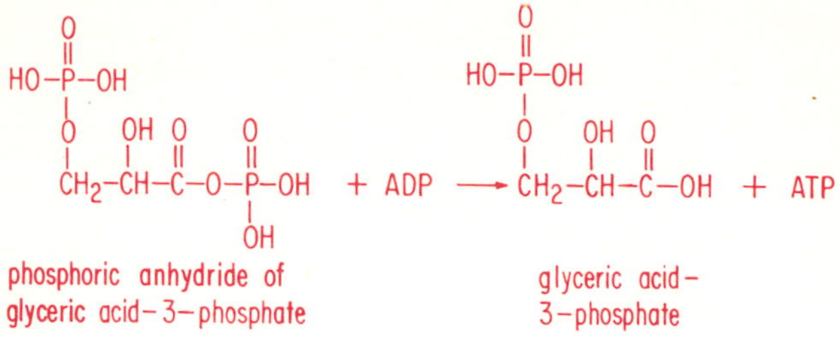

phosphoric anhydride of
glyceric acid–3–phosphate

glyceric acid–
3–phosphate

Since each molecule of glucose produces two molecules of glyceric acid phosphate, it also produces two molecules of ATP. So far, then, the metabolic pathway has used up two ATP's and produced two ATP's, for a net energy-storage change of zero. However, the next two reactions convert the 3-phosphate into an energy-rich phosphate, which goes on to make two more ATP's for an overall energy-storage gain of two ATP's per molecule of glucose. The energy is, of course, stored in the highly reactive phosphate anhydride bonds of ATP.

In the first of these two reactions, the phosphate migrates from the 3-hydroxyl to the 2-hydroxyl.

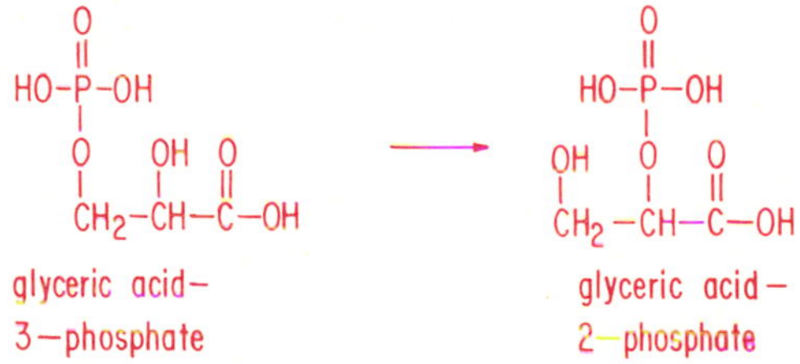

glyceric acid–
3–phosphate

glyceric acid–
2–phosphate

In the second reaction, the molecule loses water, forming a double bond between carbon atom-2 and carbon atom-3.

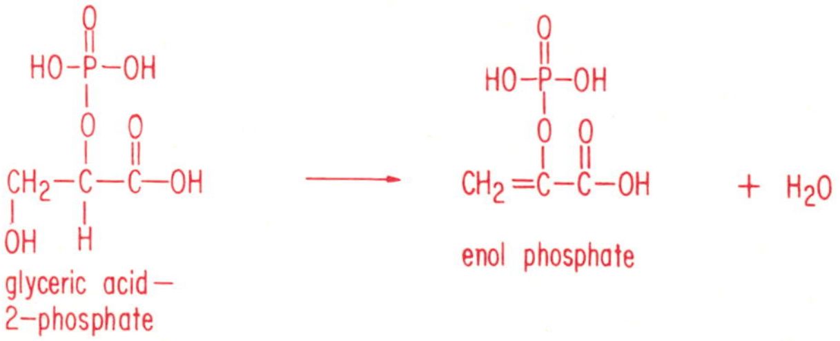

glyceric acid–
2–phosphate

enol phosphate

The unsaturated phosphate formed by this dehydration is an energy-rich phosphate, since its electronic structure is now similar to that of an anhydride. It also has the power to convert ADP into ATP.

$$
\underset{\text{enol phosphate}}{\overset{\displaystyle HO-\overset{\overset{O}{\|}}{\underset{\underset{CH_2=C-C-OH}{O \quad O}}{P}}-OH}{}}
\;+\; ADP \;\longrightarrow\;
\left[\underset{\text{enol form of}}{CH_2=\overset{OH}{C}-\overset{\overset{O}{\|}}{C}-OH}\right] \;+\; ATP
$$

$$
\underset{\text{pyruvic acid}}{CH_3-\overset{\overset{O}{\|}}{C}-\overset{\overset{O}{\|}}{C}-OH}
$$

Overall, then, this metabolic sequence converts one molecule of glucose, two molecules of phosphoric acid, and two molecules of ADP into two molecules of pyruvic acid and two molecules of ATP. It also reduces two molecules of NAD^+.

$$
\underset{+\;2\;NAD^+}{CH_2-CH-CH-CH-CH-CH}\overset{OH\ OH\ OH\ OH\ OH\ O}{}
\;+\;2\;\;HO-\overset{\overset{O}{\|}}{\underset{OH}{P}}-OH \;+\;2\;ADP
$$

$$
\downarrow
$$

$$
2\;\;CH_3-\overset{\overset{O}{\|}}{C}-\overset{\overset{O}{\|}}{C}-OH \;+\;2\;NADH \;+\;2\;ATP+2H^+
$$

Net Reaction for Anaerobic Metabolism

The glucose molecule contains five carbon–oxygen single bonds and one carbon–oxygen double bond. The two pyruvic acid molecules contain a total of ten carbon–oxygen bonds, five for each molecule. The transformation of glucose into pyruvic acid, then, is an oxidation since the product molecules contain more carbon–oxygen bonds than the starting molecule. This oxidation produces energy which the organism traps and stores chemically in the form of energy-rich ATP molecules. The organism then has this storehouse of energy available to use as needed for its energy-consuming activities of staying alive, growing, and reproducing.

The sustenance of life requires some source of energy, and the primitive organisms of the earth depended for this energy on the availability of pre-formed organic molecules, such as glucose, synthesized abiotically in the environment. Thus the situation was one in which these organisms were gradually diminishing their food supply, with the only source of replenishment being ultimately the limited amount of methane left in the atmosphere. If life were to continue on earth, some other source of these nutrient molecules must be provided. Ideally, of course, this new source should be inexhaustible. Since the nutrient molecules were being used up oxidatively, they would have to be replaced reductively. Otherwise all the organic molecules in the world would eventually be oxidized beyond the nutriment stage. Atmospheric hydrogen would be of little long-range use as a reducing agent, since it was continuously being lost to space due to its small mass and since its generation from methane photolysis would eventually cease. The development of a complementary reductive metabolism involving, for example, the conversion of pyruvic acid back into glucose, would be impossible because a reduction of this type absorbs energy. A metabolism must provide energy to sustain life.

The problem then was to find an inexhaustible energy source and to use this source somehow to reduce highly oxidized carbon compounds back to the nutrient level. The most obvious and the most widely available energy source for such purposes was the sun, the same source used to split the molecules of the primitive atmosphere into the reactive fragments which combined to produce formaldehyde and hydrogen cyanide. What was needed was a means of capturing some of the radiant energy of the sun and converting it into chemical form, for example, by using it to convert ADP into ATP. Then the ATP could be used to drive the required reduction reactions. This is exactly what happened.

The process which organisms developed to capture the sun's radiant energy, use it to make ATP, and then use the ATP to drive reduction reactions is called photosynthesis. The first step in this process—the absorption of sunlight—is accomplished by a molecule called chlorophyll, a derivative of porphyrin.

Porphyrin derivatives of prebiotic origin were probably present in the primitive oceans. They have been found in petroleum deposits, which may have formed non-biogenically, and recent simulated prebiotic synthesis experiments have produced porphyrin-like molecules. Therefore, molecules of this type were probably available to the first organisms.

Porphyrin and its derivatives contain an extended sequence of alternate single and double bonds, called conjugated double bonds. When light strikes such a molecule, it energizes the molecule, driving electrons into orbitals of higher energy. The molecule is then said to be photo-excited.

When the electrons fall back to their original orbitals, energy is released by the molecule, and under the proper conditions, this energy can be used to

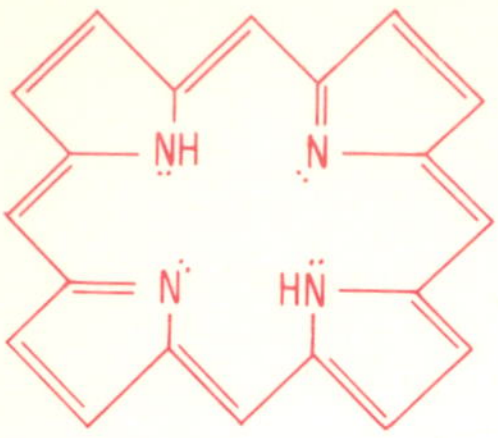

porphyrin

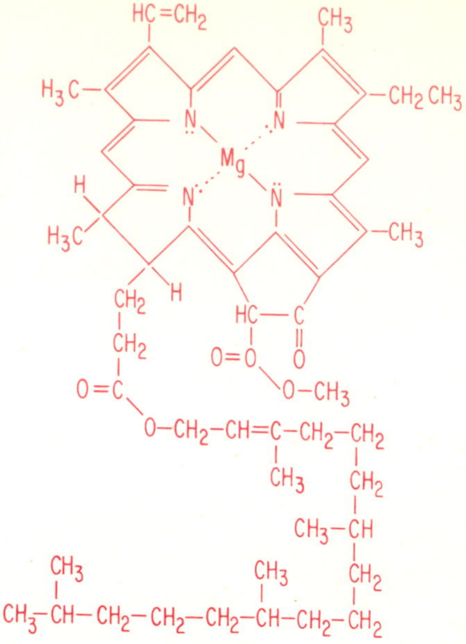

chlorophyll

drive a chemical reaction. Photo-excited chlorophyll does just this. It provides the energy needed to convert ADP into ATP.

$$\text{ADP} + \text{HO}-\overset{\overset{\displaystyle O}{\|}}{\underset{\underset{\displaystyle OH}{|}}{P}}-\text{OH} \xrightarrow[\text{chlorophyll}]{\text{sunlight}} \text{ATP}$$

The organism can use the energy from this ATP molecule and a reducing system to reduce an oxidized molecule, such as phosphoglyceric acid, to a molecule on the carbohydrate oxidation level, such as phosphoglyceraldehyde. Both these molecules are intermediates in the metabolic pathway for the anaerobic oxidation of glucose to pyruvic acid. The reducing agent is a reduced nicotinamide, also seen earlier in the anaerobic glucose oxidation scheme.

The significance of all this is that the organism can now use energy from the sun to make a molecule of food for itself. This represents a tremendous advance beyond the non-photosynthesizing organism, which must depend upon the environment to supply preformed food molecules.

This ability to reduce carbon compounds using solar energy allowed the organism to take in carbon dioxide, probably as carbonate or bicarbonate ion,

$$\underset{\text{phosphoglyceric acid}}{\overset{\text{O}}{\underset{\underset{CH_2-CH-C-OH}{\overset{\mid \quad \mid \quad \parallel}{\underset{\quad OH \quad O}{\mid}}}}{HO-P-OH}}} \quad + \quad \underset{\substack{\text{reduced} \\ \text{nicotinamide}}}{\left[\text{reduced nicotinamide ring, } CONH_2 \right]} + H^\oplus + ATP \longrightarrow$$

$$\underset{\text{phosphoglyceraldehyde}}{\overset{\text{O}}{\underset{\underset{CH_2-CH-CH}{\overset{\mid \quad \mid \quad \parallel}{\underset{\quad OH \quad O}{\mid}}}}{HO-P-OH}}} \quad + \quad \left[\text{nicotinamide ring } N^\oplus, CONH_2\right] \quad + \quad ADP \quad + \quad \underset{\overset{OH}{\mid}}{\overset{O}{\overset{\parallel}{HO-P-OH}}}$$

from the primitive ocean environment and to use the carbon dioxide as a source of carbon atoms for making sugars and other nutrient molecules. This reaction is really the heart of photosynthesis. In green plants, it involves the reduction of carbon dioxide to the aldehyde (carbohydrate) level, using water as the primary source of hydrogen atoms, and liberating free oxygen to the atmosphere. As in the case of anaerobic metabolism (see p. 21), each of the reactions involved in photosynthesis requires a very specific enzyme.

$$\underset{\text{water}}{\overset{H}{\underset{H}{\diagdown}}O} \quad + \quad \underset{\text{carbon dioxide}}{O=C=O} \quad \xrightarrow[\text{synthesis}]{\text{photo} -} \quad \left[\underset{H}{\overset{H}{\diagdown}}C=O\right] + \quad \underset{\text{oxygen}}{O=O}$$
$$\text{carbohydrates}$$

This process sounds and looks simple, but in practice is extremely complex. Only recently, and after many years of investigation, has the carbon-atom pathway finally been pieced together completely. Some details are still not known, for example, how photo-excited chlorophyll makes ATP from ADP and phosphate.

When a molecule of carbon dioxide (or bicarbonate ion) enters a photosynthetic cell, it reacts with a five-carbon sugar (ribulose diphosphate) to produce two molecules of phosphoglyceric acid. This takes only a few seconds.

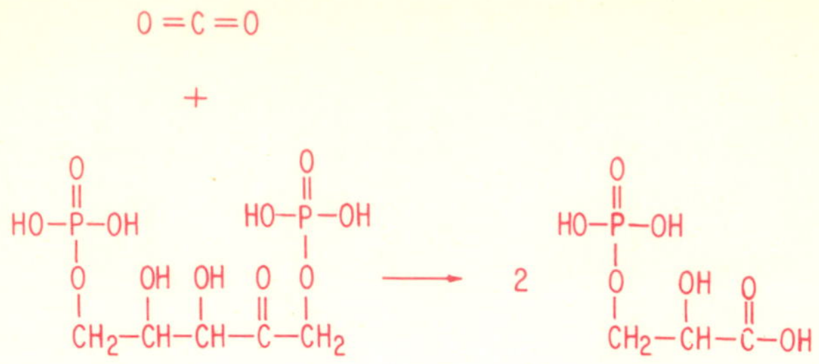

$$O = C = O$$

$$+$$

$$
\begin{array}{c}
\text{HO–P(=O)–OH} \qquad \text{HO–P(=O)–OH} \\
\text{O} \quad \text{OH OH O O} \\
\text{CH}_2\text{–CH–CH–C–CH}_2
\end{array}
\longrightarrow
2
\begin{array}{c}
\text{HO–P(=O)–OH} \\
\text{O} \quad \text{OH O} \\
\text{CH}_2\text{–CH–C–OH}
\end{array}
$$

ribulose diphosphate   phosphoglyceric acid

Just how this happens is not entirely clear, but it may involve several steps.

(1) Isomerization of ribulose diphosphate:

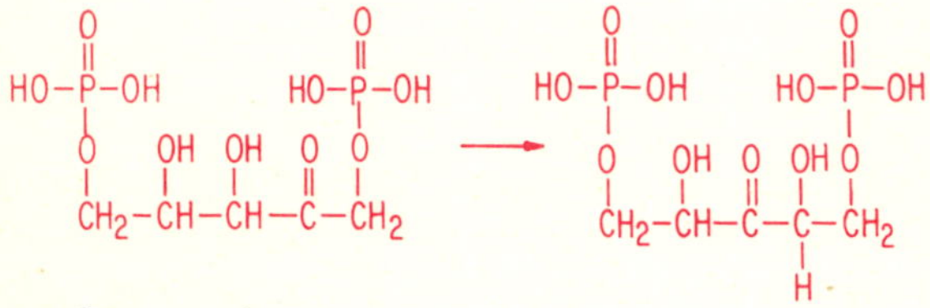

$$
\begin{array}{c}
\text{HO–P(=O)–OH} \qquad \text{HO–P(=O)–OH} \\
\text{O} \quad \text{OH OH O O} \\
\text{CH}_2\text{–CH–CH–C–CH}_2
\end{array}
\longrightarrow
\begin{array}{c}
\text{HO–P(=O)–OH} \qquad \text{HO–P(=O)–OH} \\
\text{O} \quad \text{OH O OH O} \\
\text{CH}_2\text{–CH–C–C–CH}_2 \\
\text{H}
\end{array}
$$

(2) Formation of sugar anion:

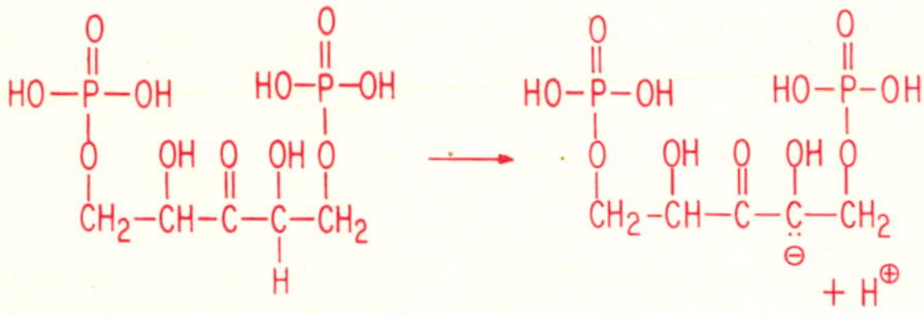

$$
\begin{array}{c}
\text{HO–P(=O)–OH} \qquad \text{HO–P(=O)–OH} \\
\text{O} \quad \text{OH O OH O} \\
\text{CH}_2\text{–CH–C–C–CH}_2 \\
\text{H}
\end{array}
\longrightarrow
\begin{array}{c}
\text{HO–P(=O)–OH} \qquad \text{HO–P(=O)–OH} \\
\text{O} \quad \text{OH O OH O} \\
\text{CH}_2\text{–CH–C–C–CH}_2 \\
\ominus
\end{array}
+ H^{\oplus}
$$

This carbanion is stabilized by enolization (see appendix).

(3) Nucleophilic attack by anion on carbon dioxide:

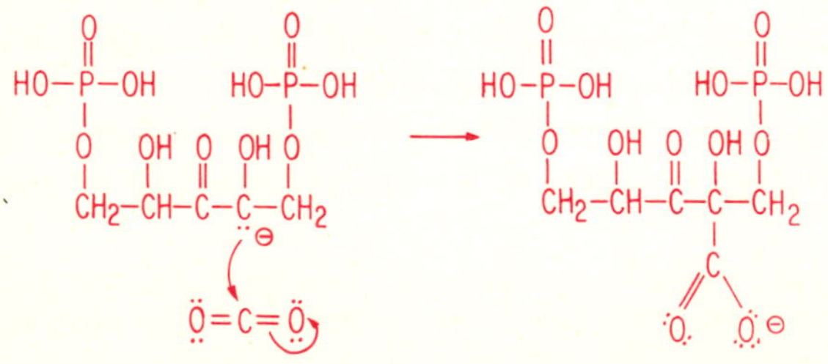

$$
\begin{array}{c}
\text{HO–P(=O)–OH} \qquad \text{HO–P(=O)–OH} \\
\text{O} \quad \text{OH O OH O} \\
\text{CH}_2\text{–CH–C–C–CH}_2 \\
\ominus \\
\ddot{O}=C=\ddot{O}
\end{array}
\longrightarrow
\begin{array}{c}
\text{HO–P(=O)–OH} \qquad \text{HO–P(=O)–OH} \\
\text{O} \quad \text{OH O OH O} \\
\text{CH}_2\text{–CH–C–C–CH}_2 \\
\text{C} \\
\ddot{O} \quad \ddot{O}^{\ominus}
\end{array}
$$

(4) Hydrolytic cleavage (nucleophilic attack by water):

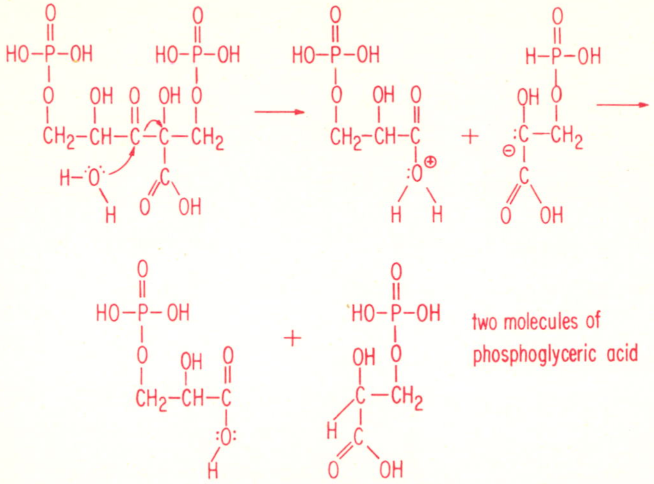

This hypothetical scheme is consistent with an observation made in the laboratories of Melvin Calvin† that the carbon atom of radioactive carbon dioxide ($C^{14}O_2$) appears first in the carboxyl group of phosphoglyceric acid.

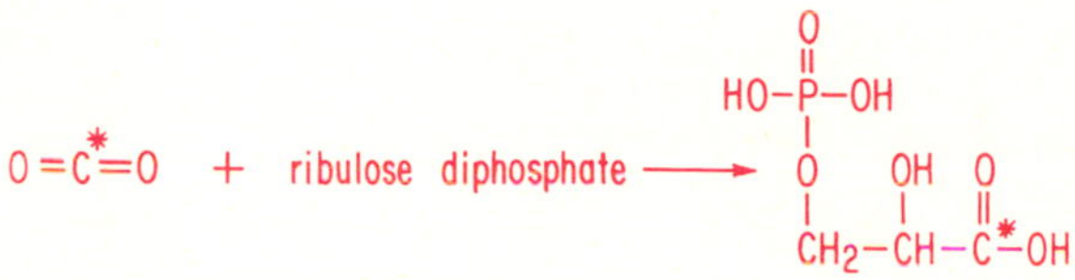

As seen earlier, the phosphoglyceric acid is reduced to phosphoglyceraldehyde by ATP and reduced nicotinamide. The glyceraldehyde may then continue to back up the anaerobic metabolism pathway through fructose and glucose and end up as food storage in starch or in a cell wall as cellulose, both of which are polysaccharides.

Photosynthesis is actually a cyclic process. Powered by sunlight, each revolution of the cycle absorbs three molecules of carbon dioxide and produces one molecule of phosphoglyceric acid (Figure 2.2).

The cycle begins with three ribulose diphosphate molecules. These pick up three carbon dioxide molecules to produce six molecules of phosphoglyceric

---

† For his pioneering work in the elucidation of the path of carbon in photosynthesis, Professor Calvin was awarded the Nobel Prize in Chemistry for the year 1961.

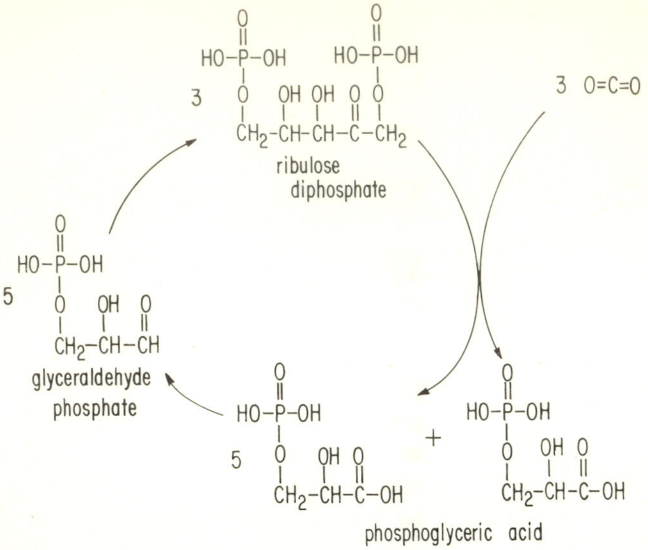

FIGURE 2.2 The photosynthetic cycle. Each revolution absorbs three carbon dioxide molecules and produces one phosphoglyceric acid molecule

C–C–C (triose)	+	C–C–C (triose)	⟶	C–C–C–C–C–C (hexose)	
C–C–C (triose)	+	C–C–C–C–C–C (hexose)	⟶	C–C–C–C–C pentose + C–C–C–C tetrose	
C–C–C (triose)	+	C–C–C–C (tetrose)	⟶	C–C–C–C–C–C–C (heptose)	
C–C–C (triose)	+	C–C–C–C–C–C–C (heptose)	⟶	2 C–C–C–C–C (pentose)	

Overall conversion: 5 C–C–C (triose) ⟶ 3 C–C–C–C–C (pentose)

Five glyceraldehyde molecules (trioses) are converted into three ribulose molecules (pentoses).

FIGURE 2.3 Summary of reactions involved in interconversion of carbohydrates in part of photosynthetic cycle

acid. Five of these remain in the cycle and are reduced to glyceraldehyde phosphate. The five glyceraldehyde phosphate molecules then interact to regenerate the original three ribulose diphosphate molecules. This completes one revolution of the cycle, and the three ribulose diphosphate molecules start another revolution by absorbing three more carbon dioxide molecules, and so on.

The first two-thirds of the cycle (now considered without mention of the phosphate groups)—from ribulose to glyceraldehyde—has already been discussed. The final third—from glyceraldehyde back to ribulose—shows even in skeleton form the remarkable intricacy of the carbon-atom pathway. The carbohydrates involved are referred to as trioses, tetroses, pentoses, etc., according to their carbon-atom content (Figure 2.3).

It should be reemphasized that the reactions in the photosynthetic cycle are all enzyme-catalyzed. The cycle requires a net energy input because of its reductive nature, and this energy is supplied by ATP formed by photo-excited chlorophyll. Photo-excited chlorophyll also supplies the energy needed to reduce NAD^+ to NADH for use as reducing agent in the cycle. In green plants, the hydrogen atoms transferred to NAD^+ in this reduction come from water.

Besides providing the reduced nicotinamide needed to make food molecules, this reaction produces oxygen. It is the source of our present oxygen-rich atmosphere. The development of photosynthesis in the primitive oceans, then, not only introduced a new food supply, it also made oxygen available to the early life forms. The presence of oxygen in turn led to the next stage in the improvement of life on earth—the development of aerobic metabolism.

The new aerobic metabolism did not displace the old anaerobic metabolism. Rather, the new was superimposed upon the old to complete and perfect it. Aerobic metabolism picks up where anaerobic metabolism leaves off. It starts with the end-product of anaerobic metabolism—pyruvic acid—and continues the oxidative degradation all the way to the end—carbon dioxide and water. The long-range significance of this is that it completes the third and final leg of the life cycle, theoretically making it possible for life to continue indefinitely on earth, at least until the sun burns out.

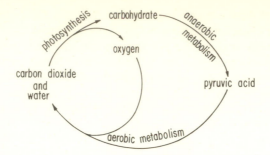

FIGURE 2.4   The life cycle

The development of aerobic metabolism also provided valuable short-range benefits. For one thing, the new metabolism produces many more ATP molecules than does the old anaerobic metabolism. Anaerobic metabolism oxidizes a molecule of glucose to two molecules of pyruvic acid and produces

Anaerobic metabolism

$$glucose \longrightarrow 2 \quad pyruvic \ acid$$
$$2 \ NAD^+ \longrightarrow 2 \quad NADH + 2H^+$$
$$2 \ (ADP + H_3PO_4) \longrightarrow 2 \quad ATP$$

Aerobic metabolism

$$6 \ O_2 + 2 \ pyruvic \ acid \longrightarrow 6 \quad CO_2 + 6 \ H_2O$$
$$30 \ (ADP + H_3PO_4) \longrightarrow 30 \quad ATP$$
$$2 \ NADH + 2H^+ + 6 \ (ADP + H_3PO_4) \longrightarrow 2 \ NAD^+ + 6 \ ATP$$

Net reaction for combined metabolisms

$$6 \ O_2 + \ glucose \ + \ 38 \ (ADP + H_3PO_4) \longrightarrow 6 \ CO_2 + 6 \ H_2O + 38 \ ATP$$

FIGURE 2.5   Summary of cellular respiration

two molecules of ATP. Aerobic metabolism further oxidizes the two pyruvic acid molecules to carbon dioxide and water, and produces thirty ATP molecules. The aerobic scheme also provides for the oxidation of the two NADH molecules produced in the anaerobic pathway. This produces six more molecules of ATP, for a grand total of 38 ATP molecules for every glucose molecule oxidized all the way to carbon dioxide and water.

Aerobic metabolism is a two-part process. The first part oxidizes pyruvic acid to carbon dioxide and reduces NAD^+ to NADH and H^+. This is called the citric acid cycle, because the process is cyclic and citric acid is an intermediate. The second part oxidizes NADH back to NAD^+, reduces oxygen to water, and produces ATP. This is called the respiratory chain.

Before entering the citric acid cycle, pyruvic acid undergoes an oxidative decarboxylation. This reaction is enzyme catalyzed, as most biological

$$\underset{\text{pyruvic acid}}{CH_3-\overset{\overset{O}{\|}}{C}-\overset{\overset{O}{\|}}{C}-OH} \quad \xrightarrow[\substack{\text{decarboxylation} \\ H_2O}]{\text{oxidative}} \quad \underset{\text{acetic acid}}{CH_3-\overset{\overset{O}{\|}}{C}-OH} \quad + \quad O=C=O$$

reactions are (see Chapter III). It does not actually produce free acetic acid. It produces a sulfur-atom derivative (a thio-ester) of acetic acid called acetyl coenzyme A (see p. 191). This thio-ester enters the citric-acid cycle by combining with oxaloacetic acid to form citric acid.

$$\begin{array}{c} \text{acetyl coenzyme A} \\ \overset{O}{\overset{\|}{}} \\ H-CH_2-C-S-R \\ + \\ O=C-CO_2H \\ | \\ CH_2-CO_2H \\ \text{oxaloacetic acid} \end{array} \quad \longrightarrow \quad \begin{array}{c} \overset{O}{\overset{\|}{}} \\ CH_2-C-OH \;+\; HS-R \\ | \\ HO-C-CO_2H \qquad \text{coenzyme A} \\ | \\ CH_2-CO_2H \\ \text{citric acid} \end{array}$$

Citric acid then isomerizes enzymatically to isocitric acid.

$$\begin{array}{c} CH_2-CO_2H \\ | \\ HO-C-CO_2H \\ | \\ H-CH-CO_2H \\ \text{citric acid} \end{array} \quad \longrightarrow \quad \begin{array}{c} CH_2-CO_2H \\ | \\ H-C-CO_2H \\ | \\ HO-CH-CO_2H \\ \text{isocitric acid} \end{array}$$

Next, isocitric acid is simultaneously oxidized and decarboxylated to produce α-ketoglutaric acid and the second molecule of carbon dioxide.

$$\begin{array}{c} CH_2-CO_2H \\ | \\ CH-CO_2H \\ | \\ HO-CH-CO_2H \\ \text{isocitric acid} \end{array} \longrightarrow \left[\begin{array}{c} CH_2-CO_2H \\ | \\ CH-CO_2H \\ | \\ O=C-CO_2H \end{array} \right] \longrightarrow \begin{array}{c} CH_2-CO_2H \\ | \\ CH-H \\ | \\ O=C-CO_2H \\ \text{α-ketoglutaric acid} \end{array} \quad + \quad CO_2$$

Only one more molecule of carbon dioxide remains to be formed to account for all three of the carbon atoms in pyruvic acid. The α-ketoglutaric acid molecule accomplishes this by undergoing an oxidative decarboxylation similar to the oxidative decarboxylation of pyruvic acid.

$$
\begin{array}{c}
CO_2H \\
| \\
CH_2 \\
| \\
CH_2 \\
| \\
O=C \\
| \\
CO_2H
\end{array}
\quad
\xrightarrow[\text{decarboxylation}]{\text{oxidative}}
\quad
\begin{array}{c}
CO_2H \\
| \\
CH_2 \\
| \\
CH_2 \\
| \\
O=C-OH
\end{array}
\quad + \quad CO_2
$$

α–ketoglutaric acid   succinic acid

The remainder of the cycle simply regenerates oxaloacetic acid by oxidizing succinic acid.

$$
\begin{array}{c}
CO_2H \\
| \\
CH_2 \\
| \\
CH_2 \\
| \\
CO_2H
\end{array}
\quad \xrightarrow{-2H} \quad
\begin{array}{c}
H \diagdown \ \diagup CO_2H \\
C \\
\| \\
C \\
HO_2C \diagup \ \diagdown H
\end{array}
$$

succinic acid H_2O fumaric acid

$$
\begin{array}{c}
CO_2H \\
| \\
HO-CH \\
| \\
CH_2 \\
| \\
CO_2H
\end{array}
\quad \xrightarrow{-2H} \quad
\begin{array}{c}
CO_2H \\
| \\
O=C \\
| \\
CH_2 \\
| \\
CO_2H
\end{array}
$$

malic acid   oxaloacetic acid

The citric acid cycle produces only one molecule of ATP directly. This happens during the oxidative decarboxylation of α-ketoglutaric acid. All the rest of the ATP molecules provided by aerobic metabolism are produced by the respiratory chain. The principal function of the citric acid cycle is to feed hydrogen atoms to the respiratory chain. The hydrogen atoms are usually in the form of NADH and H^+. Each revolution of the citric acid cycle reduces three molecules of NAD^+ to NADH. The first of these comes from the oxidative decarboxylation of pyruvic acid. The second comes from the oxidative decarboxylation of α-ketoglutaric acid. The third comes from the dehydrogenation of malic acid (Figure 2.6).

In addition, the oxidation and decarboxylation of isocitric acid reduces a molecule of nicotinamide adenine dinucleotide phosphate ($NADP^+$). This molecule differs from NAD^+ in that it contains three phosphoric acid residues

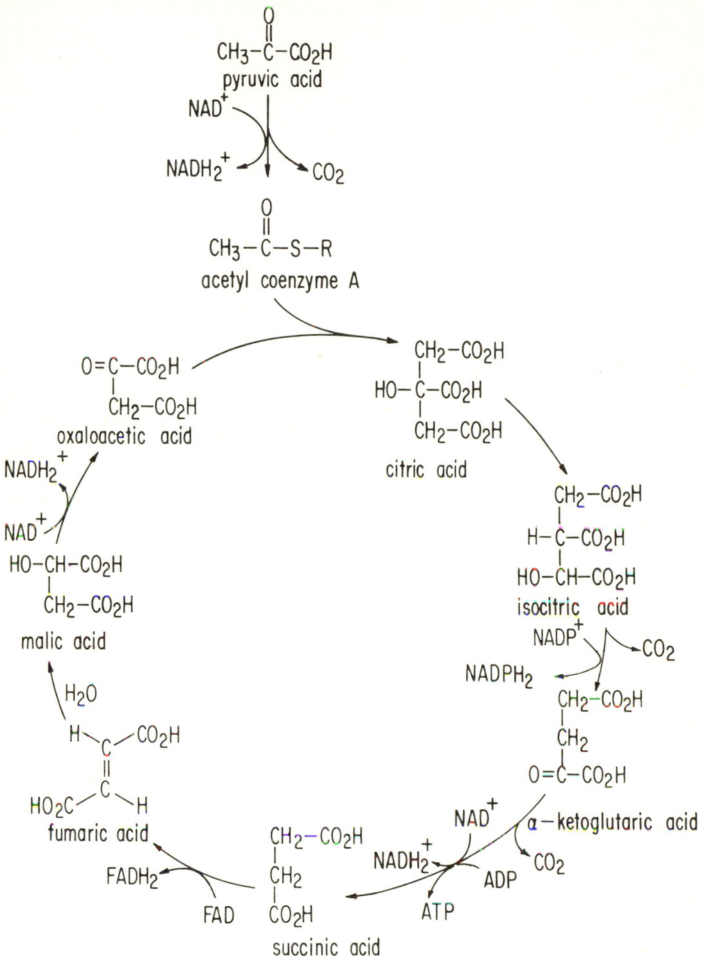

FIGURE 2.6 The citric acid cycle

instead of two. The oxidation–reduction functions of the nicotinamide portions of both molecules work the same way.

When succinic acid loses two hydrogen atoms to become fumaric acid, the oxidizing agent is not NAD^+, it is FAD (flavin adenine dinucleotide). FAD is similar to NAD^+ except that it contains the flavin oxidation–reduction system instead of the nicotinamide system. In both systems, oxidation–reduction occurs at nitrogen atoms, and in both molecules, the pendant group R is the same.

Reduction of flavin adenine dinucleotide

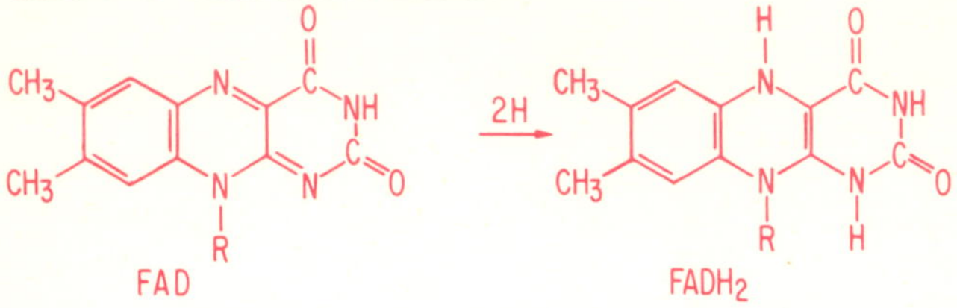

FAD   2H →   FADH₂

R-flavin is adenine–ribose–phosphate–phosphate–ribose–flavin.
R-nicotinamide is adenine–ribose–phosphate–phosphate–ribose–nicotinamide.

Each revolution of the citric acid cycle, then, produces four reduced nicotinamides and one reduced flavin. Each of these reduced molecules feeds one pair of hydrogen atoms to the respiratory chain (Figure 2.7).

The respiratory chain consists of a series of coupled oxidation–reduction reactions. It begins with the oxidation of NADH and ends with the reduction of oxygen, with molecules called cytochromes involved in intermediate steps. Each traversal of the chain produces three molecules of ATP.

The cytochromes in the respiratory chain are heme-containing proteins. Heme is a porphyrin (see p. 28) derivative which contains an iron ion.

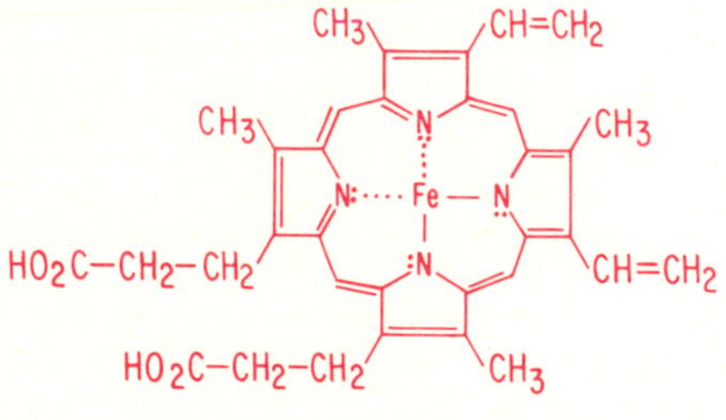

HEME

When a cytochrome is in the oxidized state, it contains ferric ion (Fe^{+++}). In the reduced state, it contains ferrous ion (Fe^{++}). The difference between these two states is only one electron. Since each hydrogen atom contains one electron, two heme units containing ferric ion are required to oxidize one molecule of $FADH_2$ to FAD.

The molecule of $FADH_2$ produced in the citric acid cycle by the oxidation of succinic acid enters the respiratory chain by reducing cytochrome b^{+++} (see Figure 2.7). This reaction sequence bypasses NAD^+ and therefore the pair of hydrogen atoms passed into the respiratory chain by $FADH_2$ results in the production of only two ATP molecules. This deficit is made up for, however, by the "extra" ATP molecule formed in the oxidative decarboxylation of α-ketoglutaric acid. Thus, each turn of the citric acid cycle generates four

$$NADH + H^+ + FAD \xrightarrow[]{ADP \quad ATP} NAD^+ + FADH_2$$

$$FADH_2 + 2\ \text{cytochrome}\ b^{+++} \longrightarrow FAD + 2\ \text{cytochrome}\ b^{++} + 2\ H^+$$

$$2\ \text{cytochrome}\ b^{++} + 2\ \text{cytochrome}\ c^{+++} \xrightarrow[]{ADP \quad ATP}$$
$$2\ \text{cytochrome}\ b^{+++} + 2\ \text{cytochrome}\ c^{++}$$

$$2\ \text{cytochrome}\ c^{++} + 2\ \text{cytochrome}\ a^{+++} \longrightarrow$$
$$2\ \text{cytochrome}\ c^{+++} + 2\ \text{cytochrome}\ a^{++}$$

$$2\ \text{cytochrome}\ a^{++} + 2\ H^+ + 1/2\ O_2 \xrightarrow[]{ADP \quad ATP}$$
$$2\ \text{cytochrome}\ a^{+++} + H_2O$$

Net reaction

$$NADH + H^+ + 1/2\ O_2 + 3\ ADP \longrightarrow NAD^+ + H_2O + 3\ ATP$$

FIGURE 2.7 The respiratory chain

reduced nicotinamides, one reduced flavin, and one ATP. The four pairs of hydrogen atoms from the reduced nicotinamides make twelve ATP molecules in the respiratory chain and the pair of hydrogen atoms from the reduced flavin makes two more ATP molecules. These plus the single ATP molecule produced directly by the citric acid cycle amount to fifteen molecules of ATP for every molecule of pyruvic acid oxidized to carbon dioxide and water.

The anaerobic oxidation of a molecule of glucose to two molecules of pyruvic acid generates two reduced nicotinamide molecules and two ATP molecules. With the advent of aerobic metabolism, the two reduced nicotin-amide molecules can pass their hydrogen atoms into the respiratory chain to manufacture six more molecules of ATP. The combination of anaerobic and aerobic metabolisms thereby makes a total of thirty-eight molecules of ATP for every molecule of glucose oxidized to carbon dioxide and water.

But the real marvel of this invention of nature is the high degree of efficiency attained in trapping and storing the oxidative energy. It takes about twelve kilocalories to convert a mole of ADP into ATP. The combustion of glucose to carbon dioxide and water releases about 690 kilocalories per mole. Thirty-eight moles of ATP represents 456 kilocalories of energy trapped and stored, amounting to an efficiency of 66%. Conventional steam-driven power plants are only about half this efficient.

SUMMARY

In this chapter we have seen that the appearance of the first life forms on the primitive earth must have occurred through highly complex organizations of individual abiotically generated organic molecules. This statistically unlikely process may have begun with the formation and separation of non-living cell-like structures resembling coacervate droplets. However it happened, the first living things developed a method for obtaining biological energy by oxidizing pre-existing organic molecules in the absence of oxygen. The next milestone in the history of life was the development of photosynthesis, whereby an organism became able to synthesize its own food by trapping and using the energy of sunlight. This caused a profound change in the earthly environment because it slowly generated an atmosphere containing free oxygen. The availability of free oxygen paved the way for the ultimate development on the molecular level—aerobic metabolism. This final breakthrough set up the dynamic balance of nature we know today as the life cycle—the process in which plants produce carbohydrates and other food molecules from carbon dioxide and water, and in which animals and other aerobes live by the energy obtained in oxidizing carbohydrates and other food molecules back to carbon dioxide and water. And so life goes on.

SOURCE MATERIALS AND SUGGESTED READING

Bassham, J. A., and M. Calvin, 1957. *The Path of Carbon in Photosynthesis*, Prentice Hall, Englewood, New Jersey.

Calvin, M., and M. Jorgenson, intro. by, 1968. *Bio-organic Chemistry* (Readings from *Scientific American*), W. H. Freeman and Co., San Francisco and London.

Fox, S. W., ed., 1965. *The Origins of Prebiological Systems and of their Molecular Matrices*, Academic Press, New York.

Margulis, L., ed., 1970. *Organic Chemistry: the Fossil Record in Origins of Life*, Gordon and Breach.

CHAPTER III

NUCLEIC ACIDS AND PROTEIN SYNTHESIS
The Code of Life

WE HAVE seen how an organism obtains and stores the energy it needs for its life activities. These energy-producing and energy-trapping reactions are catalyzed by enzymes, as are the hundreds of other reactions that take place within any living thing. To catalyze these reactions as needed, the organism must synthesize the appropriate enzymes.*

In general, each individual reaction requires its own specific enzyme. For example, the enzyme which catalyzes the oxidative decarboxylation of pyruvic acid is different from the enzyme which catalyzes the oxidative decarboxylation of α-ketoglutaric acid, even though the reactions are very similar.

Enzymes, then, are extremely specific or selective in their catalytic action. Since enzymes are proteins, this high degree of specificity arises primarily from the sequence of amino-acid residues in the protein chain. Therefore, an organism must be able to synthesize hundreds of specific protein molecules, each with its own unvarying sequence of amino-acid residues. Each enzyme may contain hundreds of amino-acid residues, and each cell must contain all the information required to replenish its supply of any of its enzymes as needed. It seems a lot to ask of a "simple" single-celled organism such as a bacterium, even in fact a lot to ask of a complex many-celled creature such as man, but nature has provided an ingenious scheme to enable even the lowest forms of life to accomplish this wondrous feat.

How is it done? Every cell contains a complete set of "blueprints" for making its own enzymes. These blueprints are stored in the nucleus of the cell in the form of molecules of deoxyribonucleic acid (DNA). The DNA molecules contain the plans for making particular protein molecules. When the cell needs a particular protein molecule, the appropriate molecule of DNA makes a copy of the plan in the form of a molecule of ribonucleic acid (RNA), called

---

* Enzymes are proteins, each of which is characterized by very specific chemical and geometric properties. Not all proteins are enzymes. Some proteins, such as those of muscle and skin, perform structural functions and possess no enzymatic ability.

messenger RNA. The messenger RNA leaves the nucleus and proceeds to pick up small molecules of another kind of RNA called transfer RNA. These small transfer-RNA molecules attach themselves to a specific site along the long molecule of messenger RNA. Each small transfer-RNA molecule carries a specific amino acid with it. As adjacent molecules of transfer RNA line up along the messenger RNA, adjacent amino acids couple with each other to form peptide bonds. As the peptide bonds form, the amino acids release the transfer RNA, and the transfer RNA in turn releases the messenger RNA. After every site along the messenger RNA molecule has been occupied by a molecule of transfer RNA, beginning at one end and continuing stepwise to the other end, the complete molecule of protein will have been synthesized. Furthermore, and most important, the molecule of protein will have an amino-acid sequence dictated precisely by the base sequence (see below) of the original molecule of DNA back in the cell nucleus (Figure 3.3, p. 51).

How does this happen? Many details of the process are still not known, but much of the mystery has been solved since Francis Crick and James Watson proposed the double-helix structure of DNA in 1953.

The deoxyribonucleic acids are long-chain polynucleotides containing hundreds of nucleotides. Each nucleotide contains a purine or pyrimidine base, a molecule of 2-deoxyribose, and phosphoric acid. If the base is a purine, it

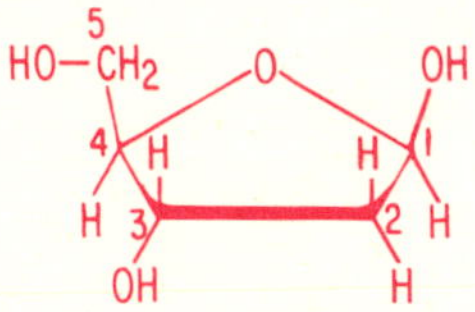

2–deoxyribose

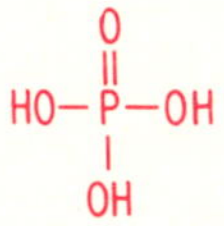

phosphoric acid

purine bases

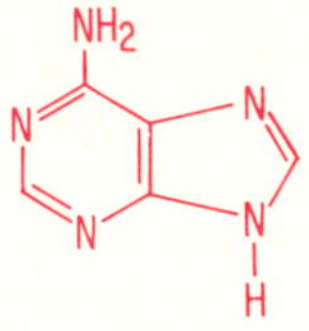

adenine

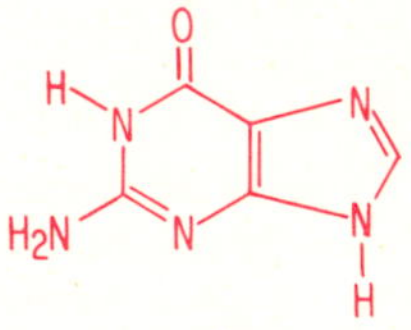

guanine

pyrimidine bases

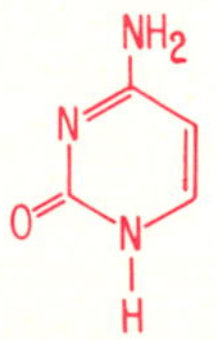

cytosine

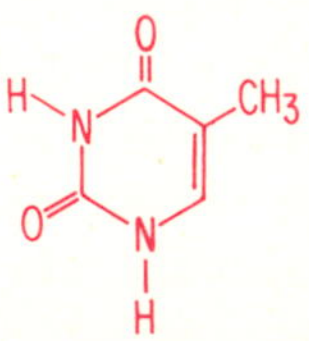

thymine

must be either adenine or guanine. If it is a pyrimidine, it must be either cytosine or thymine.

The polynucleotide has a sugar-phosphate backbone, and the purine and pyrimidine bases are attached to the sugar residue at the 1-position (the aldehydic carbon atom). Since deoxyribose does not have a hydroxyl group at the 2-position, the phosphoric acid links the 3-position of one sugar to the 5-position of the next sugar.

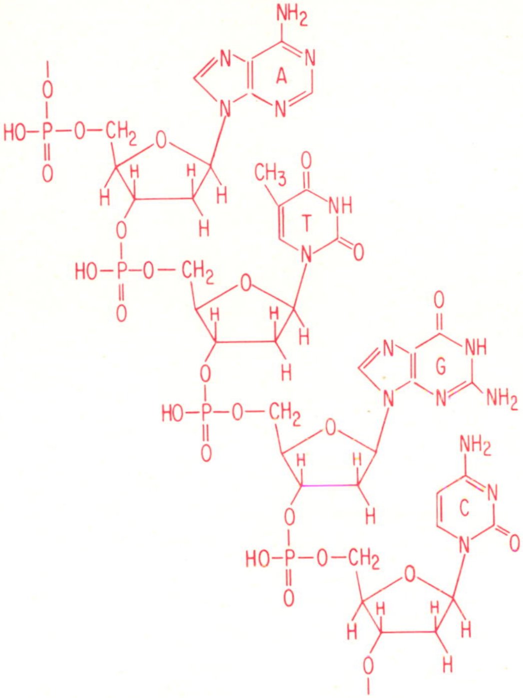

Random Portion of DNA Chain
Purine bases: Pyrimidine bases:
    A = Adenine   T = Thymine
    G = Guanine   C = Cytosine

DNA exists in the cell nucleus in the form of paired strands twisted into a double helix (Figure 3.1). Each purine or pyrimidine base is directed inwards toward the axis of the helix and is hydrogen-bonded to another purine or pyrimidine base on the other strand. For uniformity of spacing between strands, a purine base is always hydrogen-bonded to a pyrimidine base, and vice versa. Also, because of the donor-acceptor natures of the hydrogen-bonding groups, adenine is always hydrogen-bonded to thymine, and guanine

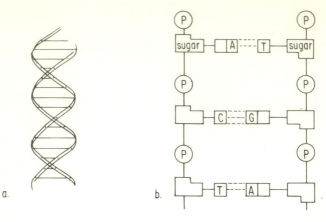

FIGURE 3.1   DNA structure.* (a) Double helix; (b) base pairing between strands

is always hydrogen-bonded to cytosine. Therefore, the number of adenine residues always equals the number of thymine residues, and the number of guanine residues always equals the number of cytosine residues.

Hydrogen bonding
between adenine (A)
and thymine (T)

Hydrogen bonding
between guanine (G)
and cytosine (C)

Base pairing in DNA is determined by hydrogen bonding. Adenine and cytosine cannot form hydrogen bonds with the same stability as adenine and thymine. In order for the adenine and cytosine to form hydrogen bonds, it is

* In Chapter XI we will consider the x-ray diffraction technique which allowed Watson and Crick to postulate the double helix.

necessary for the 6-amino group of the adenine to tautomerize* to the 6-imino group as is shown in the diagrams below. This is not as stable a configuration for the adenine to assume. Hence, A–T pairing occurs in preference to that of A–C.

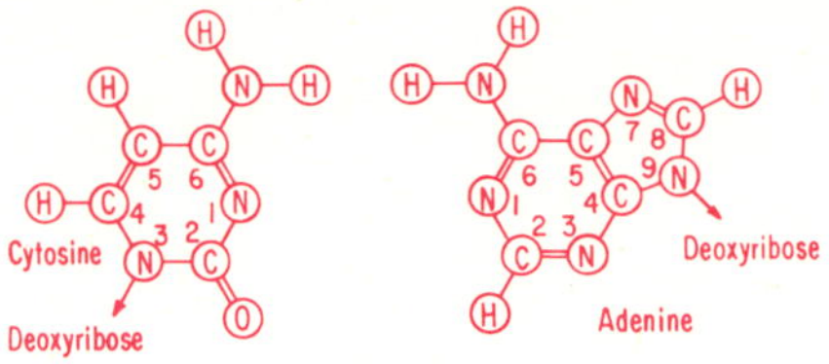

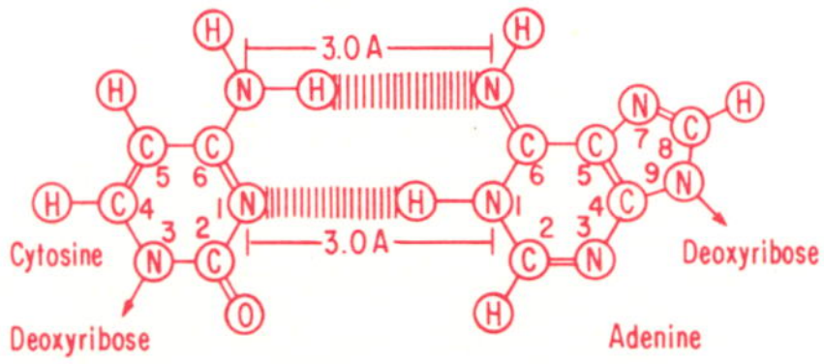

Similar considerations can be used to show why guanine and cytosine rather than guanine and thymine form the preferred base pairs.

The strict requirement of base pairing is important because it provides a mechanism for the faithful duplication of a pair of strands. DNA duplicates itself prior to cell division to provide each daughter cell with a complete set of DNA molecules. It does this by breaking the hydrogen bonds between strands and then forming new hydrogen bonds to new nucleotide partners—adenine to thymine and guanine to cytosine (Figure 3.2). The new nucleotides then form sugar-phosphate bonds between each other to form a new chain. The result is an exact reproduction of the original paired strands. This is the molecular basis of heredity. Any error in the duplication process causes a mutation.

---

* Tautomerism is a process of isomerization in which a proton shifts from atom 1 to atom 3. In the case of adenine, the following structures are involved:

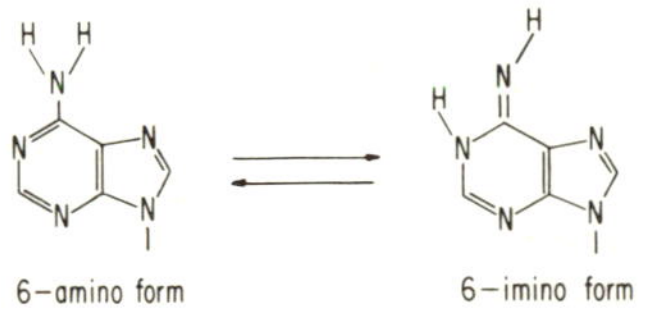

6−amino form   6−imino form

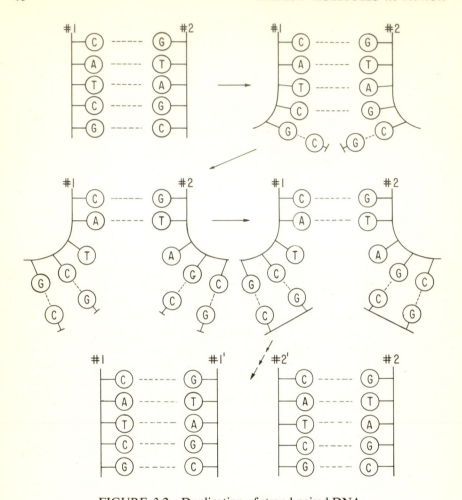

FIGURE 3.2   Duplication of strand-paired DNA

The replication of DNA is better understood than is the synthesis of a molecule of messenger RNA. The chief mystery in the messenger RNA synthesis is that only one strand of RNA is produced for each unit of double-stranded DNA. The single molecule of messenger RNA produced is an exact template of one of the paired DNA strands, but not of the other. The messenger RNA migrates out of the nucleus and becomes attached to a ribosome.*

* Ribosomes are large organized particles in the cytoplasm. These globular structures are rich in protein and RNA and are the location of protein synthesis in the cell.

An RNA molecule is similar to a DNA molecule except that RNA contains ribose instead of deoxyribose, and the base uracil instead of thymine. (Uracil is simply demethylated thymine.) Like thymine, uracil always base-pairs with

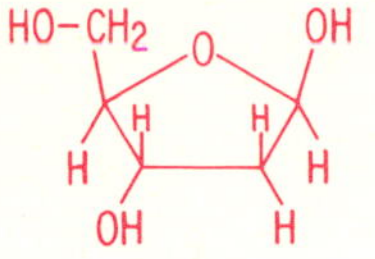

2-deoxyribose 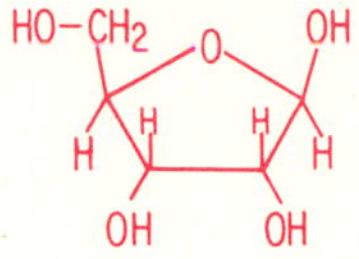   ribose

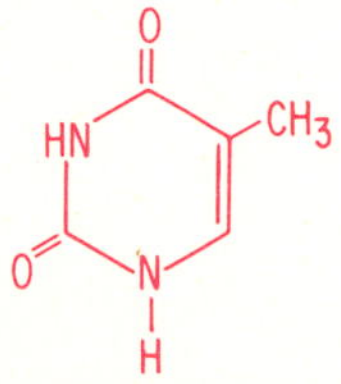

thymine 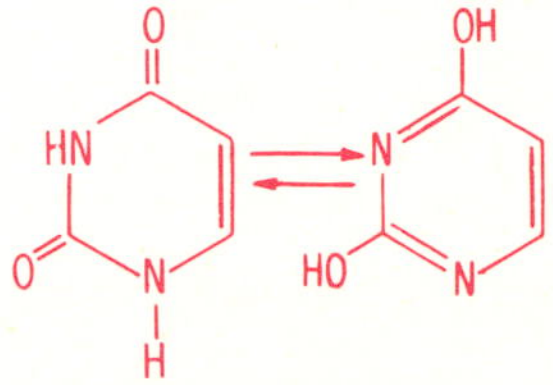   uracil

adenine. The remainder of the structures of DNA and RNA are the same, that is, the phosphoric acid residue connects the 3-position of one sugar to the 5-position of the next sugar. An important consequence of having a hydroxyl group at the 2-position in the ribose residue in RNA is that it makes RNA much more susceptible to mild alkaline hydrolysis than DNA is. The reason for this is that the 2-hydroxyl participates in the alkaline hydrolytic cleavage of RNA.

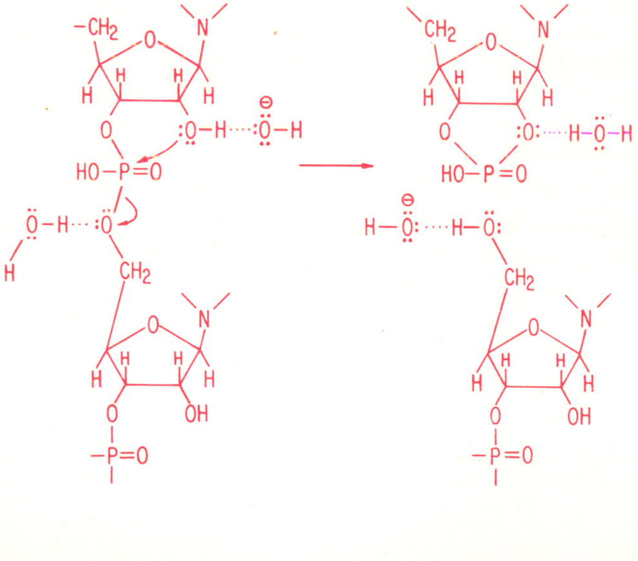

This participation leads to an intermediate cyclic 2′,3′-diester* which then hydrolyzes either to the 2-phosphate or also to the 3-phosphate. Both phosphates form from alkaline hydrolysis even though RNA itself contains a free 2-hydroxyl group.

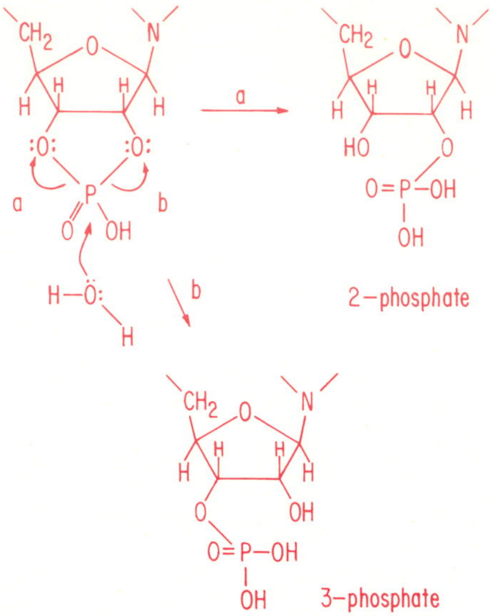

Since DNA has no 2-hydroxyl, this kind of near-neighbor assistance during hydrolysis is impossible. The practical consequence of this is that DNA is much longer lived than RNA, and RNA must be replenished as needed over the long run for protein synthesis.

The complementary base-pairing principle results in a molecule of RNA reflecting the base sequence of a DNA molecule exactly. Thus messenger RNA contains an adenine residue where DNA contains thymine, a cytosine residue where DNA contains guanine, a guanine residue where DNA contains cytosine, and a uracil residue where DNA contains adenine. But how is this translated into a specific sequence of amino acid residues in a protein molecule? This is the fascinating part of the puzzle.

The sequence of bases in messenger RNA must somehow control the sequence in which amino acids are bonded together to form a protein chain.

---

* In contrast to the 2′,3′-diester mentioned above, the 3′,5′-cyclic adenosine monophosphate has been implicated in a multiplicity of hormonal effects. It is involved with stimulation and/or retardation of the respiratory cycle (see Chapter II). For discovering the importance and ubiquitousness of this single substance, Dr. Earl Sutherland was awarded the Nobel Prize in Medicine for 1971.

Messenger RNA contains four different kinds of bases and a protein usually contains twenty different kinds of amino acids. A single base then could not control the positioning of a specific amino acid in a protein chain, because the four bases could this way control only four amino acids. Similarly, combinations of two adjacent bases could control a maximum of sixteen amino acids, since only sixteen different combinations of two adjacent bases are possible (see below). (The combination AC differs from the combination CA because of the directionality of the 3,5-diester links to the phosphoric acid residue.)

To control specifically the sequence of twenty amino acids, a combination of at least three bases on messenger RNA is required. This provides sixty-four possible combinations. These triads of bases on messenger RNA (called

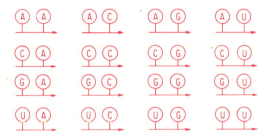

codons) act as specific docking sites for complementary triads on transfer RNA* (anticodons). Docking specificity arises from the specificity of the hydrogen bonding between adenine and uracil, and between cytosine and guanine. The complementary triad of bases on transfer RNA occurs on a loop near the middle of the transfer RNA chain. The non-looped portion of the chain folds back on itself and is double-stranded like DNA. One of the chain

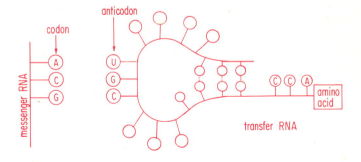

* In the course of protein synthesis within the cell, the transfer RNA becomes bonded to messenger RNA which is temporarily attached to a ribosome (see Figure 3.3, p. 51).

ends is always slightly longer than the other, however, and this loose end carries the amino acid. Before becoming attached to a specific tRNA molecule, an amino acid is first activated by enzymatically reacting with ATP to form AA–AMP. Surprisingly enough, the loose end always has the same final base sequence (CCA) regardless of which amino acid is at the end. (The amino acid is attached to the terminal ribose by an ester bond.) Clearly, the docking triad must somehow control the choice of amino acid at the loose end of the molecule, but how it does this remains one of nature's secrets.

Another secret which remained hidden until recently has to do with the problem of choosing the single correct triad from a given sequence of bases. For example, the sequence ACGU contains two triads—ACG and CGU. Only one of these can be the "true" triad corresponding to the specific amino acid scheduled to enter the particular protein for which the messenger RNA is programmed. Choice of the "wrong" triad to begin with would lead to an entire sequence of "wrong" triads, and consequently all the "wrong" amino acids, which would mean a completely "wrong" protein.

This kind of error does not occur in nature, and an ingenious explanation for this was proposed by Francis Crick, J. S. Griffith, and L. E. Orgel in the late 1950s. Assuming that the amino-acid code is based on consecutive triads (resembling three-letter words), they proposed that the code is non-overlapping. This means that in the hypothetical repeating polynucleotide UGAUGAUGA, only one of the three triads (UGA, GAU, and AUG) has any "meaning." The other two are "meaningless" in that they do not correspond to any complementary triad on transfer RNA, and therefore no transfer RNA would ever dock at that triad.

If this were true, they reasoned, the triads AAA, CCC, GGG, and UUU could not be real triads because a repetition of any one of them could produce overlapping and hence a false start to protein synthesis. Thus, the allowable combinations of the four RNA bases drops from sixty-four to sixty. Of these sixty combinations, they reasoned further, two-thirds must be meaningless (to avoid overlapping). Therefore, only one-third of the sixty will be real triads. This number (twenty) exactly matches the number of different kinds of amino acids found in proteins. As neatly as this proposal fitted in with what was

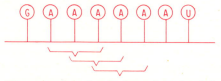

Three overlapping choices of AAA triad
would lead to different amino acid
sequences in protein.

known at the time, it is wrong. Subsequent work has shown that there are more than twenty meaningful triads. Present evidence indicates that choice of the "wrong" triad is avoided by beginning transfer-RNA docking at one of the ends of messenger RNA rather than at a random site along the chain.

How then does one find out which triad is a code word for which amino acid? The most direct method is to make a synthetic polynucleotide with a known base sequence, allow this molecule to act as messenger RNA in a protein synthesis, and then determine the amino-acid sequence of the protein.

For example, a polynucleotide containing only one kind of base can be made from the nucleotide (as a diphosphate) and an enzyme called polynucleotide phosphorylase, isolated by S. Ochoa and H. Grunberg-Manago. If the base is uracil, the synthetic polynucleotide is called poly U (UUUUUUUU...). In the presence of a mixture of transfer RNA molecules and enzymes and other cell constituents, poly U incites the synthesis of a polypeptide containing only one kind of amino-acid residue—that of phenylalanine. Clearly then, the triad UUU is the code word for phenylalanine.

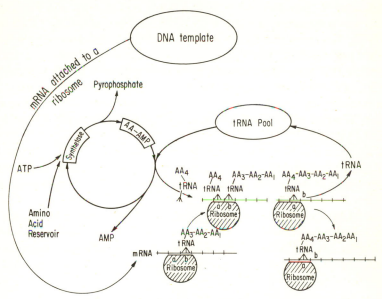

FIGURE 3.3   Representation of stepwise protein synthesis. In this figure AA refers to amino acid; ATP denotes adenosine triphosphate; AMP represents adenosine monophosphate; AA-AMP is amino acid adenylate; tRNA refers to transfer ribonucleic acid; mRNA represents the messenger ribonucleic acid and a, b, etc. represent attachment sites on the ribosomes for peptide bond formation. [Adapted from M. Yčas, *The Biological Code*, North-Holland Publishing Co., Amsterdam, 1969, p. 9]

This method can be extended (and has been by Ochoa and by Nirenberg) to mixed-base code words. For example, the polymerization of uracil nucleotide can be initiated with the dinucleotide AUUUUUU... This polynucleotide promotes the synthesis of polyphenylalanine with a single tyrosine residue at one end. Thus, the code word for the amino acid tyrosine must be AUU. This work has led to the tabulation of code words for each of the twenty amino acids. Most amino acids appear to have more than one code word.

Present thought proposes, therefore, that a particular enzyme is synthesized in a linear stepwise sequence of amino-acid coupling reactions beginning at one end of a messenger RNA molecule and continuing from triad to triad to the other end of the messenger RNA molecule, where the completed protein chain is finally released. As each peptide bond forms, the "spent" transfer RNA is released by messenger RNA. This allows fresh transfer RNAs carrying amino acids to begin synthesizing a second molecule of protein before the first protein molecule is finished (Figure 3.3).

SUMMARY

In this chapter we have covered the composition, structure and function of nucleic acids. We have seen how these important molecules are involved in genetics and protein synthesis. The code of life was explained in its simple elegance by the steps involved in protein synthesis. We saw that the code is made up on a DNA template. The message is carried by a messenger RNA which reads one strand of the DNA template. Messenger RNA then migrates to and becomes attached on the surface of a ribosome. Transfer RNA molecules, carrying specific amino acids, bond to codon sites of the messenger RNA. Thus, amino acids are arranged in a specific linear sequence which is determined by the messenger NRA. These amino acids join together to form a specific protein.

With the above system, it is possible to understand mutations and errors in protein synthesis. We can also marvel at the efficiency and reproducibility of the myriad of reactions controlled by the genes.

SOURCE MATERIALS AND SUGGESTED READING

Borek, E., 1965. *The Code of Life*, Columbia University Press, New York and London.

Calvin, M., and M. Jorgenson, intro. by, 1968. *Bio-organic Chemistry* (Readings from Scientific American), W. H. Freeman and Co., San Francisco and London.

Jukes, T. H., 1966. *Molecules and Evolution*, Columbia University Press, New York and London.

Watson, J. D., 1970. *Molecular Biology of the Gene*, Benjamin, New York.

Yčas, M., 1969. *The Biological Code*, North-Holland Publishing Co., Amsterdam.

CHAPTER IV

GIANT MOLECULES
The Magic Materials

THERE ARE three kinds of giant molecules which have been around for billions of years, as long as life has existed on earth. These are the nucleic acids, the proteins, and the polysaccharides. Each kind has a dual function in living things.

The nucleic acids, especially the deoxyribonucleic acids, are really the master molecules of life. They are responsible for heredity—the faithful continuation and proliferation of the many species of living things. They accomplish this by their ability to replicate themselves exactly within the cell nucleus and so, when a cell divides to form two daughter cells, each daughter cell contains the same set of nucleic acids which the parent cell contained. The second function of these master molecules is to direct protein synthesis within the cell, as has already been seen.

The proteins have either of two functions as another set of giant molecules of biological action. Some can act as catalysts to run the chemical reactions that keep a cell alive. Others make up muscle fiber which converts the chemical energy of adenosine triphosphate (ATP) into the mechanical energy of muscle contraction. This provides an organism with the ability to pump blood, to catch food, and to move from one place to another. Protein molecules also play an important role in the animal kingdom—that of providing protection from the environment. Skin and hide, fur and feathers are all composed of protein molecules. Man has long made use of these animal materials to add to his own comfort.

Polysaccharides are sugar polymers. As might be expected from what has been said about the glucose molecule being a primary food molecule, one of the functions of polysaccharides is to act as a food storehouse. Plants polymerize glucose into giant molecules of starch, and pack this into their seeds to provide food for tiny seedlings to grow until they can begin photosynthesizing their own food. Animals polymerize glucose into giant molecules of glycogen and store these in the liver, thus maintaining a reserve against lean times. The second principal function of polysaccharides, seen most abundantly in the plant kingdom, is a structural one. All plant cells polymerize glucose into giant molecules of cellulose to build cell walls. This protects the liquid cell contents, and in the aggregate, provides the mighty oak with the strength to stand against hurricane winds. Besides wood, the plant product cotton is also cellulose. Again, man has long availed himself of these materials to further his own well-being.

Nature held the secret of the giant molecule until only recently. Try as he might, man in the role of organic-molecule builder could do nothing to match the materials, such as wool, silk, cotton, and wood, which nature provided in profusion. Finally, inroads were made, few and slowly at first, and mostly as results of accidents, until today, when synthetic giant molecules are produced in daily tonnage quantities, tailor-made for specific properties and end-uses, surpassing by far the capabilities of natural materials. How did this marvellous technology come about? It happened mainly through tedious, patient trial and error, and through keen observation of the unexpected, by men dedicated to improving the lot of life while satisfying their natural curiosities.

The synthetic giant molecule story goes back more than 100 years to a kitchen "laboratory" in Basel, Switzerland. Christian Schoenbein, a chemistry professor, was boiling nitric and sulfuric acids, and accidentally spilled some on the floor. He quickly wiped it up with a cotton apron belonging to his wife. Realizing that the highly corrosive acids would soon decompose the apron, he sought to avoid that misfortune by rinsing the apron acid-free with water. He then hung the apparently undamaged apron near the hot stove to dry. Very shortly, the apron burst into flame and disappeared. Christian Schoenbein had discovered guncotton.

He had unwittingly converted the hydroxyl groups of the cellulose molecules in the cotton apron into nitrate esters. The cellulose had become nitrocellulose, a hitherto unknown giant molecule three times as powerful an explosive as gunpowder.

cellulose
(polyglucose)

guncotton
(cellulose tri-nitrate)

The presence of the sulfuric acid in the spilled acid mixture was fortuitous in that sulfuric acid is an excellent catalyst for esterification reactions (see appendix). This catalytic power arises from the exceptional strength of sulfuric acid. As a result, it acts as a protonating agent even for nitric acid. A nitration mixture of nitric and sulfuric acids contains nitronium ions, which form when protonated nitric acid molecules lose water. The nitrogen atom in the nitronium ion is rapidly attacked by a nucleophilic agent such as an alcohol.

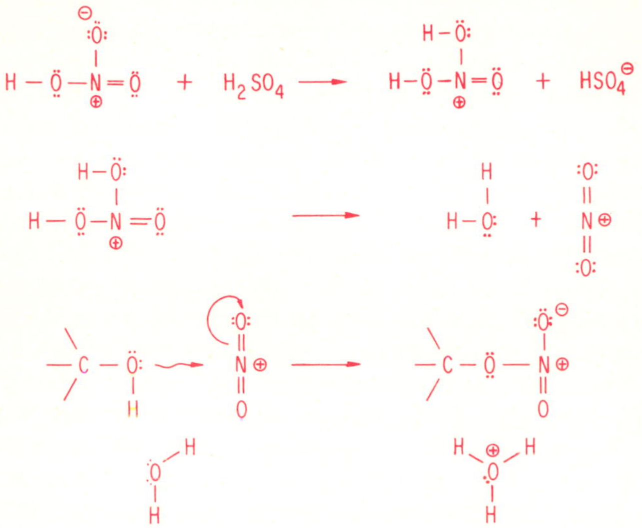

The guncotton which Christian Schoenbein stumbled upon 125 years ago was also better than gunpowder in that it produced much less smoke. Air quality on the battlefield improved which allowed contending parties, for better or worse, to have a clearer view of each other.

As useful as the discovery was, however, it strictly speaking did not represent a man-made giant molecule. It was merely a synthetic modification of a pre-existing giant molecule made by nature. Nevertheless, it led to discoveries of several very useful cellulose-based new substances.

The first widely used material of this type was celluloid. This was probably the first man-made material worthy of the name "plastic". But the way was paved for the appearance of celluloid by an earlier discovery. Almost immediately after the first preparation of guncotton, a young Frenchman named Louis Menard found that alcohol and ether dissolve guncotton and produce a non-fibrous clear material after evaporation. This material was called collodion, and it found little use except as a protective coating for minor wounds.

Louis Menard's transformation of guncotton into collodion made nitrocellulose much easier to work with experimentally. During the 1860s a printer in Albany, New York, named John Wesley Hyatt began compounding collodion with other materials in the hope of making a new material to substitute for ivory in billiard balls. Because ivory was scarce, an American company had offered $10,000 for an ivory substitute, and John Hyatt had his eye on this prize. He found that a combination of collodion and camphor produced a horn-like material which softened on warming and could be molded or extruded into many different shapes. (Camphor is a bicyclic ketone produced by the camphor tree and used in "moth balls.")

$$CH_3 \quad CH_3$$
$$CH_3$$
$$=0$$

camphor

This first synthetic plastic was called celluloid. Unfortunately, it was too brittle to use in billiard balls, but it became popular in combs, brush handles, ping-pong balls, photographic film, men's detachable collars, and many other items. John Hyatt continued his search for an inexpensive billiard-ball material.

In 1875, in Sweden, Alfred Nobel invented blasting gelatin by mixing collodion and nitroglycerin. Nitroglycerin is like nitrocellulose in that both are nitrate esters of alcohols, but nitroglycerin is not a polymer. Glycerin (or glycerol) is simply a trihydroxypropane.

$$CH_2-OH \qquad\qquad CH_2-O-NO_2$$
$$| \qquad\qquad\qquad\qquad |$$
$$CH-OH \qquad\qquad CH-O-NO_2$$
$$| \qquad\qquad\qquad\qquad |$$
$$CH_2-OH \qquad\qquad CH_2-O-NO_2$$

glycerin   nitroglycerin

Blasting gelatin is also called solidified nitroglycerin (S.N.G.). It is a rubbery elastic material—the most powerful explosive sold commercially—used for blasting rock. Nitroglycerin itself is an extremely hazardous liquid. It detonates with violence at the slightest shock. The number of oxygen atoms per molecule is more than enough to convert the carbon and hydrogen atoms to gaseous carbon dioxide and water. The simultaneous release of elemental nitrogen and

$$CH_2-ONO_2$$
$$|$$
$$2 \quad CH-ONO_2 \longrightarrow 6\ CO_2 \;+\; 5\ H_2O \;+\; 3\ N_2 \;+\; \tfrac{1}{2}\ O_2$$
$$|$$
$$CH_2-ONO_2$$

oxygen gases adds to the sudden volume increase and produces a tremendous shock wave.

The extreme sensitivity of nitroglycerin was successfully moderated by Alfred Nobel in 1866, 20 years after nitroglycerin was first prepared. He found that kiesulguhr, a diatomaceous earth, could adsorb three times its weight of nitroglycerin and still appear dry. The absorbed nitroglycerin is still explosive, but much less sensitive. This mixture is called dynamite.

Then in 1889, a French physiologist named Hilaire de Chardonnet extruded collodion into a fine thread and rewove it into a fabric. This new form of nitro-cellulose resembled silk which is a protein fabric made from silkworm cocoons. Chardonnet's artificial silk was received warmly, but then rejected heatedly, since a small spark would send it the way of Schoenbein's apron. It was still guncotton.

Success in making an acceptable synthetic fabric came in the form of a less explosive ester of cellulose—the acetate instead of the nitrate. When cotton is heated with acetic anhydride, its hydroxyl groups are acetylated, and the giant cellulose molecules become giant cellulose acetate molecules. Like cellulose

$$
\begin{array}{c}
\diagdown \\
-\,C\,-\,\ddot{O}\, + \quad CH_3\,-\,\overset{\displaystyle O}{\overset{\|}{C}}\,-\,O\,-\,\overset{\displaystyle O}{\overset{\|}{C}}\,-\,CH_3 \quad \longrightarrow \\
\diagup \qquad | \\
H \qquad\qquad \text{acetic anhydride}
\end{array}
$$

$$
\begin{array}{c}
\diagdown \qquad\qquad \overset{\displaystyle O}{\overset{\|}{}} \qquad\qquad\qquad \overset{\displaystyle O}{\overset{\|}{}} \\
-\,C\,-\,O\,-\,C\,-\,CH_3 \;+\; CH_3\,-\,C\,-\,OH \\
\diagup \qquad \text{acetate} \qquad\qquad \text{acetic acid}
\end{array}
$$

nitrate, cellulose acetate is not a fully synthetic polymer, but rather a synthetic modification of a natural polymer. Cellulose acetate, also known as acetate rayon, is obtained as a white powder from the acetic anhydride treatment. The dry material is dissolved in acetone, extruded* into a fiber, and then woven into a fabric.

Phenolic Resins

The first fully synthetic polymer made deliberately by man for a purpose— not made accidentally as a tarry product from a reaction that "went wrong"— was made in 1907 by Leo Baekeland. He knew that chemists in the past had

---

* The concentrated polymer solution is forced through a fine orifice (called a spinaret). This process is called extrusion. The extrudate is pushed through a solvent in which the acetone is soluble but not the polymer. A fiber results.

produced unwanted resinous materials—materials impervious to acids and
bases and which could not be dissolved out of the laboratory apparatus in
which they formed. The German chemist Adolf von Baeyer in 1871 described
the formation of a troublesome gray resinous material from phenol and formaldehyde (of prebiotic synthesis fame; see Chapter I). Baekeland repeated this
work, modifying the reaction conditions to encourage rather than discourage
polymerization, until he came up with what he wanted—a clear, hard, amber-
like resin molded in the shape of the reaction vessel. He named it Bakelite
and received a patent for the invention in 1909. This new, completely man-made
material achieved instant success and is used in a wide variety of everyday
articles, including buttons, electrical insulators, cameras, distributor heads,
radio and telephone equipment, etc.

Another scientist, Dr. L. V. Redman claimed to have produced the same
polymeric system independently and naturally called his material Redmanol.
Bitter competition and long, costly litigation ensued. In 1922 the courts
finally awarded the patents to Baekeland who licensed many companies to
produce Bakelite. The basic patents ran out in 1926. Union Carbide absorbed
the Bakelite Company in 1939.

Bakelite is a giant cross-linked molecule containing phenol units linked
and locked in a three-dimensional array by methylene ($-CH_2-$) groups. How

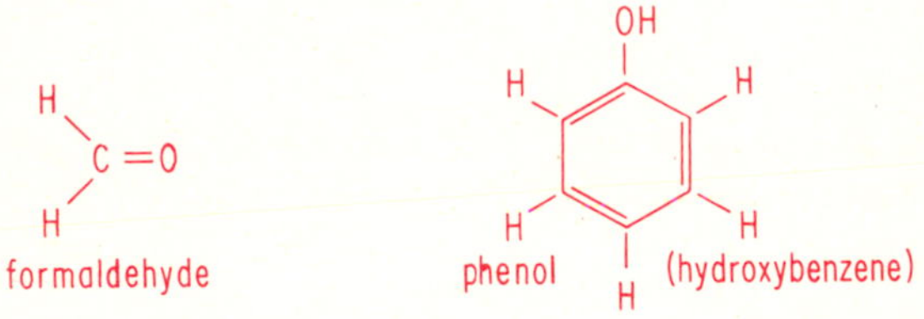

formaldehyde   phenol   (hydroxybenzene)

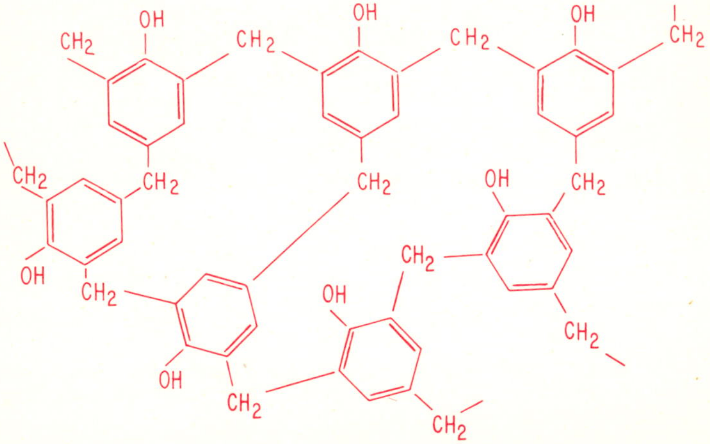

Bakelite

does this insoluble, infusible mass of covalently bonded atoms arise from a simple mixture of tiny phenol and formaldehyde molecules?

First, the phenol molecule contains three "activated" carbon atoms—the so-called *ortho*- and *para*-carbon atoms. These three carbon atoms are

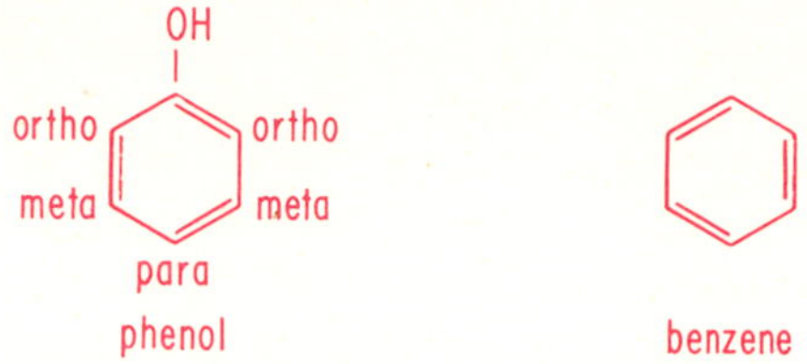

more nucleophilic than the ordinary carbon atoms in benzene (the parent hydrocarbon of phenol). They are more nucleophilic because the oxygen atom in phenol delocalizes some of its high electron density to these three positions.

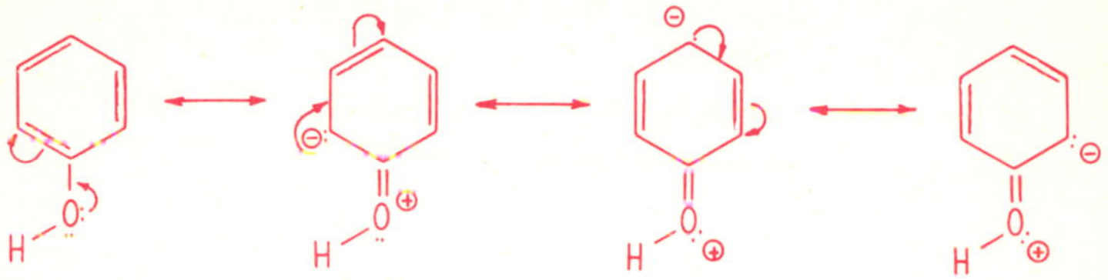

Because of this, any of these three carbon atoms can attack the formaldehyde carbon atom nucleophilically.

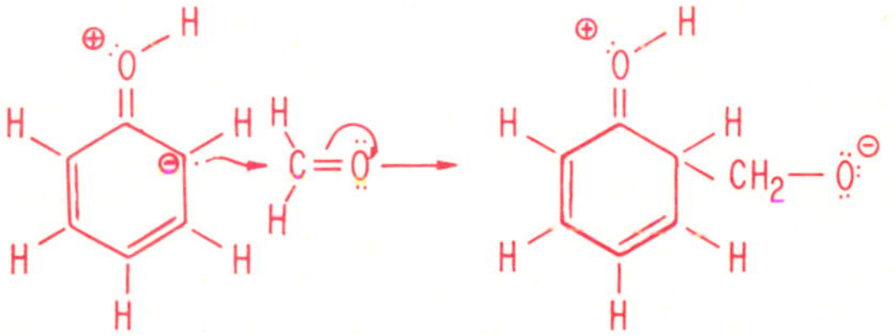

A proton shifts to neutralize the charges.

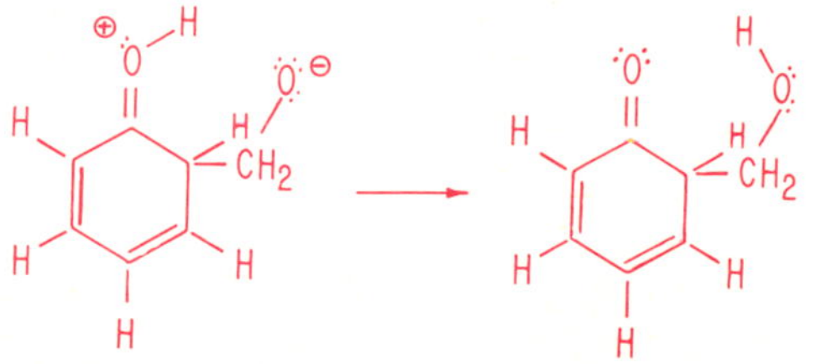

Another proton shift restores the benzene structure of the molecule. This over-all process is called substitution.

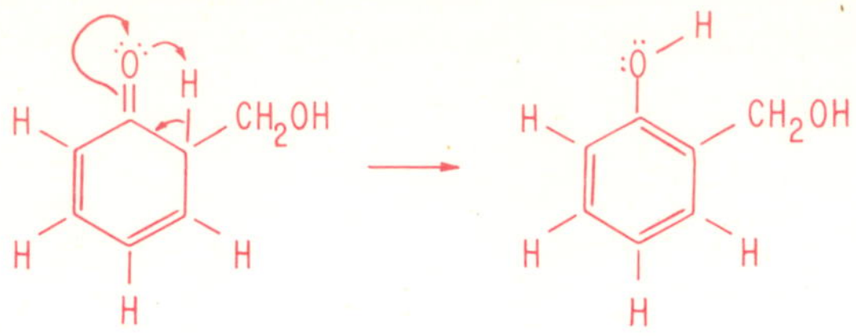

In this way, as many as three formaldehyde molecules can be substituted on a single phenol molecule.

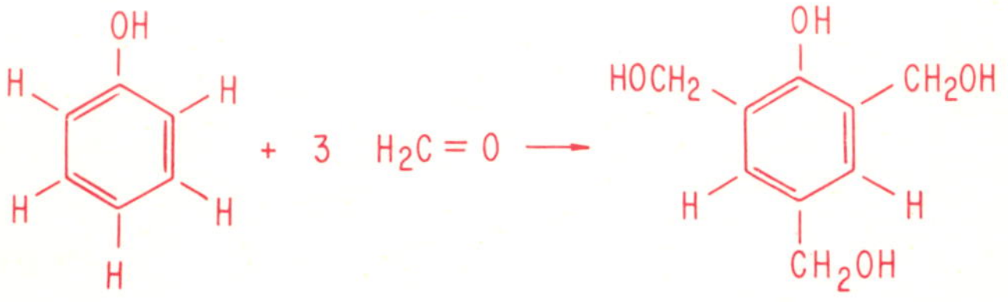

When a hydroxymethylene group (—CH$_2$OH) is bonded to a benzene ring, the hydroxyl group is extremely reactive. This high reactivity arises from the ease with which the parent molecule (benzyl alcohol) forms a carbonium ion.

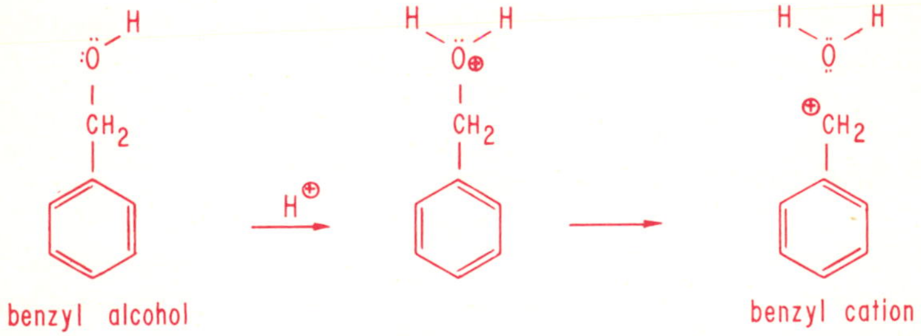

benzyl alcohol   benzyl cation

This carbonium ion (the benzyl cation) forms easily because it is stabilized by resonance which allows charge delocalization.

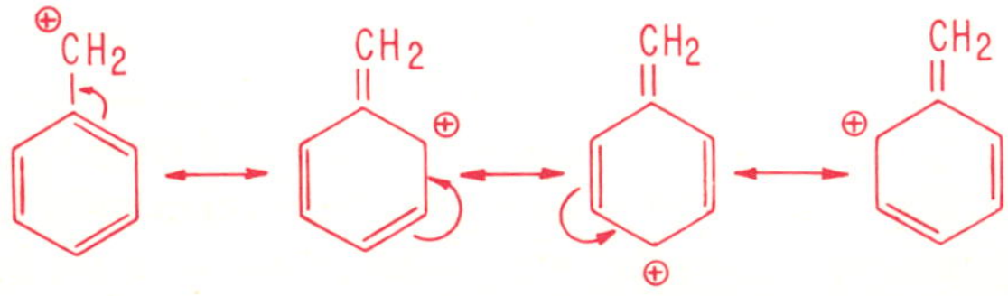

The benzyl cation will of course be extremely susceptible to nucleophilic attack by any of the three activated *ortho-* and *para-*carbon atoms of the phenol molecule. This is how the phenol units become linked by methylene groups. The process of joining the phenol units together via methylene bridges is called condensation.

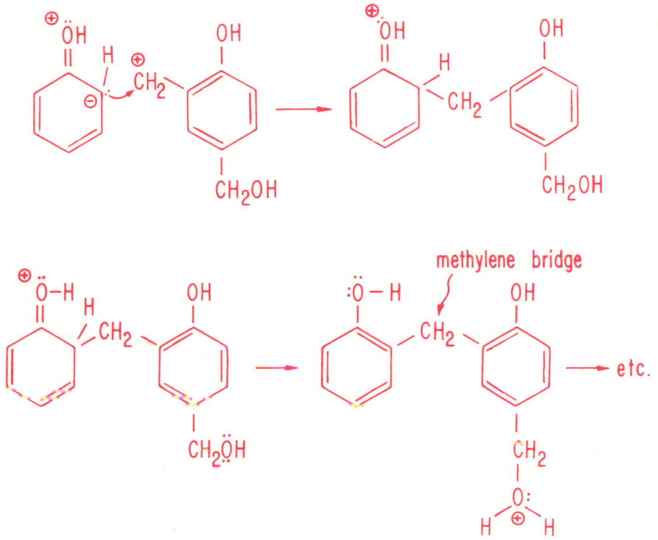

By continued substitution and condensation reactions, the giant Bakelite molecules are built up (see p. 58).

Vinyl Addition Polymerization

About 20 years later, during which time a great deal was learned about molecular structure, a research team at the I. G. Farben Company in Germany began making a series of different types of giant molecules. These new families of polymers differed from Bakelite in that the new giant molecules were not branched and cross-linked like the three-dimensional molecular mass of Bakelite. The backbone of these chains is said to be linear (i.e., unbranched) since the chain forms when each unit going into the chain is bifunctional, that is, can bond to only two other units. The chains in Bakelite can branch and cross-link because the phenol unit is trifunctional (three places for substitution on each phenol unit). This bonding to three other units leads to dense, solid three-dimensional molecular chunks. Bifunctional units can only form long string-like molecules. Linear-chain polymers tend to soften when heated and are called thermoplastics. Cross-linked polymers remain hard and are thus called thermosets.

The family of linear polymers developed at I. G. Farben comprised mainly polyolefins. The first important one of these was polystyrene, a clear, colorless hard, and shiny plastic, which can be molded at temperatures only slightly higher than the boiling of water.

An olefin is a molecule containing a carbon–carbon double bond. The simplest olefin is ethylene. Styrene is phenylethylene.

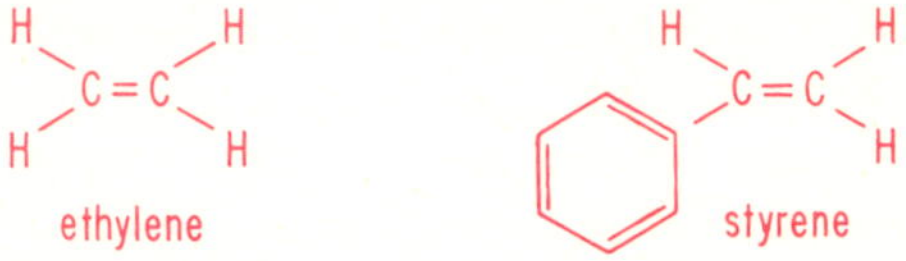

ethylene   styrene

A polystyrene molecule is a long chain of styrene units linked to each other with regular repeating bonding arrangement.

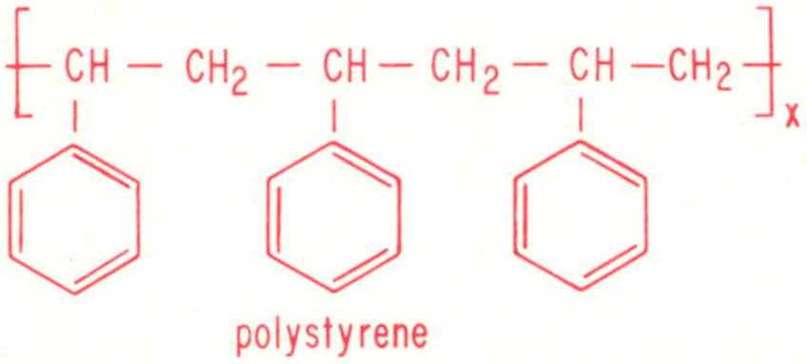

polystyrene

Since the chain is usually produced by a series of free-radical addition reactions, the regular arrangement results from the greater stability of one of the two possible intermediate radicals. For example, the polymerization can be initiated by the generation of a free radical in a system containing styrene. One way to do this is to decompose an azo compound, usually done by heating. We will illustrate initiation propagation and terminations of polymerization using an azo initiator, azoisobutyronitrile. It is common to use peroxides as initiators in olefin free radical polymerizations.

azoisobutyronitrile

Now the isobutyronitrile radical could presumably attack the olefin in two ways to produce two different new radicals.

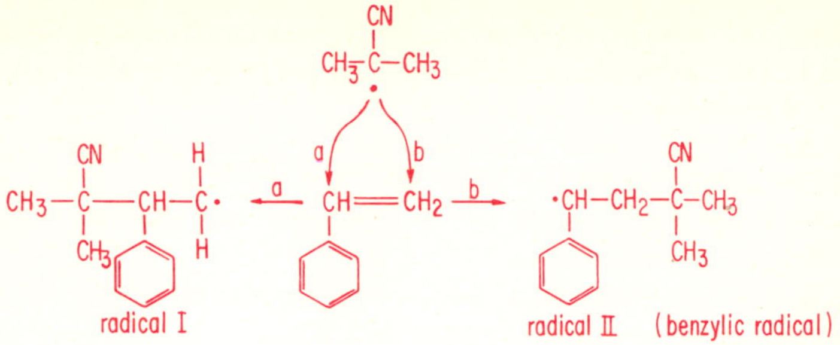

$$CH_3-\overset{CN}{\underset{\bullet}{C}}-CH_3$$

$$CH_3-\overset{CN}{\underset{CH_3}{C}}-CH-\overset{H}{\underset{H}{C}}\bullet \quad \xleftarrow{a} \quad \overset{a}{\underset{}{}}\,CH\!\!=\!\!CH_2\,\overset{b}{\underset{}{}} \quad \xrightarrow{b} \quad \bullet CH-CH_2-\overset{CN}{\underset{CH_3}{C}}-CH_3$$

radical I   radical II   (benzylic radical)

The new radical II is considerably more stable than the new radical I, because radical II has the benzylic structure seen earlier in the stable benzyl cation. Therefore, radical II forms almost exclusively in preference to radical I. Radical II can now attack another styrene molecule, and when it does, it forms another stable benzylic radical.

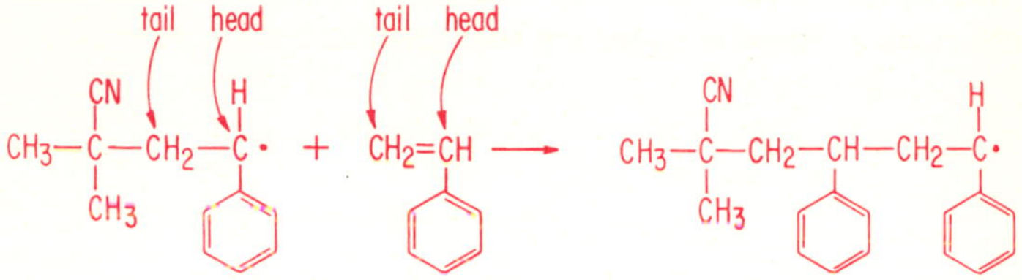

tail   head   tail   head

$$CH_3-\overset{CN}{\underset{CH_3}{C}}-CH_2-\overset{H}{\underset{}{C}}\bullet \;+\; CH_2\!\!=\!\!CH \longrightarrow CH_3-\overset{CN}{\underset{CH_3}{C}}-CH_2-CH-CH_2-\overset{H}{\underset{}{C}}\bullet$$

Repetition of this step leads eventually to an extremely long chain of styrene units linked in a regular head-to-tail pattern. Occasionally, however, two radicals collide and combine to form a simple covalent bond. This terminates the growth of both chains, and produces a head-to-head juncture.

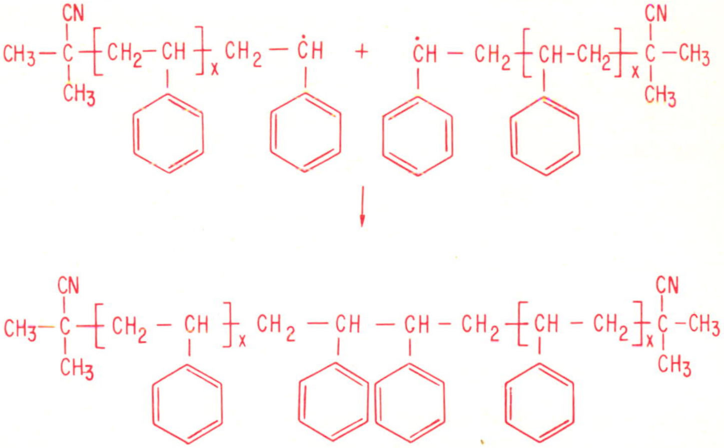

$$CH_3-\overset{CN}{\underset{CH_3}{C}}\!\!\left[CH_2\!-\!CH\right]_x\!\!CH_2-\overset{\bullet}{C}H \;+\; \overset{\bullet}{C}H-CH_2\!\!\left[CH\!-\!CH_2\right]_x\!\!\overset{CN}{\underset{CH_3}{C}}-CH_3$$

$$\downarrow$$

$$CH_3-\overset{CN}{\underset{CH_3}{C}}\!\!\left[CH_2-CH\right]_x\!\!CH_2-CH-CH-CH_2\!\!\left[CH-CH_2\right]_x\!\!\overset{CN}{\underset{CH_3}{C}}-CH_3$$

A growing chain can also terminate its growth by abstracting an atom from some other molecule.

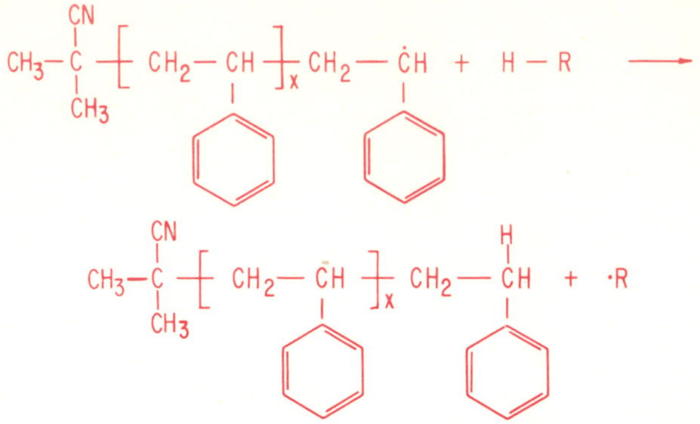

This terminates the chain but does not kill the polymerization since the new radical (R·) can carry on.

A third type of termination occurs when a growing chain abstracts a hydrogen atom adjacent to another radical.

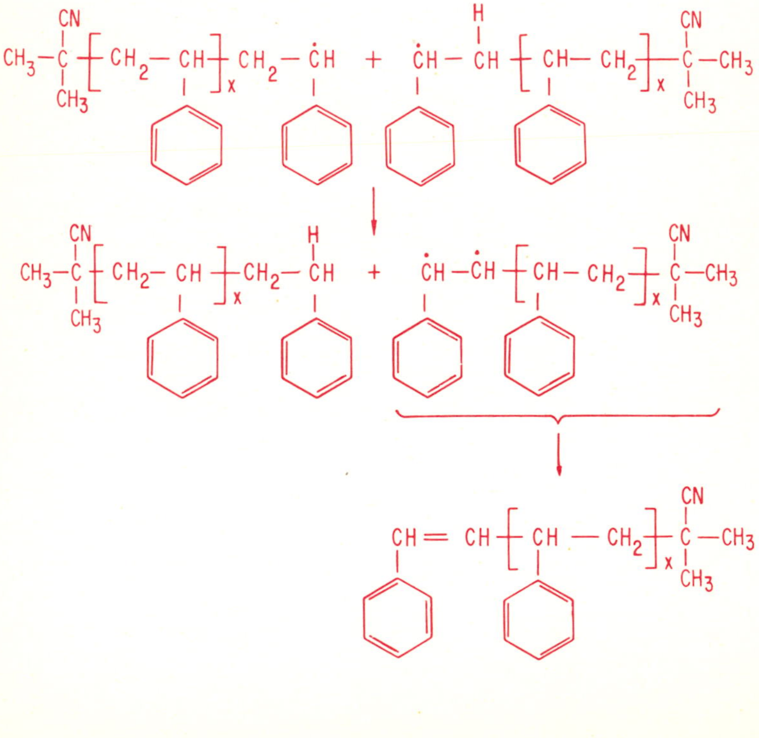

The new olefin formed in this disproportionation reaction cannot carry on the polymerization because an olefin must contain an unsubstituted carbon (CH_2=C⟨) as part of the double bond to be polymerizable (under the conditions usually used to polymerize simple olefins).

The usual goal in most polymerizations of the styrene type is to obtain maximum chain length for best material strength. This requires keeping the chances for termination reactions at a minimum. Termination by collision of two radicals, resulting either in combination or in disproportionation, can be minimized by minimizing the number of radicals in the system. Theoretically havingly only one radical in the system would provide the highest molecular-weight product possible. This ideal situation has been approached by an ingenious method of polymerization called emulsion polymerization, in which single chains grow within single droplets of emulsified monomer molecules.

The first step in emulsion polymerization is to add an emulsifying agent (soap, for example) to water. A soap molecule has a hydrophilic head and a hydrophobic tail. Consequently, soap has a limited solubility in water. The first portions of soap added to the water dissolve, but once a critical amount has been added, further additions to not dissolve. The soap molecules form

$$\text{hydrophobic} \qquad CH_3 \left(CH_2 \right)_{14} \; C \underset{O^{\ominus} \, Na^{\oplus}}{\overset{O}{\Bigl\|}} \qquad \text{hydrophilic}$$

soap molecule

tiny droplets called micelles. The surface of the micelles is composed of the hydrophilic heads of the soap molecules, and the hydrophobic tails are directed into the interior of the micelles. Then a hydrophobic monomer, such as styrene, is added. The hydrophobic monomer molecules diffuse into the hydrophobic interiors of the micelles. Finally, a small quantity of hydrophilic initiator, a source of free radicals, is added. Occasionally, a hydrophilic free radical will make its way from the aqueous medium into the hydrophobic interior of a micelle. Once in, it initiates a single growing chain. As the growing chain consumes the monomer molecules in the micelle, more monomer diffuses into the micelle from the aqueous medium, feeding the single growing chain. As the aqueous medium is depleted of monomer, more monomer enters the aqueous medium from the micelles which do not contain a growing chain. Thus, the non-polymerizing micelles shrink and disappear as the polymerizing micelles swell. The relative concentrations of the various components of the system are such that, should a second free radical enter a polymerizing micelle, termination would be faster than polymerization. Thus, any micelle at any instant would contain either one free radical or no free radical. This optimizes the ultimate chain length of the product.

Besides styrene, many other olefins have been polymerized and have found extensive application in a wide variety of molded, extruded, or woven fibrous products. These monomers include vinyl chloride, vinyl acetate, methyl methacrylate, and acrylonitrile.

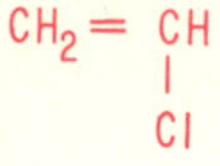

$$CH_2 = CH$$
$$| $$
$$Cl$$

vinyl chloride

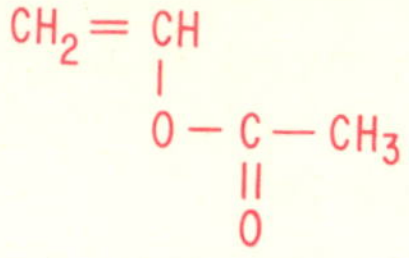

$$CH_2 = CH$$
$$|$$
$$O - C - CH_3$$
$$\|$$
$$O$$

vinyl acetate

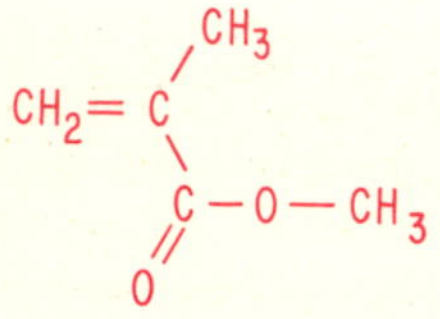

$$CH_2 = C \stackrel{CH_3}{\underset{C - O - CH_3}{}}$$
$$\|$$
$$O$$

methyl methacrylate

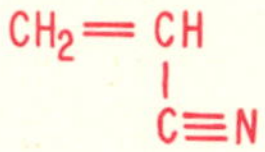

$$CH_2 = CH$$
$$|$$
$$C \equiv N$$

acrylonitrile

Ethylene (the simplest olefin monomer) is not polymerizable by the free radical techniques described above. The molecule is too stable for such processes to be effective. E. W. Fawcett, R. O. Gibson and M. W. Perrin, working in England for the Imperial Chemical Industries (ICI) in the early 1930s, made the revolutionary discovery that ethylene at 230°C and 1500 atm. pressure with a trace of oxygen and a peroxide initiator gives polyethylene in a rapid and exothermic reaction. It is obvious that ethylene polymerization required the development of chemical and engineering technologies.

The polyethylene that resulted is highly branched, slightly crosslinked and softens under 100°C. As we shall shortly see, another revolution took place in the early 1950s that led to linear polyethylene at much lower temperatures and pressures (see section on Ziegler–Natta polymerization, pp. 74–77).

Nylons

Another fabulously successful kind of man-made giant molecule was developed during the 1930s at Du Pont by a research team under the direction of Wallace Hume Carothers. It is said that Dr. Carothers and his associates worked years trying to perfect the technique to raise the molecular weight of these synthetic polymers to the range where they would be expected to form strong fibers. A story has arisen which may be apocryphal that Dr. Julien Hill, a collaborator of Dr. Carothers, left a stirring rod overnight in a flask that contained the heated components of nylon. The next morning he tried to remove the stirring rod from the cooled viscous mass. In the process, he pulled a long fiber which became tougher with elongation. And so the age of nylon, a new artificial silk was born. Du Pont introduced the fiber in the early 1940s for ladies' hosiery. It was an instant success. During World War II, it was employed to make parachutes.

Nylon molecules are not addition polymers as are the polyolefins. They are condensation polymers as are the proteins. They further differ from all polyolefins (except polyethylene) and from proteins in that they contain no groups pendant to the main polymer chain. Finally, they are unlike polyolefins, but similar to proteins, in that their monomer units are linked to each other by amide groups (—C(=O)—NH—). The most famous member of the family is nylon 66, so named because it is made from a six-carbon diamine and a six-carbon diacid.

$$H_2N-CH_2CH_2CH_2CH_2CH_2CH_2-NH_2 + \underset{\text{O}}{HO-\overset{\text{O}}{\overset{\|}{C}}-CH_2CH_2CH_2CH_2-\overset{\text{O}}{\overset{\|}{C}}-OH}$$

$$\text{hexamethylenediamine} \qquad\qquad \text{adipic acid}$$

$$\downarrow$$

$$H-\left[NH-CH_2-CH_2-CH_2-CH_2-CH_2CH_2-NH-\overset{\text{O}}{\overset{\|}{C}}-CH_2-CH_2-CH_2CH_2-\overset{\text{O}}{\overset{\|}{C}}\right]_x OH$$

$$\text{nylon 66}$$

Simply mixing an amine with a carboxylic acid does not produce an amide. Because of the basicity of amines, mixing with an acid forms an ammonium salt.

$$CH_3CH_2-NH_2 + HO-\overset{\text{O}}{\overset{\|}{C}}-CH_3 \longrightarrow CH_3CH_2-\overset{\oplus}{NH_3} + \overset{\ominus}{O}-\overset{\text{O}}{\overset{\|}{C}}-CH_3$$

$$\text{ethylamine} \qquad \text{acetic acid} \qquad\qquad \text{ethylammonium acetate}$$

However, heating an ammonium carboxylate salt to its melting point causes dehydration and formation of the covalent amide linkage.

$$CH_3CH_2 - \overset{\oplus}{N}H_3 + \overset{\ominus}{O} - \overset{\overset{\displaystyle O}{\|}}{C} - CH_3 \xrightarrow{heat} CH_3-CH_2 - NH - \overset{\overset{\displaystyle O}{\|}}{C} - CH_3 + H_2O$$

ethylacetamide

This salt formation, in making nylon 66, is a help rather than a hindrance. It allows exactly equivalent amounts of diamine and diacid to be obtained and used without ridiculously meticulous weighing and measuring. Formation of salt from approximately equivalent amounts of hexamethylenediamine and adipic acid, followed by recrystallization to remove the excess of either component, provides pure salt containing an exact one-to-one ratio of reactants.

An exact equivalence of the two reactants is necessary for making high-molecular-weight nylon 66. If one of the reactants is present even in small excess, the final stages of the reaction will see all molecules with the same end group, either all carboxyl ends or all amine ends. This completely destroys any possibility of chains joining in amide bonds to form longer chains. Hence, the more nearly equivalent the amounts of reactants, the longer the ultimate chain length.

The fact that the nylon molecule can grow at both ends shows an interesting contrast to the single-ended growth of polyolefin molecules. The immediate consequence of the double-ended growth of nylons is that two long chains can connect without terminating the polymerization. This means that chain length increases rapidly toward the end of a nylon polymerization.

The immediate consequence of the single-ended growth of polyolefin molecules is that combination of two chains terminates polymerization. Obviously, then, a polyolefin chain can grow only one unit at a time if it is to continue growing. Thus molecular weight builds more linearly with time in the polyolefin case.

Another well-known member of the nylon family—nylon 6—does not have the problem of equivalent amounts of reactants because it forms from a

$$\xrightarrow{\hspace{1.5cm}} \left[NH - \overset{\epsilon}{C}H_2\overset{\delta}{C}H_2\overset{\gamma}{C}H_2\overset{\beta}{C}H_2\overset{\alpha}{C}H_2\overset{\overset{\displaystyle O}{\|}}{C} \right]_x$$

caprolactam

nylon 6

single reactant—caprolactam. Caprolactam is the cyclic amide of ε-amino-caproic acid. Caproic acid is the straight-chain six-carbon carboxylic acid. The terminal carbon atom is called the epsilon carbon atom since carbon-atom-2 is called the alpha-carbon atom.

The polymerization of caprolactam is quite a different thing from the building of nylon 66 chains. First, it is not a condensation polymerization; it is an addition polymerization, as are the olefin polymerizations. Also, like the olefin polymerizations, it is single-ended, but unlike the olefin polymerizations, two chains cannot combine to terminate the reaction. How does this ring-opening type of addition polymerization work?

Initiation is by a nucleophile. The nucleophile may be the amine end of an ε-aminocaproic acid molecule.

$$HO-\overset{\overset{O}{\|}}{C}-CH_2\text{-}CH_2\text{-}CH_2\text{-}CH_2\text{-}CH_2-\ddot{N}H_2 \; + \quad \underset{\underset{\underset{CH_2}{\diagdown}CH_2}{CH_2}}{\overset{\overset{O}{\|}}{C}}\diagup^{NH}\diagdown\underset{\underset{CH_2}{|}}{\overset{CH_2}{|}} \longrightarrow$$

$$HO-\overset{\overset{O}{\|}}{C}-CH_2\text{-}CH_2\text{-}CH_2\text{-}CH_2\text{-}CH_2-NH-\overset{\overset{O}{\|}}{C}-CH_2CH_2\text{-}CH_2\text{-}CH_2\text{-}CH_2-NH_2$$

This generates a new amino group which can go on to attack another molecule of caprolactam, which adds another monomer unit to the chain as above, and regenerates a new amine end, and so on. In this way, the nylon 6 chain grows at only one end—the amine end—and grows by only one monomer unit at a time. The end result, however, is a giant molecule very similar in structure and therefore in properties to the nylon 66 molecule. The main structural difference between the two molecules is that all the amide linkages in nylon 6 "face" in the same direction, while adjacent amide linkages in

$$-NH-\overset{\overset{O}{\|}}{C}-CH_2CH_2\,CH_2\,CH_2\,CH_2-NH\text{-}\overset{\overset{O}{\|}}{C}-CH_2CH_2CH_2CH_2CH_2-NH-\overset{\overset{O}{\|}}{C}-$$

section of nylon 6

$$-NH-\overset{\overset{O}{\|}}{C}-CH_2CH_2\,CH_2\,CH_2-\overset{\overset{O}{\|}}{C}-NH-CH_2CH_2CH_2CH_2\,CH_2\,CH_2-NH-\overset{\overset{O}{\|}}{C}-$$

section of nylon 66

nylon 66 "face" in opposite directions. Also, the amide groups in nylon 6 are evenly spaced, and the amide group in nylon 66 are not.

Synthetic Elastomers

At the same time that Du Pont was scrambling to supply our country with enough parachute nylon, another wartime panic call went out from Washington to the polymer-chemist community to develop a new giant molecule which could substitute for natural rubber. Our supply of rubber had been suddenly and unexpectedly cut off by the rapid conquest of Southeast Asia by the Japanese. The situation was extremely critical and could well have determined the outcome of the war. U.S. stockpiles contained less than one year's peacetime supply of rubber—about half-a-million tons. Our previous leisurely attempt to develop a decent synthetic rubber had resulted only in a few pilot-sized plants which provided less than 7500 tons in 1941. In contrast, the Germans had been working hard to develop a man-made rubber industry for their war effort and had succeeded. By 1942, they were making more than 90,000 tons per year. Japan, of course, had more natural rubber in its conquered territory than it could use. American chemists and engineers had to multiply our existing synthetic-rubber capacity at least one-hundred-fold to meet the crisis.

Natural rubber consists of giant molecules of polyisoprene. Isoprene is a branched-chain hydrocarbon containing five carbon atoms and two double bonds (called a diene).

isoprene   natural rubber

Theoretically, it should be possible to make "natural" rubber in the laboratory simply by polymerizing isoprene. Prior to World War II, however, attempts to do this were unsuccessful. Isoprene could be polymerized, but the synthetic product was different in some way from the natural product, and was unsatisfactory as a rubber substitute.

German chemists had been experimenting during the 1920s with another diene—a simpler four-carbon straight-chain diene called butadiene. This molecule also polymerized into a rubbery polymer, but once again, the product

butadiene

was inferior to natural rubber and unsuitable for use in automobile tires. In 1928, however, they discovered that a mixture of butadiene and styrene forms a copolymer—a giant molecule containing two kinds of monomer units—

$$CH_2= CH - CH = CH_2 \quad + \quad CH_2 = CH \longrightarrow$$

butadiene

styrene

$$\left(CH_2 - CH = CH - CH_2 - CH_2 - CH \right)_x$$

butadiene—styrene rubber

which serves as an excellent and economical all-purpose replacement for natural rubber. The success of this butadiene-styrene rubber made Germany independent from imported rubber during the war years.

The first decent rubber substitute made in the U.S. came from Carother's research group at Du Pont in 1931. This new material, called neoprene, was a polymer of chloroprene, a chlorinated butadiene. The chlorine atoms give the polymer better resistance to oils, acids, sunlight, and oxidation than natural rubber has. Thus neoprene was used in preference to natural rubber in many low-volume applications, such as life rafts and self-sealing fuel tanks, but it was too expensive for the big need—tires.

The most efficient solution to the urgent need for synthetic tire material seemed to be to follow the path already blazed by the Germans. This was accomplished by obtaining the butadiene-styrene rubber patents from a German-headed cartel which had possession of them. The only thing remaining to do was to build a giant industry based on this technology from scratch. More than fifty plants quickly sprang up from California to Connecticut, making butadiene, styrene, and finally the synthetic rubber. By 1943, the U.S. was turning out more than 200,000 tons per year, and the following year 700,000 tons—enough to keep the wheels of national defense up off the rims.

But not until the late 1950s did man exactly duplicate a natural rubber molecule in the laboratory and then produce it on a commercial scale. This in fact is still the only natural giant molecule which man has ever been able to produce artificially on a large scale. The key to the success is stereo-control.

Nature, in all its molecular transformations, has the unerring knack of completely specifying the three-dimensional orientation of the atoms within the molecules it builds. This facility arises primarily from the three-dimensional specificity of the enzymes which catalyze the "natural" molecular transformations. Such strict stereochemical control is difficult to attain in artificial systems.

For example, without stereochemical control, isoprene can polymerize in four different ways. It can polymerize to form 1,2-polyisoprene.

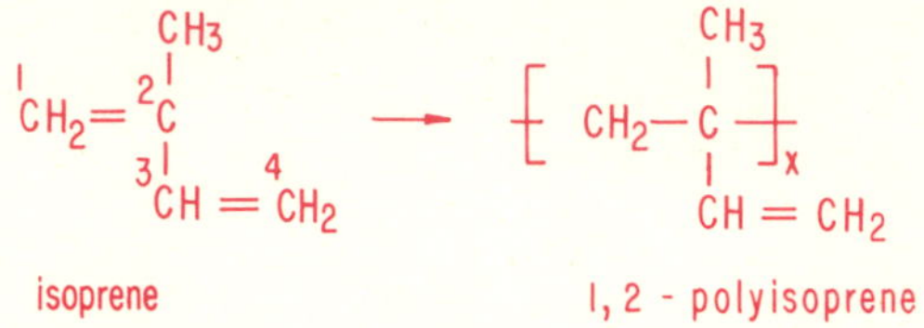

isoprene   1, 2 - polyisoprene

It can polymerize to form 3,4-polyisoprene.

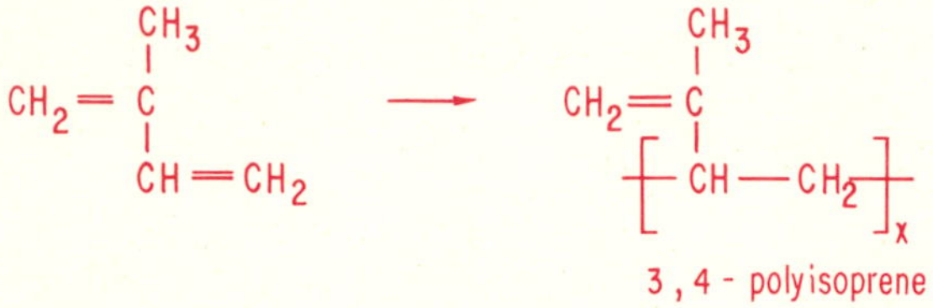

3 , 4 - polyisoprene

It can polymerize to form *cis*-1,4-polyisoprene.

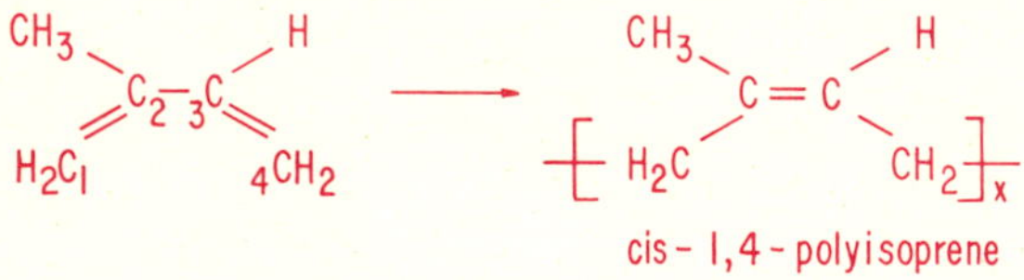

cis - 1,4 - polyisoprene

And it can polymerize to form *trans*-1,4-polyisoprene.

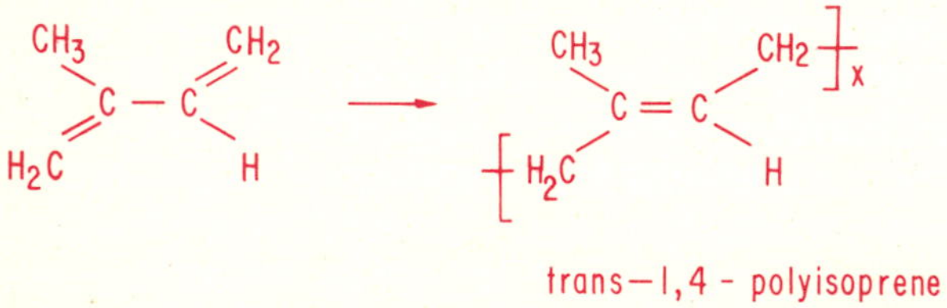

trans—1,4 - polyisoprene

In the first two cases, isoprene polymerizes like a simple mono-olefin. Only one of the double bonds in the molecule is involved in each case.

In the second two cases, however, isoprene polymerizes with migration of its second double bond. The polymerization is initiated by a cation.

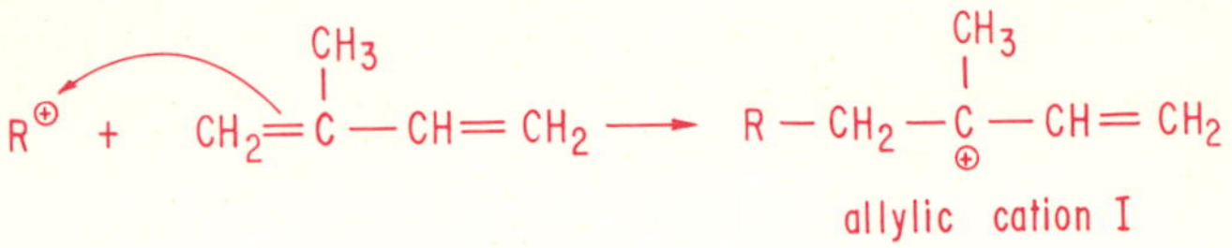

$$R^{\oplus} + CH_2 = \underset{\underset{CH_3}{|}}{C} - CH = CH_2 \longrightarrow R - CH_2 - \underset{\underset{\oplus}{|}}{\underset{CH_3}{|}}{C} - CH = CH_2$$

allylic cation I

The intermediate is an allylic cation which can rearrange to another allylic cation by double-bond migration.

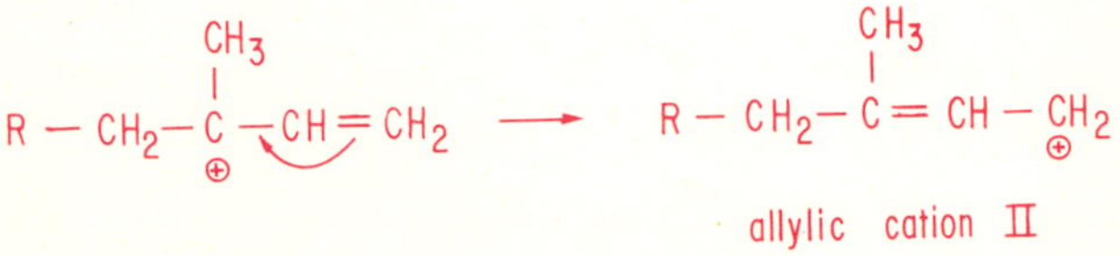

$$R - CH_2 - \underset{\underset{\oplus}{|}}{\underset{CH_3}{|}}{C} - CH = CH_2 \longrightarrow R - CH_2 - \underset{\underset{CH_3}{|}}{C} = CH - \underset{\oplus}{CH_2}$$

allylic cation II

Repetition of cationic attack and double-bond migration causes the 1,4-structure to grow into a long chain.

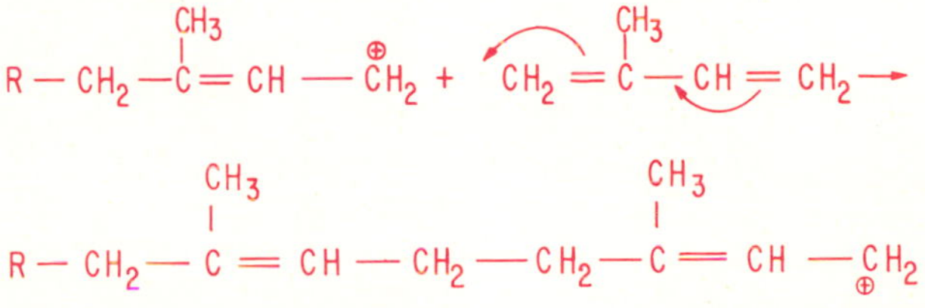

$$R - CH_2 - \underset{\underset{CH_3}{|}}{C} = CH - \overset{\oplus}{CH_2} + CH_2 = \underset{\underset{CH_3}{|}}{C} - CH = CH_2 \longrightarrow$$

$$R - CH_2 - \underset{\underset{CH_3}{|}}{C} = CH - CH_2 - CH_2 - \underset{\underset{CH_3}{|}}{C} = CH - \underset{\oplus}{CH_2}$$

Without double-bond migration, the product will contain the 1,2-structure or the 3,4-structure (or both).

Stereoregular Diene Polymerization

How can the structure of a polymer be controlled? In particular, how can the structure of natural rubber (*cis*-1,4-polyisoprene) be obtained artificially?

Experiments using the alkali metals (lithium, sodium, potassium, etc.) as cationic initiators have provided interesting insight into the problem. First, in a polar solvent such as tetrahydrofuran, all the alkali metals cause the formation of polyisoprene containing only the 1,2-structure, the 3,4-structure, and the *trans*-structure. None of the desired *cis*-structure is obtained. In a nonpolar solvent, all the alkali metals except lithium lead to mixed polymers containing all four possible structures. Lithium, however, produces 94% *cis*-structure—a material very similar to natural rubber. The exceptional behavior of lithium is explained on the basis of the exceptionally small size of the atom.

To initiate the *cis*-polymerization, two lithium atoms attach themselves to the ends of an isoprene molecule.

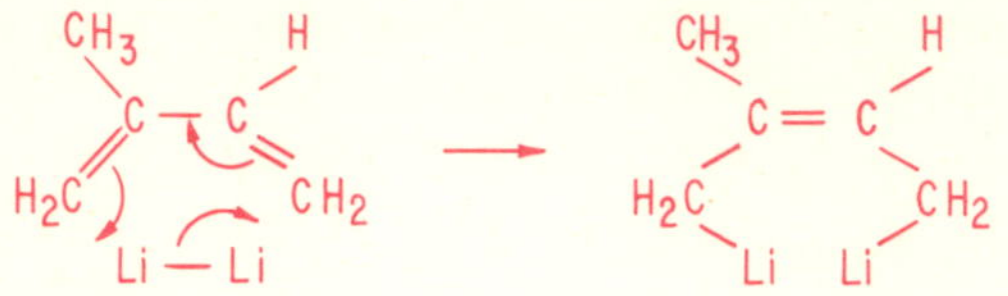

$$CH_3\ C{-}C\ H,\ H_2C{-}CH_2,\ Li{-}Li \longrightarrow CH_3\ C{=}C\ H,\ H_2C{-}CH_2,\ Li\ Li$$

This forms the first *cis*-unit in the chain. The second and the remainder of the *cis*-units form through an analogous six-atom cyclic configuration.

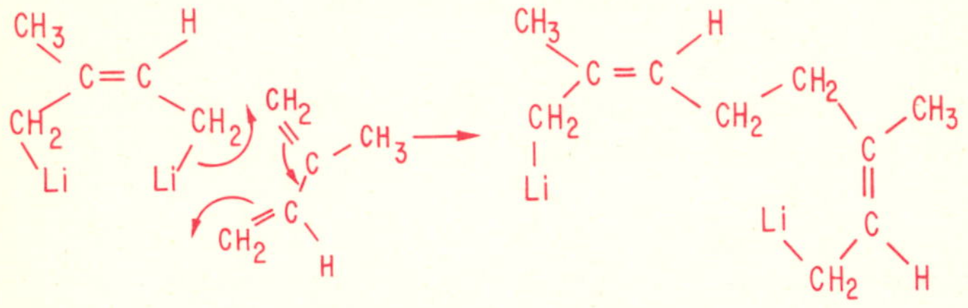

$$CH_3\ C{=}C\ H,\ CH_2,\ Li,\ Li,\ CH_2,\ CH_2{=}C{-}CH_3,\ CH_2{=}C{-}H \longrightarrow CH_3\ C{=}C\ H,\ CH_2,\ Li,\ CH_2,\ CH_2,\ C{-}CH_3,\ Li,\ C,\ CH_2{-}H$$

The cyclic arrangement of the reacting atomic centers always results in *cis*-structure for the added unit. The other alkali metals cannot do this as efficiently as lithium because they are too large to fit into the required cyclic arrangement of reacting atoms. Polar solvents completely prevent the formation of the *cis*-structure by increasing the sizes of the alkali metal ions through solvation.

Linear and Stereoregular Vinyl Polymerization

Another, more subtle, kind of structural variation is encountered in polymers of most simple olefins. As noted earlier (p. 66) ethylene polymers produced at high temperatures and pressures are branched. Dr. Karl Ziegler showed that it is possible to prepare linear polyethylene using a complex catalyst made from titanium tetrachloride and aluminum triethyl in heptane under very dry conditions. Dr. Giulio Natta in a related series of experiments with related catalysts showed that propylene could be polymerized to a high melting, crystalline polypropylene. When propylene polymerizes, it forms a normal head-to-tail polymer as does styrene.

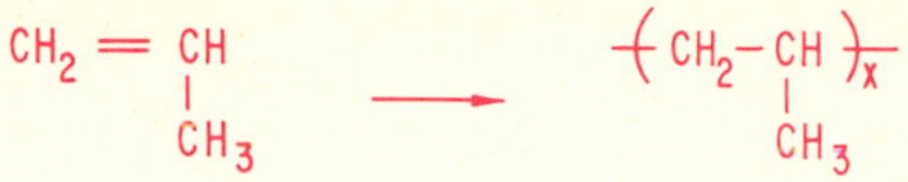

$$CH_2 = \underset{\underset{CH_3}{|}}{CH} \longrightarrow \left(CH_2 - \underset{\underset{CH_3}{|}}{CH} \right)_x$$

propylene   polypropylene

The subtlety of structural variation shows up with a consideration of the relationship between adjacent methyl groups. Adjacent methyl groups may be on the same side of the chain or on opposite sides of the chain.

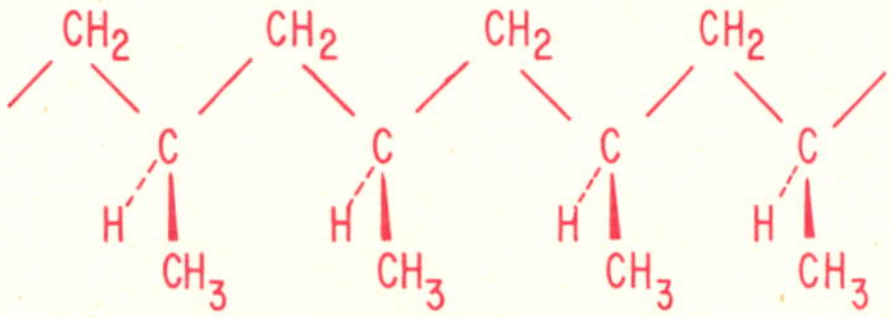

isotactic polypropylene

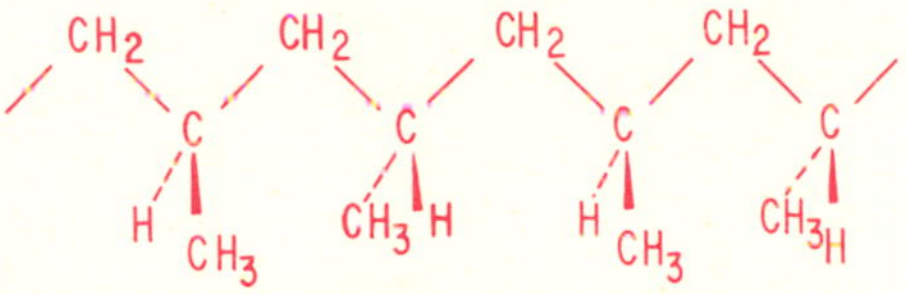

syndiotactic polypropylene

If the methyl groups are all on the same side of the chain, the arrangement is called isotactic. If adjacent methyl groups alternate positions, the arrangement is called syndiotactic. Both arrangements, however, are rather special in that they are stereoregular—their three-dimensional structures can be described in terms of a repeating unit. An ordinary free-radical polymerization of propylene produces a polymer containing random, rather than regular, orientations of the methyl groups. Such an arrangement is described as atactic.

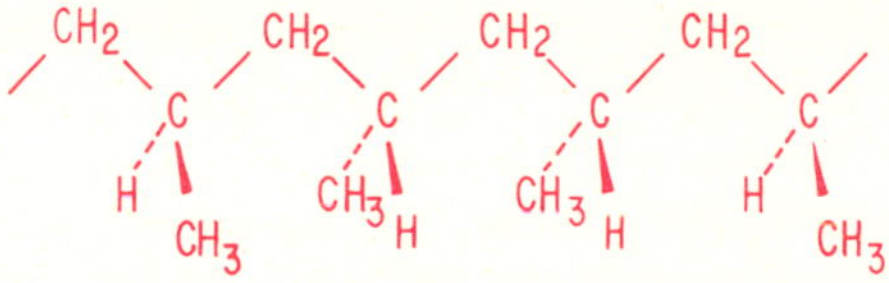

atactic polypropylene (stereo – nonregular)

As noted above, special catalyst systems were required to produce linear polyethylene and stereoregular polypropylene. How were these catalysts discovered? Were they a fortuitous development? A component of the unexpected was most certainly involved. Karl Ziegler and his associates were working on organometallic compounds. They were examining the reaction of aluminum hydride with ethylene under modest pressure. In one stainless steel reactor an almost quantitative yield of 1-butene was obtained. Immediately, Ziegler realized that insertion and abstraction had taken place with an unknown catalyst. He knew that aluminum hydride alone does not give such results. Rather low yields of low molecular weight chains of ethylene units attached to aluminum result.

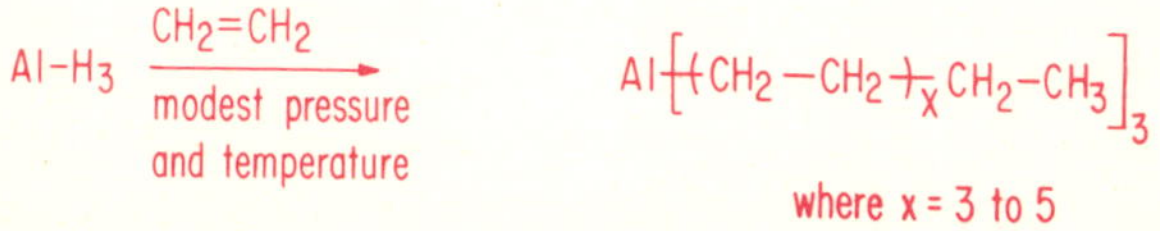

Al-H3
CH2=CH2
modest pressure
and temperature
Al[(CH2-CH2)x CH2-CH3]3
where x = 3 to 5

Careful study of the composition of the stainless steel reactor revealed trace amounts of nickel phosphate. Aluminum hydride reacts with ethylene initially to form aluminum triethyl (in the reaction above, $x = 0$), which then alkylates the nickel chloride. The alkylated nickel chloride then reacts once again with ethylene to give a four carbon chain attached to the nickel. It was postulated that ethylene then reacted with the alkyl nickel to give 1-butene and an ethylated nickel which could now insert another ethylene and repeat the process.

~Ni-CH2-CH-CH2CH3
H
H2C=CH2
abstraction
of a
hydride
CH2=CH-CH2-CH3 + ~Ni-CH2-CH3
butene-1
CH2=CH2
insertion
~Ni-CH2CH CH2CH3
H

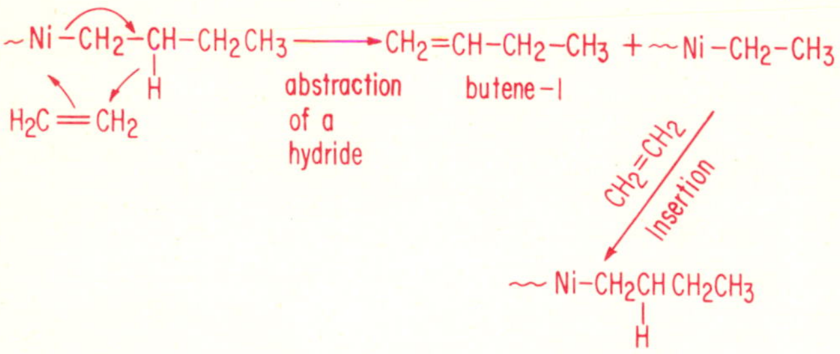

Where ~~~ denotes that the Ni is attached to a nickel chloride crystalline lattice

Ziegler and his colleagues systematically studied other transition metals in combination with aluminum triethyl and found that titanium tetrachloride as a component of this catalyst led efficiently to high yields of linear polyethylene.

About this time, Giulio Natta and his coworkers found that catalysts made from titanium trichloride and aluminum triethyl could be used to prepare isotactic polypropylene and other isotactic poly-α-olefins.* And so the age of stereoregular polymerization developed. For these momentous discoveries both Karl Ziegler and Giulio Natta received the Nobel Prize in Chemistry in 1963.

The kind of stereoregularity within a giant molecule can sometimes be varied by varying the constitution of the catalyst. For example, isotactic polypropylene forms in the presence of a modified Ziegler catalyst containing titanium trichloride and diethylaluminum chloride. Syndiotactic instead of isotactic polypropylene can be obtained by substituting vanadium tetrachloride for titanium trichloride in preparing the complex catalyst.

Sometimes even grosser variations in structure can be obtained simply by varying the bimetallic ratio. For example, with a titanium-to-aluminum ratio greater than 1.0, the bicyclic olefin norbornylene polymerizes by simple olefin addition. With a titanium-to-aluminum ratio less than 1.0, it polymerizes by ring opening.

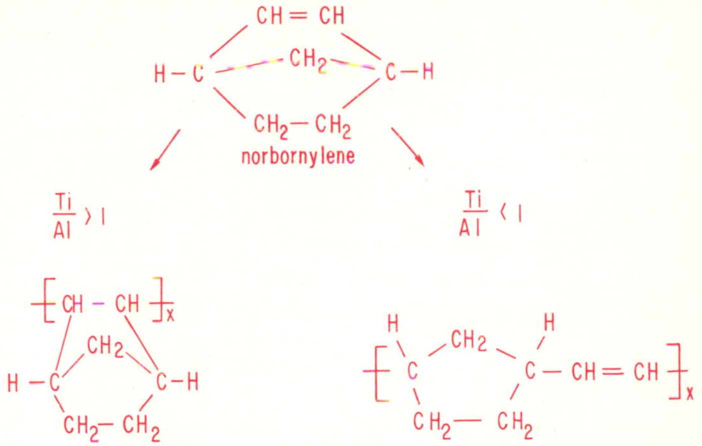

SUMMARY

In this chapter we attempted to trace the development of synthetic polymer chemistry. We saw how a kitchen apron played its part in creating guncotton and modified cellulosic materials; how a stirring rod indicated success in nylon fiber formation; and how a trace amount of nickel in a stainless steel reactor

---

* An α-olefin is a hydrocarbon compound with a terminal double bond: $CH_2{=}CH{-}R$.

led to the discovery of linear polyethylene and stereoregular polymers. Each time an apparent accident occurred, a gifted person was ready to grasp its meaning.

Today, the fruits of their labors and others surround us. Plastics, fibers, elastomers, coatings, packaging materials, etc., are integral components of our society. With the world's limited natural resources, it is clear that more and better synthetics will be required. With the anticipated world's population, it is also clear that disposal of waste materials will remain an enormous and important problem. The polymer scientists must develop new techniques to degrade and/or reclaim synthetics. New devices and materials in the medical and building areas must be created. In short, our society will require imaginative and innovative designers of giant molecules.

SOURCE MATERIALS AND SUGGESTED READINGS

Billmeyer, F. W., Jr., 1962. *Textbook of Polymer Science*, Interscience Publishers, New York.

Mark, H. F., 1966. *Giant Molecules* (Life Science Library), Time-Life Books, New York.

Melville, H., 1958. *Big Molecules*, The Macmillan Co., New York.

Sorenson, W. R., and T. W. Campbell, 1968. *Preparative Methods of Polymer Chemistry*, second edition, Interscience Publishers, New York.

CHAPTER V

MOLECULES OF MERCY
The Pain Relievers

THE MOST popular, and yet poorly understood, of the marvellous molecules used medicinally to alleviate the common aches and pains of man is acetylsalicylic acid—aspirin. Besides being an inexpensive and non-addicting pain killer, the remarkable aspirin molecule reduces fever and local inflammation, and promotes the elimination of uric acid (the thorn of gout sufferers) (see section on prostaglandins, Chapter VII, p. 145). On the negative side, the most common undesirable side effect of the drug is gastrointestinal bleeding. More rarely, a person may develop an allergic hypersensitivity to aspirin so severe that further use of the drug can cause death. Judging from the many thousands of tons of aspirin swallowed each year in the U.S., the benefits outweigh the hazards in the public eye.

Aspirin is an exceptional drug, too, in that it does not occur in nature. Almost all the other long-established medicinal organic molecules occur naturally in plants. Quinine and morphine are notable examples. Aspirin, however, is an artificial molecule, first synthesized in the laboratory by organic chemists. Still, most of the credit for the total synthesis must go to the plant kingdom, because man's contribution was only a simple modification of the salicylic-acid carbon skeleton originally obtained from plants.

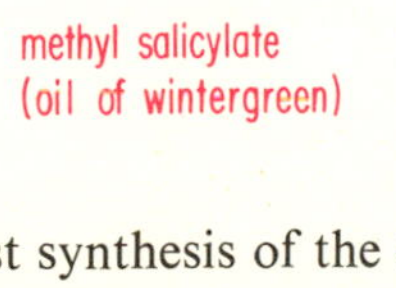

methyl salicylate
(oil of wintergreen)

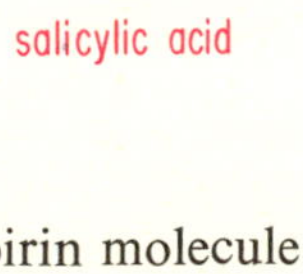

salicylic acid

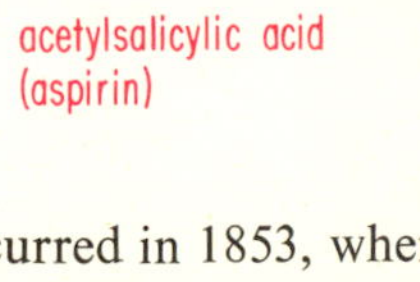

acetylsalicylic acid
(aspirin)

The first synthesis of the aspirin molecule occurred in 1853, when Charles Gerhardt of Strasbourg acetylated salicylic acid with acetic anhydride.

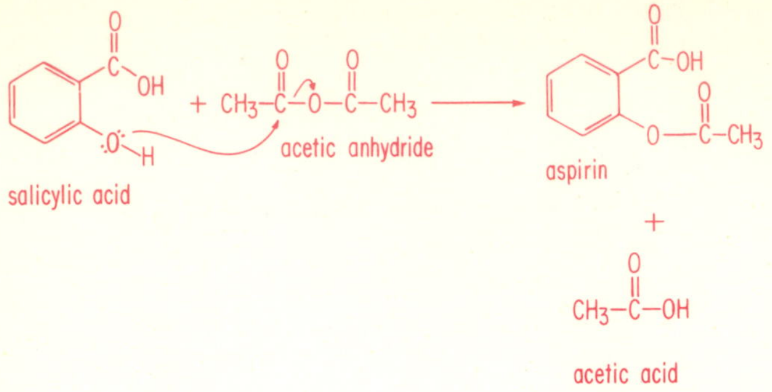

Salicylic acid had been prepared and named Spirsäure in 1835 by the German chemist Karl Löwig, who isolated it from a mixture of products formed by the action of alkali on salicylaldehyde. The pertinent reaction here is probably an oxidation-reduction disproportionation between two aldehyde molecules—a reaction later discovered (1853) by the Italian chemist Stanislao Cannizzaro and now called the Cannizzaro reaction.

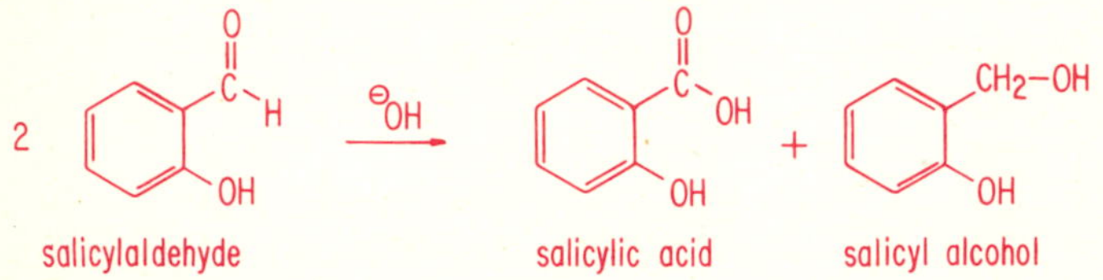

Salicylaldehyde had been obtained a few years earlier (1831) by a Swiss pharmacist, Johann Pagenstecher, who collected it from a distillation of meadowsweet flowers. One variety of meadowsweet bears the scientific name *Spiraea salicifolia*. The name "aspirin" is taken from the first of these names, with the initial letter "a" standing for acetyl.

Another early natural source of the salicylic-acid carbon skeleton was the leaf of the wintergreen plant. An extract of these leaves, called oil of wintergreen, was found to contain methyl salicylate as its principal component in 1843 by the French chemist Auguste Cahours and by an American chemist named William Procter. Oil of wintergreen is used as a flavoring and medicinally as a counter-irritant skin rub. Methyl salicylate is easily hydrolyzed to salicylic acid.

Physiologically, salicylic acid has the same four beneficial effects that aspirin has (analgesic, antipyretic, antiinflammatory, and uricosuric). Because of this, and because a convenient inexpensive method for its large-scale synthesis was shortly developed (1859) by Hermann Kolbe, salicylic acid became fairly widely used in treating arthritis and gout during the latter half of the nineteenth century. In Kolbe's method, the sodium salt of phenol and carbon dioxide are heated under several atmospheres of pressure. This causes carboxylation at the *ortho* position of the phenol molecule, which produces the sodium salt of salicylic acid.

sodium phenolate   sodium salicylate

The reaction proceeds by nucleophilic attack on carbon dioxide by the *ortho* position of phenol as explained in the section on phenolformaldehyde resins, p. 59.

Salicylic acid is more acidic than aspirin, and causes considerable irritation and tissue damage in the mouth, throat, and stomach. To avoid this, it was often administered as the sodium salt in solution, but many patients found the sweetish taste of this potion nauseating. Hence, salicylic acid never attained the popularity now enjoyed by its acetyl derivative. Salicylic acid is now used externally only, to remove local horny tissue, such as corns.

Although salicylic acid had been acetylated to aspirin by Gerhardt in 1853, the reaction was ignored for forty years. Then in 1893, a Bayer chemist named Felix Hofmann reinvestigated the acetylation reaction, in the hope that the product might provide a less obnoxious method for relieving the sufferings of his rheumatic father. His hopes were borne out, and the Bayer company had a new product.

Just how aspirin works in performing its fourfold mission of mercy remains a mystery (see Chapter VII, p. 150). Aspirin does not dissolve in the acidic aqueous fluids of the stomach, and so it passes through the stomach into the intestine largely unchanged. In the alkaline medium of the intestine, it does

aspirin   acetylsalicylate   salicylate
                                                        +

                                                acetate

dissolve, ionizing to the acetylsalicylate anion. Also, because of the alkaline conditions, it begins to hydrolyze slowly to salicylate and acetate. The dissolved anions—acetylsalicylate and salicylate—are absorbed into the blood stream and so distributed to the various tissues of the body. Hydrolysis of acetyl-salicylate to salicylate continues in the blood stream.

It was generally believed for a long time that the active species responsible for the physiological effects of aspirin was the salicylate anion, rather than the acetylsalicylate anion, and that the acetylsalicylate anion was simply a convenient vehicle for the delivery and release of salicylate anion within the body. This belief was based on early clinical studies which indicated that sodium salicylate was just as effective as aspirin in antipyretic, antiinflammatory, and uricosuric activities. Later, however, it was shown that in many cases aspirin is a more powerful analgesic than salicylate is. This cast doubt on the theory that salicylate is the active form of the drug, but this doubt was partially dispelled by the suggestion that acetylsalicylate is more mobile within the body than salicylate, and can therefore reach the affected tissues better. Once at the site, acetylsalicylate hydrolyzes to the active analgesic salicylate.

More recent findings, though, have caused a reversion to the idea that acetylsalicylate can exert its effects independently from its ability to hydrolyze to salicylate. For one thing, aspirin is very effective in inhibiting the adverse reddening reaction of the skin ordinarily caused by a nicotinate-containing cream, whereas salicylate at comparable dosages is impotent. In theory, aspirin should be able to do all the things that salicylate does, since aspirin slowly hydrolyzes to salicylate in the body. It is also conceivable that aspirin can do some things that salicylate cannot do, simply because the intact acetylsalicylate ion is structurally and functionally different from the salicylate ion.

But how does aspirin (or salicylate) really work on the molecular level? This is a complex question, one reason being that aspirin has more than one function and its mode of action in one function, say relief of pain, could hardly be expected to be the same as in another function, say promotion of uric acid excretion. Even within the same function, there may be different molecular activities, for example, it may not relieve headache the same way it relieves the pain of arthritis.

Still, within the same function, say pain relief, there should be similarities in mode of action. A simple view of the function of a pain-killing molecule is that it somehow prevents the action of a pain-causing molecule. The pain-killing molecule could do this by reacting with the pain-causing molecule to change the structure and therefore the function of the latter. Another way might be to block some kind of active site on a nerve ending where the pain-causing molecule would normally attach itself to make its painful presence felt.

One rather well-known pain molecule is a peptide called bradykinin. It is a nonapeptide, that is, it contains nine amino-acid residues. It is one of a number of molecules released by the blood at sites of disturbances, such as wounds or bacterial infections.

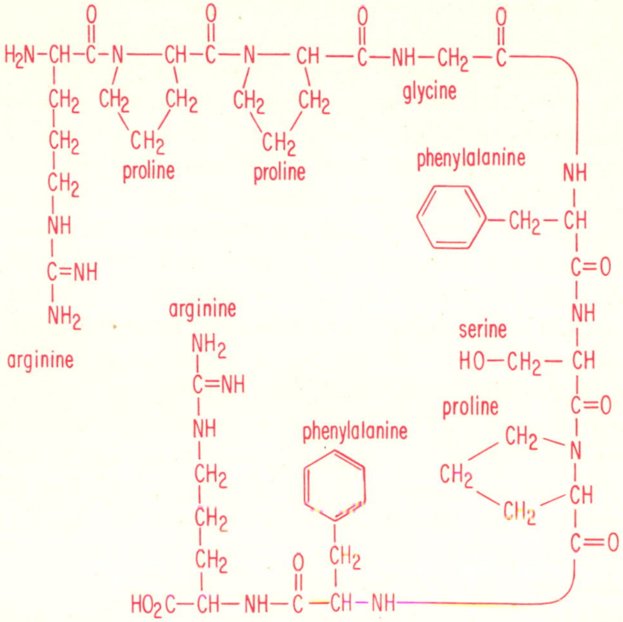

This peptide comes from the tail end of one of the giant protein molecules that make up the hemoglobin complex, which transports oxygen molecules from the lungs to all the tissues of the body.

When bradykinin is injected intravenously into a guinea pig, it causes a bronchospasm—a violent constriction of the throat muscles—which restricts breathing often to the point of suffocation. A small dose of aspirin administered a few minutes before the bradykinin injection completely prevents the bronchospasm.

But, curiously enough, aspirin does not appear to be a universal antagonist to bradykinin. For example, when bradykinin is injected into a guinea pig's skin, instead of into a vein, it causes a scratching and licking response by the animal. Aspirin does not prevent this response. Thus, it seems that aspirin does not block the action of bradykinin by directly attacking the bradykinin molecule. Instead, it may block the action of bradykinin by attaching itself to certain kinds (but not all kinds) of the receptor sites where bradykinin normally attaches itself to induce discomfort of one kind or another.

Other Analgesics—Morphine and Family

The organic molecule which for centuries has held the place of primacy in the doctor's bag for relief of severe pain is morphine. Morphine is the principal active constituent of opium, and as such has been in use since the times of the earliest written records. As an analgesic, morphine is some fifty times as potent as aspirin. Its principal disadvantages are that it is addicting and that it represses respiration.

In spite of its having been an indispensable medicinal agent for so long, morphine was not isolated in a pure state from opium until 1803. Then, more than a century passed before a not quite complete, but correct, structure for the complex molecule was proposed. Credit for this 1925 breakthrough goes to John Gulland and Robert Robinson in England.

The morphine molecule viewed in its entirety seems forbiddingly complicated, but it is simple enough when viewed as a derivative of phenanthrene. The main skeletal difference between phenanthrene and morphine is the

phenanthrene   morphine

—N—C—C— bridge between positions 9 and 13. The Gulland-Robinson structure was uncertain about the point of attachment of the carbon end of this bridge, now known to be position 13.

The presence of the phenanthrene carbon skeleton in morphine was suggested in 1881, when phenanthrene was isolated from a distillation of morphine from zinc dust. Through a series of other degradation reactions, the sites of the oxygen atoms in morphine were located. In 1949, Rudolf Grewe in Germany synthesized a phenanthrene derivative with an —N—C—C— bridge from position 9 to position 13. This molecule proved to be identical to a degradation product of morphine, strongly suggesting that the carbon end of the bridge in morphine is attached at position 13. One reason the bridge problem was so troublesome is that morphine degrades and rearranges by shifting the carbon end of the bridge in hot hydrochloric acid to form apo-

morphine. At first, the rearrangement was unsuspected, and morphine was
believed to have the same carbon–nitrogen skeleton that apomorphine has.

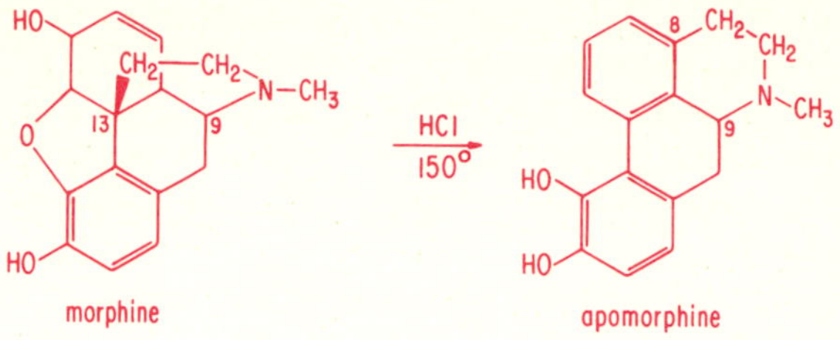

morphine   apomorphine

Total synthesis of the morphine molecule by Marshall Gates in 1952 put the
final lid of confirmation on the structure, as a total synthesis always does.
Traditionally, a proposed structure is not considered "proved" until the
proposed structure has been synthesized from simple molecules with well-
established structures, and the synthesis product has been shown to be identical
to the material in question. The key step in the synthesis of morphine was the
addition of butadiene to an appropriate derivative of 1,2-naphthoquinone.
This kind of addition reaction is called a Diels-Alder addition, so named in
honor of the co-discoverers of the reaction. The reaction is extremely useful in
synthesizing molecules containing six-membered rings, which occur in so many
natural products. It involves the addition of a diene to an olefin. The olefin is
usually activated by an adjacent carbonyl group, as in a quinone.

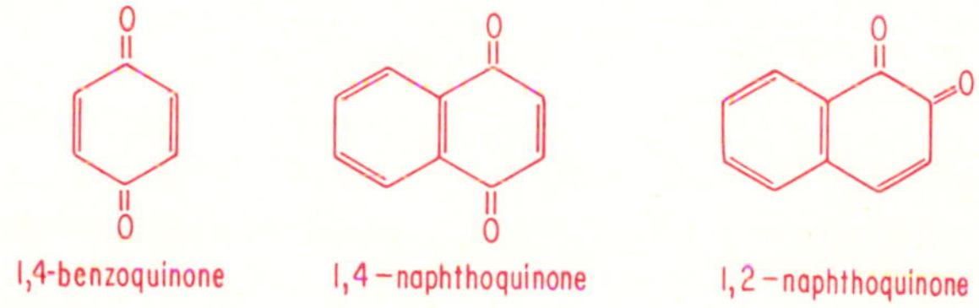

1,4-benzoquinone   1,4 –naphthoquinone   1,2 –naphthoquinone

When a diene adds to the double bond in 1,2-naphthoquinone, it forms the
phenanthrene carbon skeleton found in morphine.

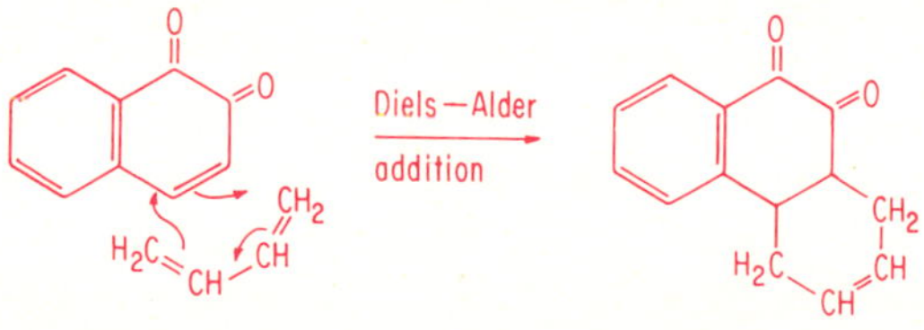

Diels—Alder
addition

The naphthoquinone derivative used in the morphine synthesis had been specially prepared with a cyanomethyl group (—CH$_2$—C≡N) in the proper position so that it could later be converted into the C—C—N morphine bridge by a series of functional-group transformations.

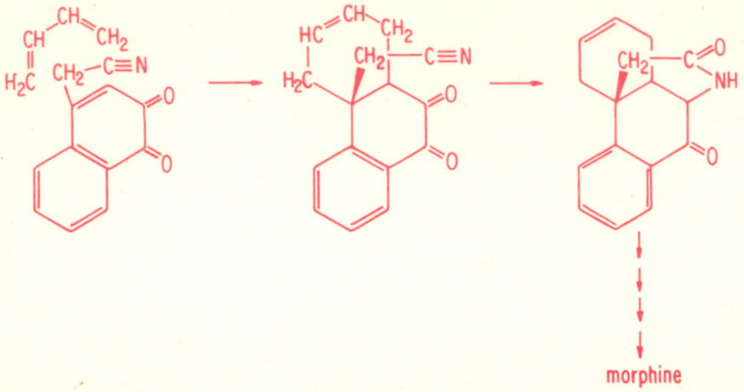

morphine

The synthesis of morphine is a lengthy, painstaking process in the laboratory. It is much easier and less expensive to grow opium poppies as a natural source of morphine.

Nonetheless, the morphine molecule has been and continues to be the focus of an enormous amount of laboratory research. The principal end of this effort is to modify the structure somehow to eliminate the unwanted properties of morphine, such as addictiveness and respiratory depression, without destroying its pain-killing potency. This primary goal has not yet been completely realized, but the work has borne side-fruit both in fundamental knowledge and in new and useful partially synthetic drugs. The study is a fascinating one in the relationship between molecular structure and function.

Two well-known simple derivatives of morphine are codeine and heroin. Codeine is methylmorphine, and heroin is diacetylmorphine.

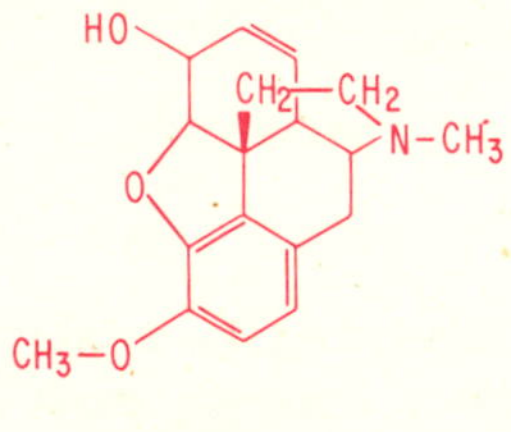

codeine 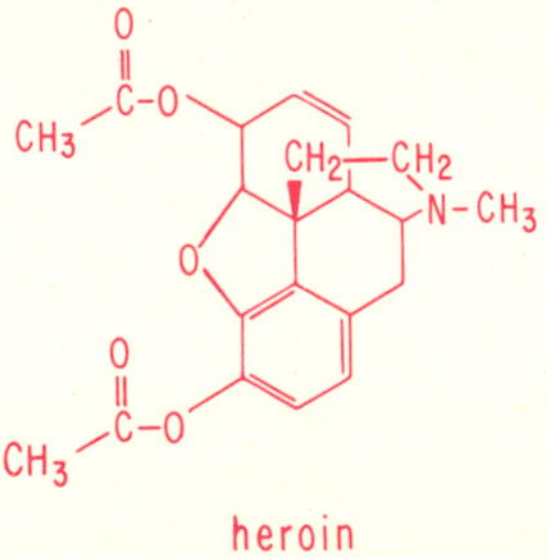   heroin

Like morphine, codeine is present in opium, but in such small amounts that it is usually prepared by methylation of the more abundant morphine. Codeine is less addicting than morphine and is used as a local anesthetic and as a narcotic in cough medicines.

Heroin does not occur naturally, but is made by acetylating morphine, as aspirin is made by acetylating salicylic acid. Heroin is so dangerously addicting that its manufacture and importation are forbidden by law in the U.S.

A striking example of a super-narcotic appeared recently in the form of a structural modification of morphine. Prepared by the British chemist K. W. Bentley, this molecule is 10,000 times as potent as morphine. A single milligram can reduce a wild elephant to easily manageable docility. Hopes that the tiny

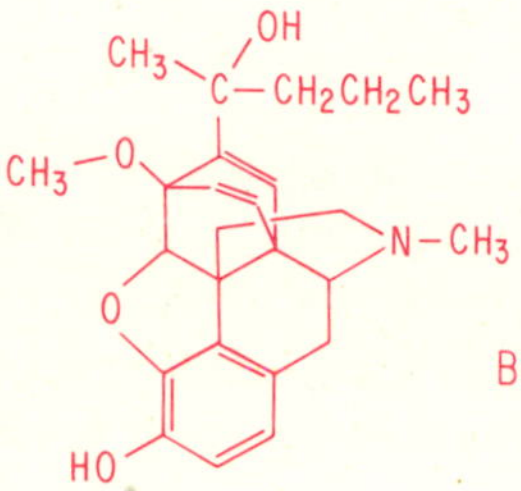

Bentley's compound,
a supernarcotic

quantities of this material, which would be effective as an analgesic in a human, might not depress respiration were dashed. Also, the material shows evidence of being highly addicting. For a long time, hopes of separating these three properties seemed futile. Only weak analgesics appeared to be free of the curse of addictiveness.

A structural analog of morphine which has found wide application as a less dangerous substitute for heroin in treating addicts is methadone. However, what relieves a heroin addict can lead to the degeneration of a normal person, since methadone is normally classified as an addicting narcotic.

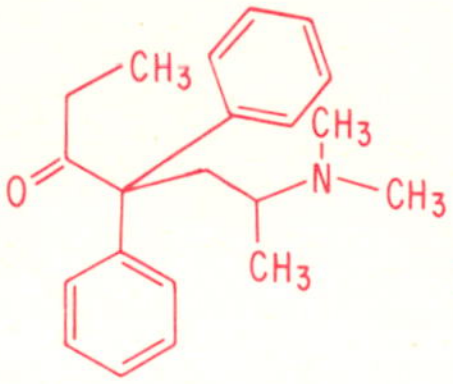

methadone

Recent successes in at least partially separating the good from the evil in morphine-like molecules have arisen from a discovery made more than 50 years ago, the implications of which went unappreciated for decades. Julius

Pohl of the University of Breslau removed the methyl group from the nitrogen atom in codeine and replaced it with an allyl group ($-CH_2-CH=CH_2$). The resultant molecule is called N-allyl-norcodeine, the prefix *nor* indicating that a carbon atom had been removed from the molecule. When this allyl derivative was injected into a morphine-drugged animal, it counteracted respiratory depression.

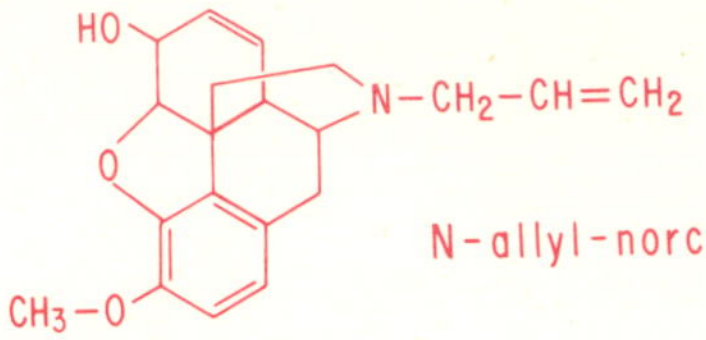

N-allyl-norcodeine

But the matter went largely ignored until 1942 when two Merck chemists did the same thing to morphine, converting it to the molecule known as nalorphine—a contraction of N-allyl-normorphine.

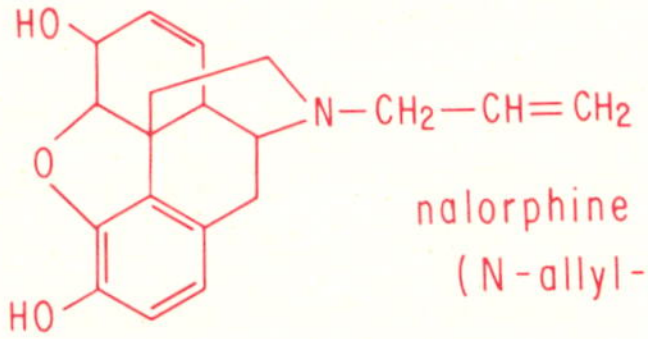

nalorphine
(N-allyl-normorphine)

As might have been expected, nalorphine counteracted respiratory depression and most of the other effects of morphine, including acute morphine poisoning. When administered to a narcotics addict, it brought on severe withdrawal symptoms. Unfortunately, animal tests showed that nalorphine did not relieve pain.

Then in 1954 at Massachusetts General Hospital, Henry Beecher and Louis Lasagna tried an experiment to find out whether nalorphine could reduce morphine's respiration-depressing effect without reducing its analgesic effect. They gave morphine and nalorphine to one group of patients and only nalorphine to another group for comparison. Surprisingly, and in contrast to the results of earlier animal tests, nalorphine itself relieved pain as well as did the combination of morphine and nalorphine. And as a bonus, further tests indicated that nalorphine is not addicting. This was the first example of a non-addicting potent analgesic. Prior to this, only weak analgesics had shown no danger of addiction. The rosy glow soon faded, however, when it was discovered that nalorphine at clinical dosages produces hallucinations.

One of the best new molecules to come out of the morphine-modification studies is pentazocine, prepared at the Sterling–Winthrop Institute.

pentazocine

This molecule is as effective as morphine in relieving post-operative pain, labor pains, cardiac pain, and other agonizing types of pain. Most importantly, actual practice has proved it non-addicting. It does, however, fall short of the ideal analgesic, since it depresses respiration. Thus the search goes on, and will continue until the ideal is in hand.

Cocaine and Novocain (Procaine)

The current drug abuse problem in the U.S., particularly that in which heroin is the demon, had its counterpart at the turn of the century in the misuse of cocaine.

The cocaine-abuse problem was somewhat different, however, in that cocaine was a legitimate medical molecule widely used as a local anesthetic. Hence, there were large legal supplies of the drug available for pilfering by those who would traffic unconscionably in its perversion.

Heroin, however, is not a legally allowable medical molecule in the United States. In those countries where its medical use is possible, the governmental authorities exert strict control over its dissemination. Certain terminal cancer patients are allowed this highly addicting, highly effective pain killer. In the vast majority of cases, however, heroin is produced illegally solely for the financial benefit of smugglers and dope peddlers who feed on the self-imposed misery and frantic crimes of the ultimate consumer.

There are other differences between the two drugs. Heroin is enormously physically and psychically addicting. Cocaine is merely psychically addicting, but it is so treacherously powerful in this respect that the habitual user is just as tightly "hooked" as the heroin addict. The only practical difference, then, is that deprivation of cocaine does not cause the strictly physical portion of the agony of withdrawal felt by the heroin addict. Physically, however, prolonged use of cocaine does cause serious mental deterioration, and withdrawal even in the least inveterate instances is accompanied by severe depression (see Chapter VIII where psychotomimetic drugs are discussed).

Another difference is that heroin is a synthetic derivative of the natural molecule morphine, while cocaine occurs naturally in the coca plant. Cocaine is a member of the tropane alkaloid family, along with atropine (like cocaine, a stimulant) and scopolamine (a sedative).

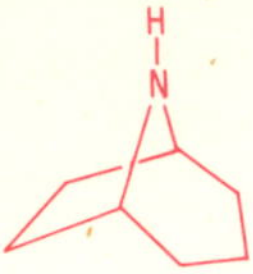

tropane 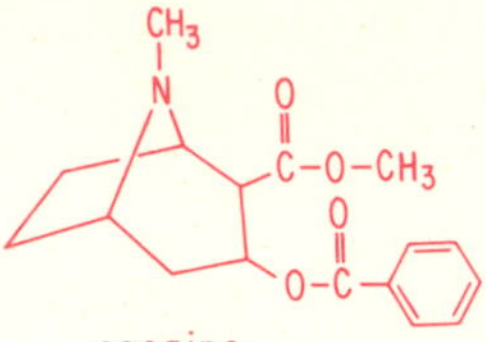   cocaine

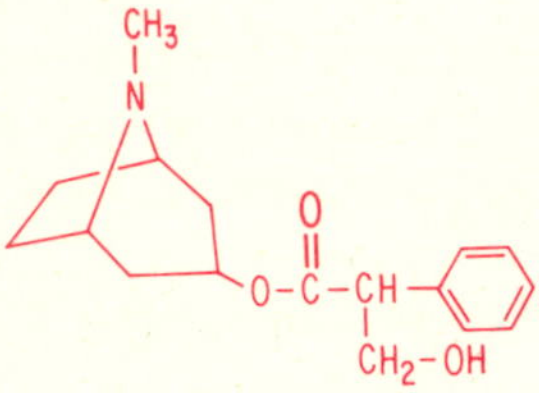

atropine 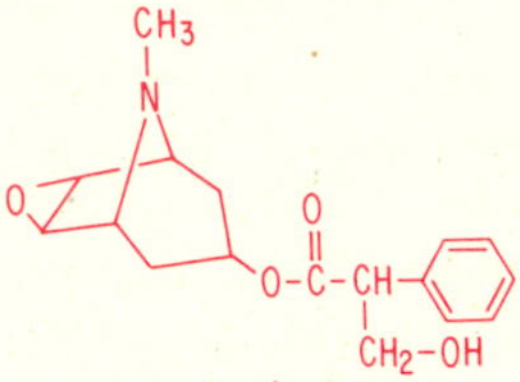   scopolamine

The parent molecule tropane is simply cycloheptane with a 1,4-bridging nitrogen atom.

The decline of the cocaine-abuse problem* resulted largely from the success of organic chemists in synthesizing structural analogs of the drug in an effort to produce a new molecule which would retain the pain-deadening properties of cocaine without retaining the demoralizing psychological effects. This they succeeded in doing in 1905 with the synthesis of procaine, trade-named Novocain. Substitution of Novocain for cocaine in medical practice made

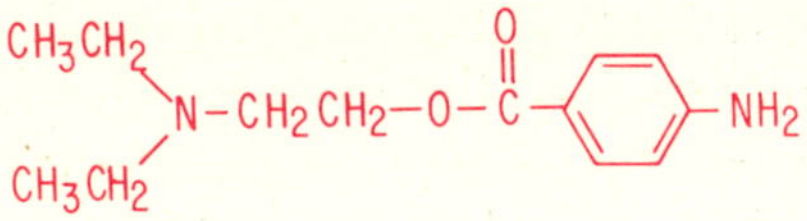

procaine (Novocain)

---

* Recently there has been a reemergence of cocaine use as a "narcotic." Since it can be produced in very large quantities and high purity, criminal elements have been distributing the material as a snuff. Numerous cases of heart seizures and stroke have been reported from deep nasal inhalation of cocaine powder.

cocaine much more difficult to obtain—a highly desirable sociological consequence of basic laboratory research.

As in the case of heroin and methadone, the useful synthetic analog is simpler than the natural drug. In fact, there appears little obvious similarity between the two molecules. However, each is a benzoic-acid ester of an alcohol which contains a fully alkylated (tertiary) amine function. This much, at least, they have in common. The dissimilarity between the two illustrates the wealth of possibilities open to the synthetic organic chemist in his continuing efforts to separate the desirable and undesirable properties of molecules by structural modification.

Tranquilizers, Sleeping Pills and Truth Serums

Feel tense? Take a pill. Can't sleep? Take a pill. By use of a proper pill or injection, the suspect had to tell the truth. How many times have we heard such refrains? People, especially in the highly industrialized countries, partake of energizers, sedatives, analgesics, and even hypnotics in order better to cope with their pressure-filled lives. Even Cleopatra supposedly took sleeping pills while Mark Antony was away on a trip.

It was probably Shakespeare who best described a man's yearning for escape from tension. Macbeth asked his physician:

> Canst thou not minister to a mind diseased,
> Pluck from my memory of rooted sorrow,
> Page out the written troubles of the brain
> And with some sweet oblivion's antidote
> Cleanse the stuffed bosom of that perilous stuff
> Which weighs upon the heart?

The chemist obliged by synthesizing many compounds apparently answering Macbeth's plea. From the laboratories have come sedatives such as the barbiturates and tranquilizers such as phenothiazine, meprobamate and chlordiazepoxide. With these drugs has come the realization that drug-taking by itself rarely provides solutions for healthy peoples' psychological difficulties. The more potent the psychoactive molecule, the more likely there are to be serious side effects. Physicians and patients should recognize that drugs always possess many simultaneous and independent biological effects. Usually they travel over the entire body causing desirable and undesirable changes in body chemistry. With this cautionary note in mind, we can consider the chemistry of some representative substances used as sleeping pills, sedatives and tranquilizers.

Barbiturates (substituted pyrimidines) have been known for a hundred years. Today the annual production of these compounds is in excess of two million pounds—enough to provide every man, woman and child with about fifteen separate typical doses per year. The parent substance, barbituric acid, was first prepared in 1864 by Professor Johann Friedrich Wilhelm Adolph von Baeyer who supposedly named the compound for a pretty lady friend named Barbara. It is most amusing to think that this powerful and formal, nineteenth-century academicer had both a sense of humor and a lady friend.

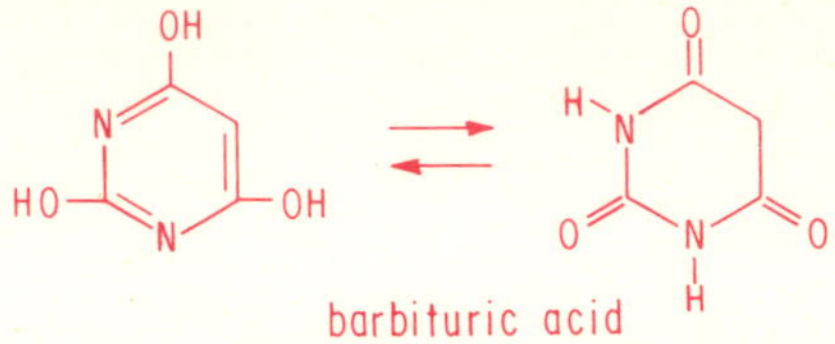

barbituric acid
(pyrimidine ring system)

Barbituric acid itself is not a sedative or soporific. As shown in the above formula, it can exist in both an enol or keto form (see appendix). A group of

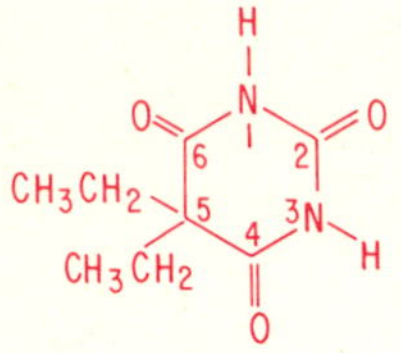

barbital
(veronal)

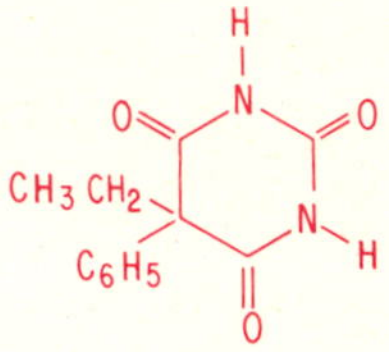

phenobarbital

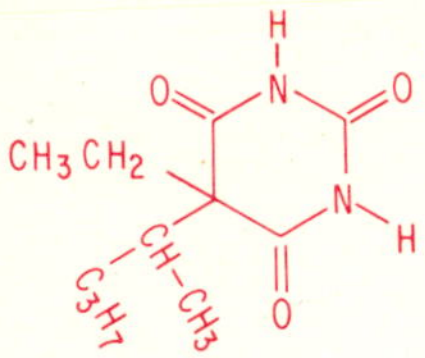

pentobarbital
(nembutal)

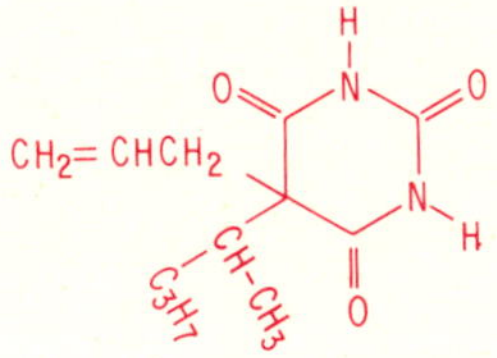

secobarbital
(seconal)

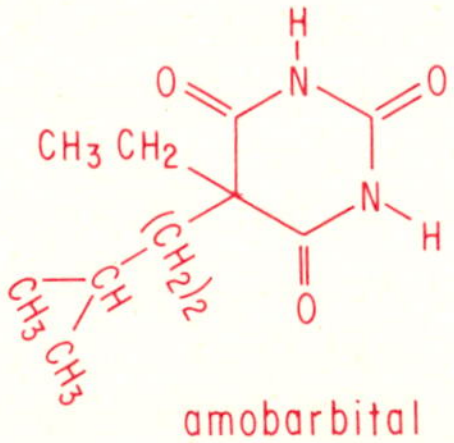

amobarbital
(amytal)

five 5-disubstituted compounds has a strong depressant action on the central nervous system (see p. 92).

Barbiturates can be prepared by a well-documented series of reactions involving the condensation of a urea and a suitably disubstituted malonate ester under strongly basic conditions.

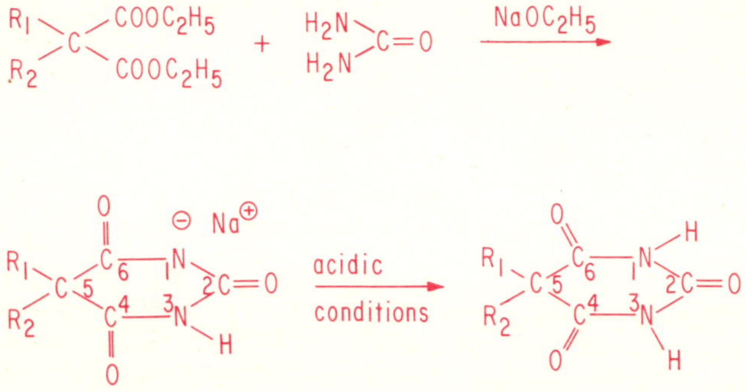

The initial reaction proceeds because the urea nitrogens can be made strongly nucleophilic by sodium ethoxide. They can attack the malonate ester groups.

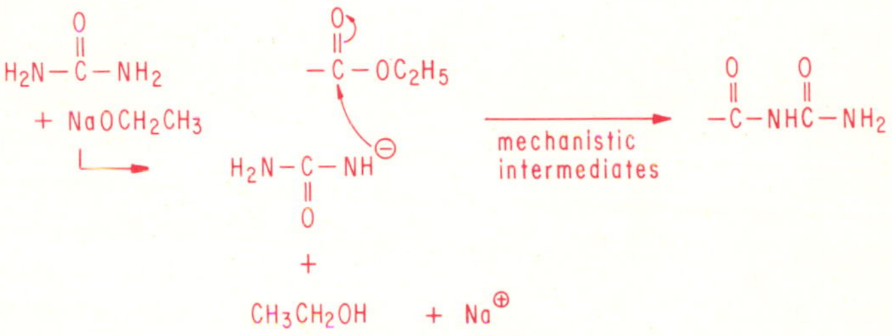

It should be noted that the pyrimidine system of the barbituric acids forms sodium salts because of the active hydrogens which enolize (see appendix on enolization). The name "barbiturates", strictly speaking, should refer to the alkali metal salts of these pyrimidine compounds. The medical profession labels the free compounds or their metal salts generically as "barbiturates".

Each barbiturate exhibits a special combination of drug effects. Typical doses vary from 10 mg to well over 100 mg. As mentioned above, all of these pyrimidine derivatives depress the central nervous system. Barbital is a long-acting sedative. Phenobarbital is more of an anticonvulsant than a soporific. Nembutal and seconal are similar to barbital in biological properties except that they act more rapidly and for shorter periods.

It has been erroneously reported that amytal is a truth serum. Rather this barbiturate has a broad sedating effect and a large therapeutic index* which allow extreme sedation to take place. Some patients react by stripping away their inhibitions in recalling painful and hidden memories. By no means does injection of amytal force subjects to tell the truth.

A related compound, pentothal or pentothal sodium is used as a general anesthetic by intravenous injection.

R = H   pentothal
R = Na   pentothal sodium

This drug is made by the condensation of thiourea with the suitably disubstituted malonate ester (mechanistically identical to the reactions shown above for the other barbiturates). Use of pentothal as a general anesthetic requires careful control because the levels necessary are almost 75% of the lethal dose. In spite of this, the relaxation inducing properties, rapid action and recovery make pentothal an important drug for modern surgical techniques.

Unfortunately, the side use of barbiturates as sedatives and sleeping pills has caused severe psychical addictions. People who routinely use these compounds find that larger and larger doses are necessary to achieve the desired effects. Overdoses are not uncommon which lead to such strong suppression of the parts of the brain controlling heart and lung function that death results. It has been shown that consumption of alcohol while under barbiturate sedation can have dire consequences. There appears to be substantially enhanced depressant effect on the central nervous system from the combination of alcohol and barbiturates than for much larger doses of each separately.

Medicinal use of rauwolfia plants and extracts by ancient Hindus have been documented. These peoples used the materials to treat epilepsy, chronic headaches and as an antidote for snake bites. (One of the species of this plant is actually called *Rauwolfia serpentina*.) Reserpine is one of the main rauwolfia active agents. It has been characterized and synthesized.

---

* The therapeutic index represents the ratio of the maximum tolerated dose to the minimum effective dose. High values indicate that extreme effects can be achieved.

The rauwolfia alkaloids have the ability to interfere with the storage of adrenaline, serotonin and other important biological amines (see Chapter VIII, p. 163). As a result, they induce a calming effect since stimulation by the amines is removed. Blood pressure and metabolic rates are reduced. Snake bite victims are helped by the slowdown in that the toxins are not as rapidly distributed throughout their bodies. They can marshall their natural defenses and perhaps destroy the invading poisons.

The phenothiazines are widely employed as tranquilizers. Chlorpromazine, the prime example of this group of molecules, is administered to mentally disturbed people including those with severe and disabling maladies.

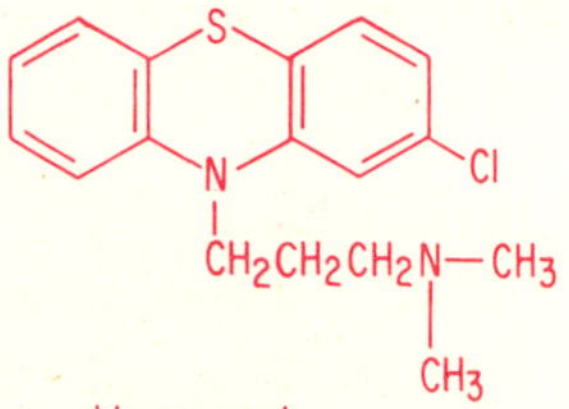

chlorpromazine

The structure of chlorpromazine is rather complex. In some ways, it is not dissimilar to some of the hallucinogens we will consider in Chapter VIII. Unlike the barbiturates, the phenothiazines induce calmness without concurrent sleepiness. Institutionalized schizophrenics have been allowed to resume nearly normal lives under medically supervised chlorpromazine treatment. The mechanism of action of these drugs remains mostly unknown. The problem can be seen to be complicated by the fact that closely related imipramine is a strong psychic energizer rather than a tranquilizer.

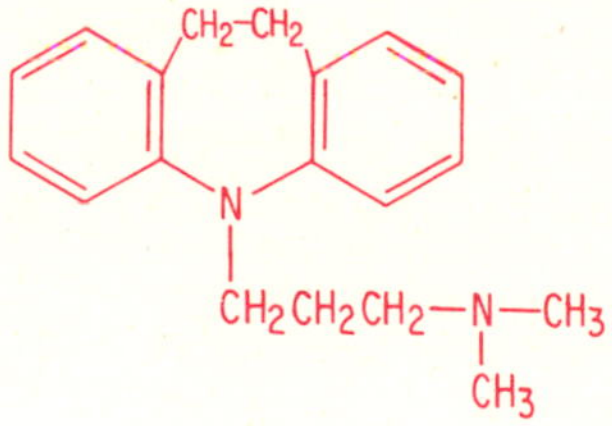

imipramine

Meprobamate, better known as Equanil or Miltown, is representative of another class of tranquilizers. This drug is widely used because it appears to relieve anxiety without outwardly impairing normal performance. It induces a mild euphoria in addition to causing tranquilizing effects. Neurotic people generally exhibit apprehension, insecurity and insomnia which may lead to debilitating psychosomatic diseases. Physicians often prescribe meprobamate to alleviate the anxiety symptoms and help the patient cope. In this aspect, tranquilizers are useful. They do not, however, provide cures or insights into the causes of the mental problems.

meprobamate

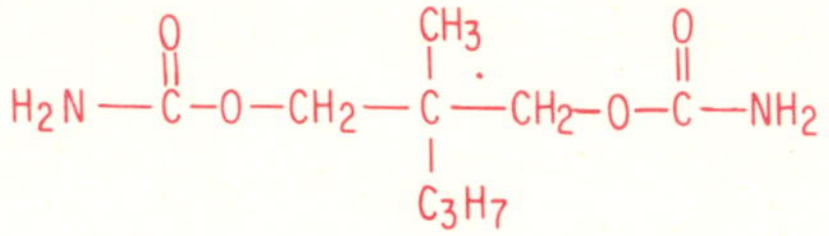

$$H_2N - \overset{\overset{O}{\|}}{C} - O - CH_2 - \overset{\overset{CH_3}{|}}{\underset{\underset{C_3H_7}{|}}{C}} - CH_2 - O - \overset{\overset{O}{\|}}{C} - NH_2$$

Lastly, let us consider benzodiazepine molecules of which chlordiazepoxide hydrochloride (librium) and diazepam (valium) are representative examples.

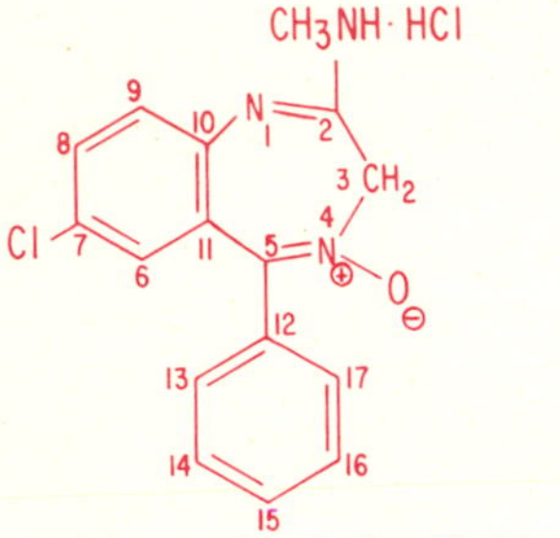

chlordiazepoxide hydrochloride

7-chloro-2-methylamino-5-phenyl-
3H-1,4-benzodiazepine-4-oxide
hydrochloride

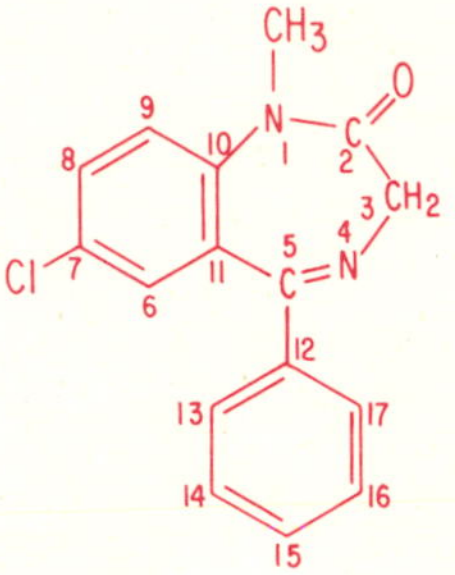

diazepam

7-chloro-1,3-dihydro-1-methyl-
5-phenyl-2H-1,4-benzodiazepine-
2-one

Both compounds are used as alternatives to meprobamate. Most tests show chlordiazepoxide and diazepam to be intermediate in potency between meprobamate and the more powerful chlorpromazine. Typical dosage for chlordiazepoxide and diazepam varies over a wide range from 5 mg to more than 100 mg a day for adult patients. The metabolic fate of these drugs is complex. A single dose of diazepam labeled with tritium administered to an adult human subject is mostly excreted in the urine. Part of the drug appears as the N-demethylated compound; part as the 3-hydroxylated material and part as a product which is N-demethylated and 3-hydroxylated.

Chlordiazepoxide and diazepam are similar to other tranquilizers in that they accomplish their desired effects without strong soporific effects. Both drugs are used for a variety of purposes including anxiety states, neuroses, tension and even alcoholism. Symptoms of alcoholic withdrawal and combativeness are often alleviated with these drugs.

Although the toxicities of chlordiazepoxide and diazepam are low, serious side reactions do occur including skin rashes, nausea, headaches and impaired sexual activity. As with barbiturates and meprobamate, physical dependence is frequently encountered. Symptoms following withdrawal are depression, loss of appetite, insomnia and even psychotic-like states.

Unquestionably, sleeping pills and tranquilizers have their place in chemotherapy. Promiscuous use of these substances is not use but abuse.

SUMMARY

In this chapter, we traced the story of aspirin and showed how extensively it is used. The details of how aspirin works continue to be obscure. Recently, reports have appeared involving aspirin with the inhibition of the prostaglandin synthesis (see Chapter VII). Much research remains to be carried out to explain the fever reduction by aspirin and its prevention of bronchospasm induced by bradykinin.

Morphine alkaloids were next considered. We saw the structural relationships among morphine, codeine and heroin. Methadone was also discussed. The mechanism of addiction for these narcotics is not understood. Our sojourn into the chemistry of alkaloids next took us to cocaine and related drugs, barbiturates and tranquilizers. Each compound has its place in medicine. It is obvious that each also represents an ever present danger in the wrong hands.

SOURCE MATERIALS AND SUGGESTED READING

Burger, A., ed., 1970. *Medicinal Chemistry*, 3rd edition, Wiley-Interscience, New York.

Poynter, F. H. L., 1963. *Chemistry in the Service of Medicine*, J. B. Lippincott Co., Philadelphia.

Russo, J. P., 1967. *Take as Directed (our Modern Medicines)*, F. E. Shideman, ed., C. R. C. Press, Chemical Rubber Co., Cleveland.

Swidler, G., 1971. *Handbook of Drug Interactions*, Wiley-Interscience, New York.

CHAPTER VI

MOLECULES OF MIGHT
The Germ Killers

AS WE SAW in the last chapter, man has accumulated a substantial armamentarium devoted to protecting himself from pain and physical suffering. The magic molecules vary from relatively innocuous physical effects to those of lasting and perhaps disastrous consequences. We now turn ourselves to the problem of disease and how man has utilized nature's miracle molecules and then how man has created many of his own specific molecules to help ward off microbial invasions.

In pre-quinine times, that is, prior to about three hundred years ago, the dreaded parasites of the chills-and-fever disease, malaria, fed without restraint within the bloodstream of humanity. Even in what now seem unlikely places, such as Rome and London, malaria was by far the most common and devastating of the afflictions of man. The disease is spread by the mosquito *Anopheles* infected by a protozoa of the genus *Plasmodium*. Spores of this protozoa are stored in the infected mosquito. Once inside the blood of man, they develop into an active form which invade, and multiply in, red blood cells.

The quinine molecule brought an almost unbelievable cure for the wracking tortures and death of malaria. It also precipitated the demise of Galenism—that false science of medicine which for 1500 years dominated physicians' thinking and practice with its purges and bloodlettings. The quinine molecule is a remarkable molecule with a fascinating history.

In view of this history, it seems incredible that even during the past decade, one-third of the world population is still threatened by malaria. About two million persons die from malaria each year, and chronic sufferers of the disease number in the hundreds of millions.

And yet the specific ability to cure malaria lies within the structure of the quinine molecule. What does it look like?

The quinine molecule consists of a quinoline portion and a quinuclidine portion attached to each other through an alcoholic carbon atom (a carbinol bridge).

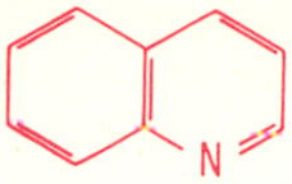

quinoline

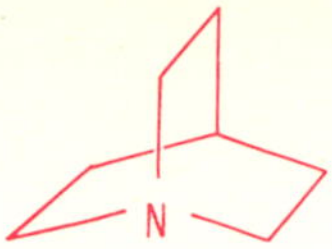

quinuclidine

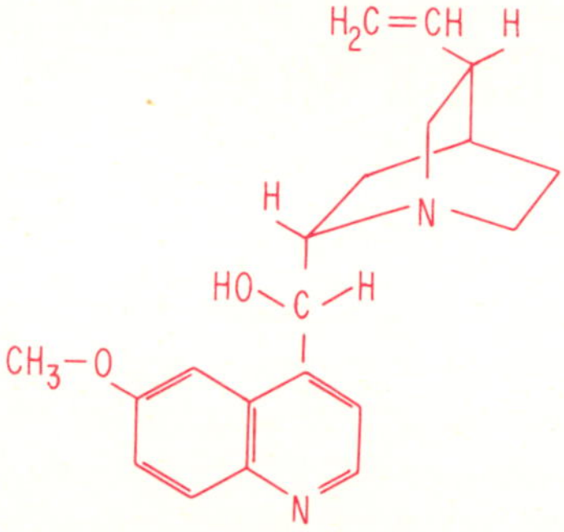

$H_2C=CH$ H

H

HO—C—H

CH_3—O

quinine

    The molecule is produced by the cinchona tree—a tree indigenous to
Peru—and is found in the bark of the tree. Why the cinchona tree produces
quinine is not known. The bark also contains three closely related alkaloids—
quinidine, cinchonine, and cinchonidine—all of which are active against
malaria. The overriding popularity of quinine arose fortuitously from its being
the first of the four to be isolated. Quinine and quinidine differ in configuration
at the asymmetric quinuclidine carbon atom attached to the carbinol bridge.
Cinchonidine and cinchonine have no methoxy group (CH_3O—) on the
quinoline portion of the molecule.

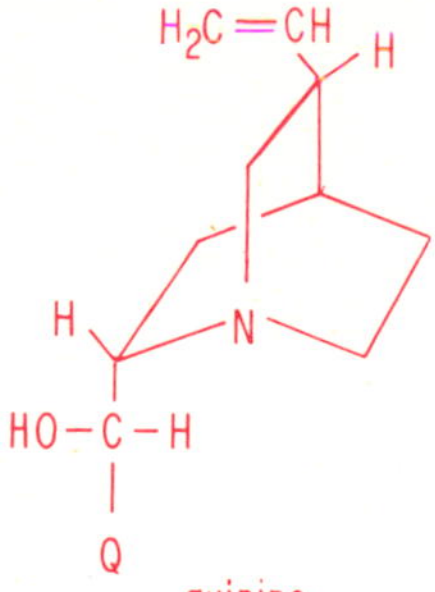

$H_2C=CH$ H

H

HO—C—H

Q

quinine
(and cinchonidine)

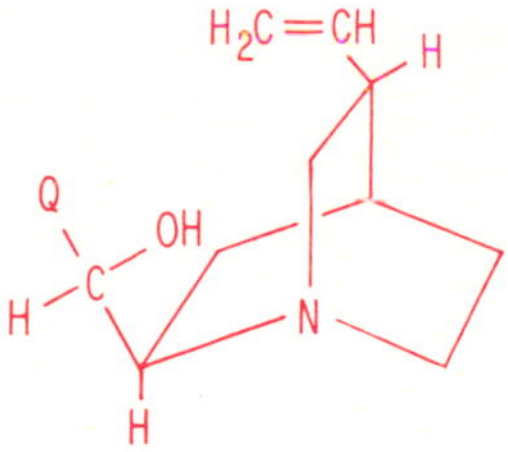

$H_2C=CH$ H

Q

OH

H—C

H

quinidine
(and cinchonine)

(Q represents quinoline portion of molecule)

How the antimalarial value of cinchona bark was actually discovered is not known. There are many versions of the story of the discovery. One version attributes the discovery to Jesuit priests, who allegedly chewed barks as an aid to differentiating between botanical species. Another credits a ship's surgeon with the discovery after he observed that feverish mountain lions seek out the cinchona tree and chew its bark. A third version holds that a malaria sufferer, smitten suddenly by burning fever and thirst, drank in desperation from a pool of water made bitter by cinchona trees felled in a recent storm and was cured.

To muddy further the murky origin of the cure, some historians believe that South American indians knew about the antimalarial properties of the bark long before the discovery of the New World, but did not share their knowledge with Europeans in the hope that the disease would wipe out the unwelcome foreigners. Others claim that the indians did not know the curative value of the bark, and in fact had no need of it, since the disease was unknown in the New World until carried in by infected Europeans.

Be that as it may, the first recorded report of the wonderful workings of the remedy appeared in 1633 in the *Chronicle of St. Augustine* by a Father Calancha in South America. He wrote that a tree called the fever tree has cinnamon-colored bark which, when powdered and made into a beverage, cures the fevers. Oddly enough, this announcement did not result immediately in the bark being brought back to the malaria-ridden cities of Europe.

Twelve long feverish years passed and finally, so the story goes, a Father Tafur brought some with him to a religious gathering in Rome, then a malarial pest hole. The cures which resulted soon inspired Jesuit priests in South America to begin shipping the bark to Rome, where it shortly became known as Jesuits' bark. Religious pilgrims to Rome obtained packets of the powder and, returning home, rapidly spread knowledge of the cure through Europe. In spite of the bark's success, physicians indoctrinated in the principles of Galenism would have nothing to do with the Jesuits' brew. According to Galen, who was a Greek physician practicing in Rome back in the second century, diseases were caused by an imbalance among the four humors of the body, namely blood, phlegm, black bile, and yellow bile. The Galen-prescribed treatment for fever, supposedly a bile-based disorder, was bleeding or purging or a combination thereof. Galen's influence, dispersed through his five-hundred-and-some books, squelched medical progress for about 1500 years, until after the discovery of Jesuits' bark. A Roman conclave held in 1655 was the first in the history of the Church to be free from death from malaria. Religious differences were also responsible for hindering the more rapid distribution and use of the life-saving powder. Protestant countries would not allow the material within their borders. At the time of the death-free Roman conclave, London was ravaged by malaria. Oliver Cromwell himself, the celebrated defender of Protestantism, fell to the fever, would not hear of succumbing to the popish cure, and so died of malaria in 1658.

Twelve years later, those well-to-do about London who could afford the fee could be cured of malaria in the fashionable offices of one Robert Talbor, until recently an obscure but enterprising druggist's apprentice. Cleverly, Talbor pubicly warned of the dangers of the insidious Jesuits' powder, while privately dispensing it under the guise of a new secret remedy of his own. His successes soon had the Royal College of Physicians down on him, however, because he neither bled nor purged. But their cries of anti-Galenism were soon stifled. King Charles II contracted malaria, called in Talbor, and was cured. Talbor was knighted for his service and named Physician to the King. After Talbor's death, the secret remedy was revealed, and the Peruvian bark gradually became accepted for what it is.

The next advance in man's still unfinished struggle against malaria came in 1820, when two French pharmacists isolated the quinine molecule from cinchona bark. These two men, Joseph Pelletier and Joseph Caventou, did not patent their method. Instead, they allowed manufacturers to make and distribute pure quinine, which for the first time provided the world with the powerful anti-malarial remedy in easily controllable dosages.

The classic unravelling of the structure of the quinine molecule began in 1854 when A. Strecker found that the formula of quinine is $C_{20}H_{24}N_2O_2$. By the end of the century, several other essential bits of information about the structure had been assembled, although the complete constitution of the molecule was not yet known. For one thing, the two nitrogen atoms were known to be tertiary (R_3N:), since each was capable of reacting with one molecule of ethyl iodide to form the quaternary ammonium salt. Therefore, the molecule contains no N—H groups.

$$R_3N: \quad + \quad CH_3CH_2 - I \longrightarrow R_3\overset{\oplus}{N} - CH_2 CH_3 + \overset{\ominus}{I}$$

tertiary amine   ethyl iodide   quaternary
                                              ammonium salt

Since the molecule forms an acetyl derivative when treated with acetic anhydride, one of the oxygen atoms must be in an O—H group, the only other kind of acetylatable group possible with the atoms available in quinine.

$$R-O-H + CH_3-\overset{\overset{O}{\|}}{C}-O-\overset{\overset{O}{\|}}{C}-CH_3 \longrightarrow R-O-\overset{\overset{O}{\|}}{C}-CH_3 + CH_3-\overset{\overset{O}{\|}}{C}-OH$$

$$R-NH_2 + CH_3-\overset{\overset{O}{\|}}{C}-O-\overset{\overset{O}{\|}}{C}-CH_3 \longrightarrow RNH-\overset{\overset{O}{\|}}{C}-CH_3 + CH_3-\overset{\overset{O}{\|}}{C}-OH$$

The presence of the double bond was indicated by the formation of addition products (with halogen acids and with bromine).

$$R-CH=CH_2 \;+\; H-Br \;\longrightarrow\; R-\underset{\underset{Br}{|}}{C}H-\underset{\underset{H}{|}}{C}H_2$$

$$R-CH=CH_2 \;+\; Br_2 \;\longrightarrow\; R-\underset{\underset{Br}{|}}{C}H-\underset{\underset{Br}{|}}{C}H_2$$

This double bond was found to be a vinyl type of double bond ($-CH=CH_2$), since oxidation produces two carboxylic acids, one of which is formic acid.

$$RCH=CH_2 \xrightarrow{\text{oxidation}} RCO_2H \;+\; \overset{\overset{O}{\|}}{H}C-OH$$

Thus, the quinine molecule was known to contain two tertiary amine functions, an alcohol group, and a vinylic double bond.

The first important information about the main skeletal framework of the molecule came in 1883 when a degradation product from potash fusion was shown to be the 6-methoxy derivative of quinoline.

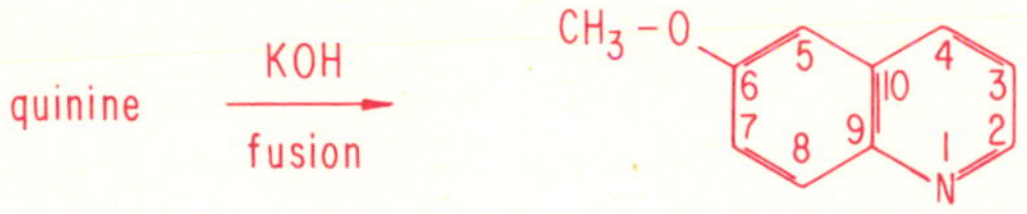

$$\text{quinine} \xrightarrow[\text{fusion}]{\text{KOH}}$$

6- methoxyquinoline

The 4-position in this fragment seemed a likely point of attachment to the rest of the molecule when an oxidation fragment of the molecule was shown to have a carboxyl group at that position. This fragment was named quininic acid.

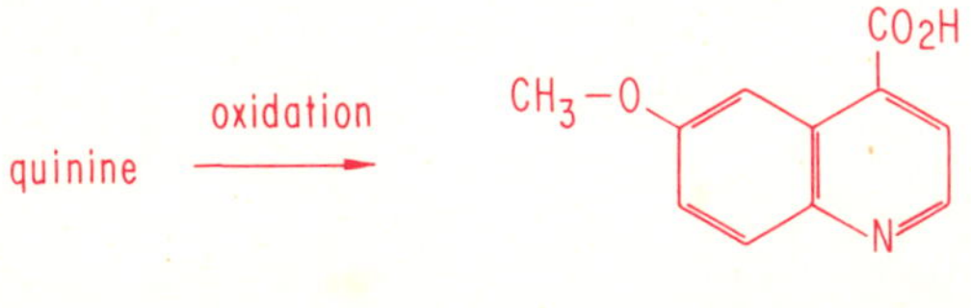

$$\text{quinine} \xrightarrow{\text{oxidation}}$$

quininic acid

Quininic acid contains eleven carbon atoms. The remaining nine carbon atoms of quinine appeared in another key degradation fragment called meroquinene. This was obtained by phosphoric acid degradation of anhydroquinine.

anhydroquinine $\xrightarrow{\text{dehydration}}$ $\xrightarrow[\text{acid}]{\text{phosphoric}}$ meroquinene $C_9H_{15}NO_2$

By a complicated series of transformations among interrelated compounds, meroquinene was shown just before the turn of the century to be a piperidine molecule substituted in the 3-position by a vinyl group and in the 4-position by an acetic acid side chain.

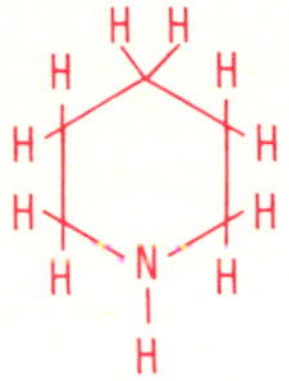

piperidine

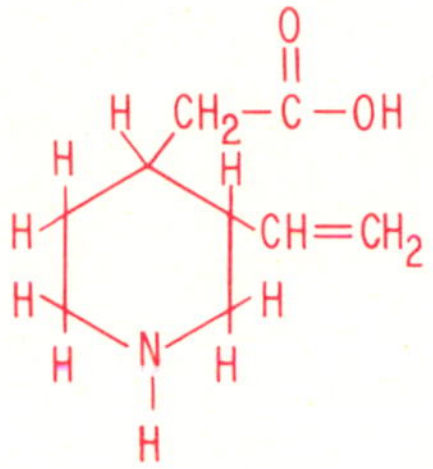

meroquinene

Looking at the various atoms in the meroquinene molecule leads one to conclude, on the basis of what is already known about the intact quinine molecule, that the carboxyl carbon atom and the nitrogen atom have been altered during the degradation. This follows from the facts that quinine contains no carboxyl group and no N—H group. The simplest way, then, to reconstruct the molecule would be to bond the carboxyl carbon to the nitrogen atom. This forms the quinuclidine portion of the molecule.

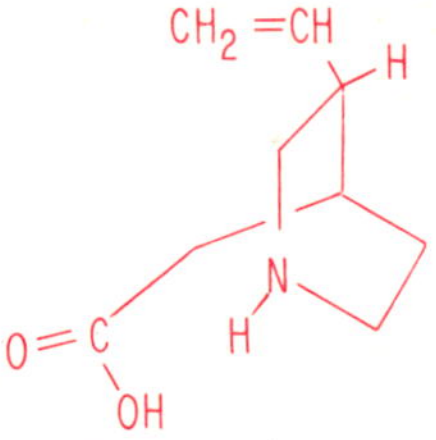

meroquinene

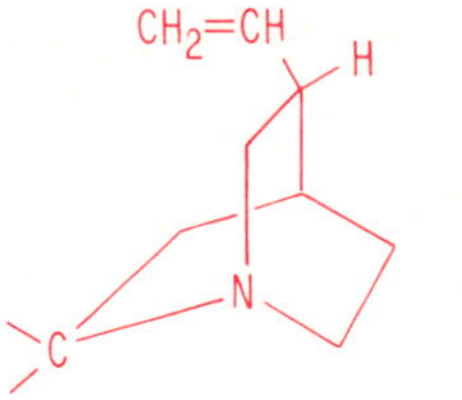

possible partial structure for quinine

If no other atoms within this nine-carbon fragment were disturbed during the phosphoric acid treatment, then the carboxyl carbon must also be the point of attachment for the rest of the molecule.

The remaining eleven carbon atoms seen earlier in the fragment called quininic acid include one carboxyl carbon atom. This atom was also obviously altered during degradation, again because quinine contains no carboxyl group. The most obvious reconstruction of the molecule, then, is to join these two carbon atoms to each other.

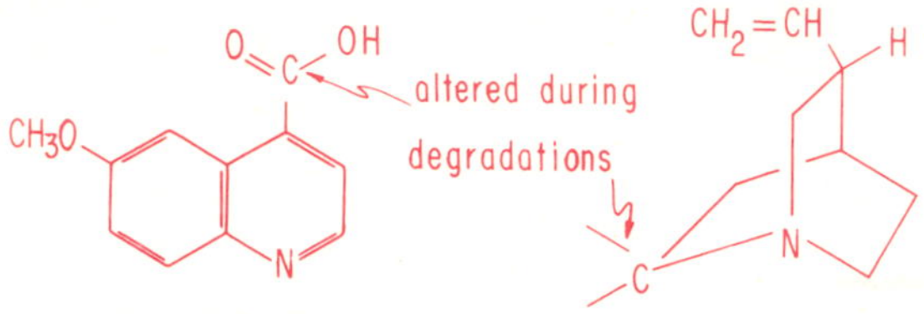

quininic acid

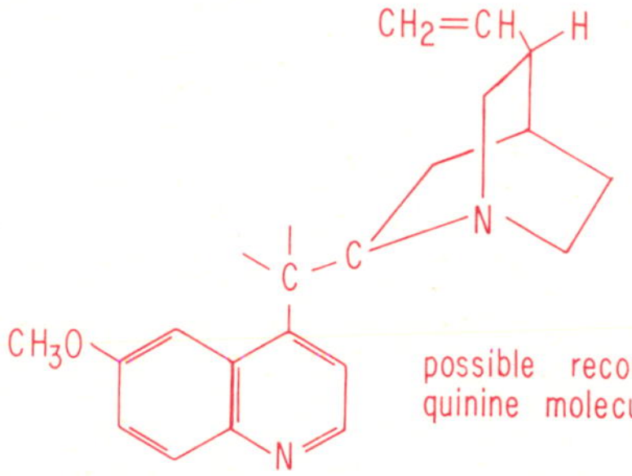

possible reconstruction of quinine molecule

The resulting structure accounts for all the atoms of the quinine molecule except for one oxygen atom, which is known to be present in a hydroxyl group. This hydroxyl group could presumably be on either of the altered carbon atoms. If it were on the quininic acid carbon atom, it would be a secondary hydroxyl group (R_2CH—OH), and should be oxidizable directly to a ketone.

secondary hydroxyl   ketone

The alternative would put the hydroxyl group on a carbon atom already bearing a nitrogen atom. This is much less likely, considering the stability of the molecule. At any rate, direct ketone formation without breaking up the molecular framework would be impossible. In 1907, the direct ketone formation was achieved, showing that the hydroxyl is indeed a secondary one.

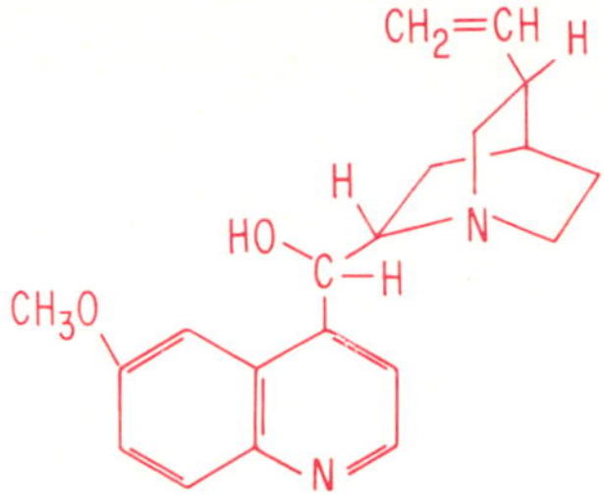

The total synthesis of this proposed structure which would ultimately prove its validity and hopefully answer some of the unsettled stereochemical aspects of the matter was stimulated by the critical shortage of quinine during the Second World War. It was brilliantly accomplished by R. B. Woodward and W. Doering during the war years and reported in 1945. The molecule is so complex, however, that the total synthesis would probably never have been a practical source of the material. Besides, in the meantime, a frantic program to develop simpler synthetic antimalarials succeeded and provided the Allied forces with an antidote for the debilitating disease.

Nevertheless, the synthesis of quinine stands as one of the truly classic organic syntheses. As is true of most "total" syntheses, it depends upon and incorporates earlier work done by other people, as Woodward and Doering point out in their report of the synthesis in the Journal of the American Chemical Society. They describe how they built their synthesis of quinine on earlier work which had culminated in the total synthesis of dihydroquinine fourteen years earlier.

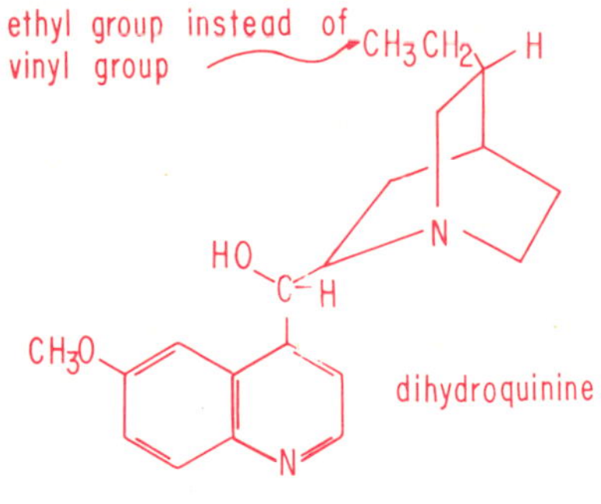

The total synthesis of dihydroquinine had been accomplished by the synthesis and the separation of the stereoisomers of dihydro-homo-meroquinene. This was done by the German chemist Rabe, who first proposed the correct structure for quinine in 1908.

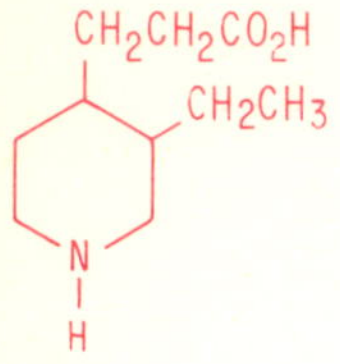

dihydro – homo – meroquinene 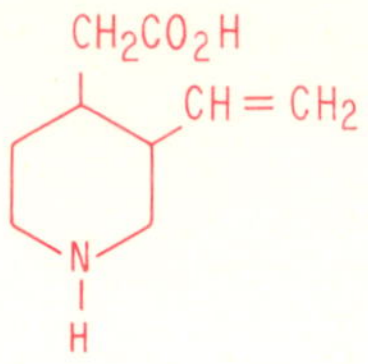   meroquinene 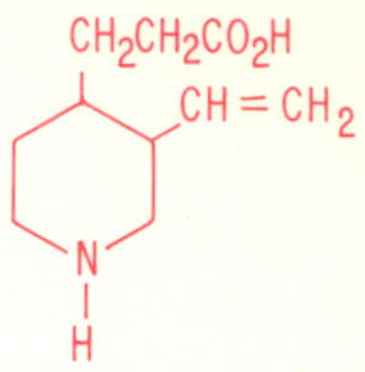   homo– meroquinene

The remainder of the work had already been done. Quininic acid had been synthesized independently by two groups in 1912. Rabe had already shown that dihydro-homo-meroquinene (obtained from natural sources) could be coupled to quininic acid esters, yielding a dihydroquinotoxine.

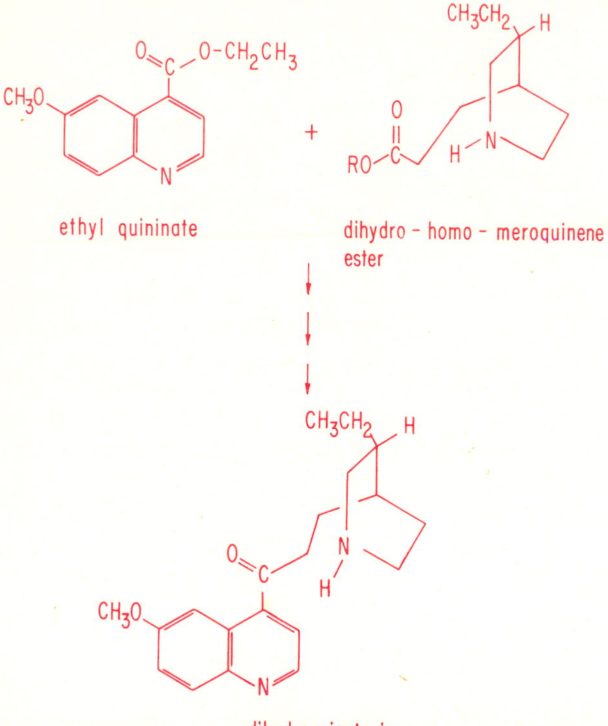

ethyl quininate   dihydro – homo – meroquinene
                                   ester

dihydroquinotoxine

Quinotoxine itself had first been obtained from quinine by Pasteur back in 1853. Rabe himself had succeeded in 1918 in converting natural quinotoxine back into quinine. Therefore, the only blank space in the total synthesis of dihydroquinine was the synthesis of dihydro-homo-meroquinene. Rabe accomplished this in 1931, starting with β-collidine, which had first been synthesized some ten years earlier.

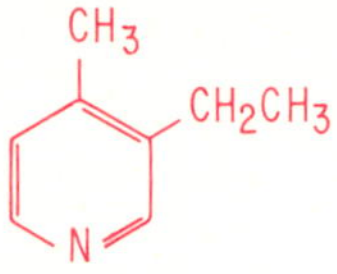

β - collidine 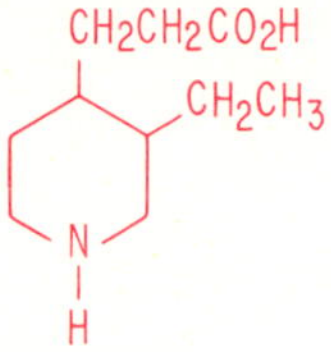   dihydro - homo - meroquinene

With all this work behind them, Woodward and Doering reasoned that the pathway to quinine itself would be smooth once they had synthesized homo-meroquinene. They therefore directed their efforts toward this objective.

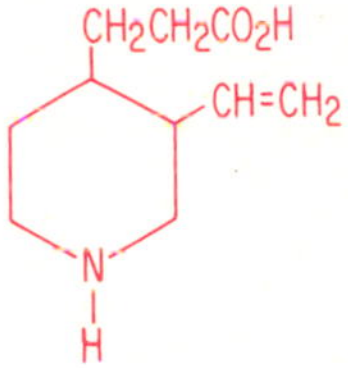

homo—meroquinene

As a starting material, they chose 7-hydroxyisoquinoline, since this molecule contains almost the complete carbon–nitrogen skeleton of homo-meroquinene and had already been synthesized by Fritsch in 1895.

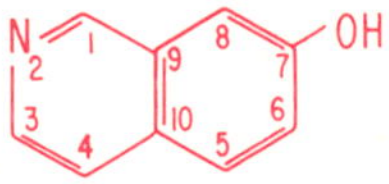

7 - hydroxyisoquinoline

Without delving into the many difficulties and complications which attended the conversion of 7-hydroxyisoquinoline into homo-meroquinene, an abbreviated outline of the synthetic pathway is shown below.

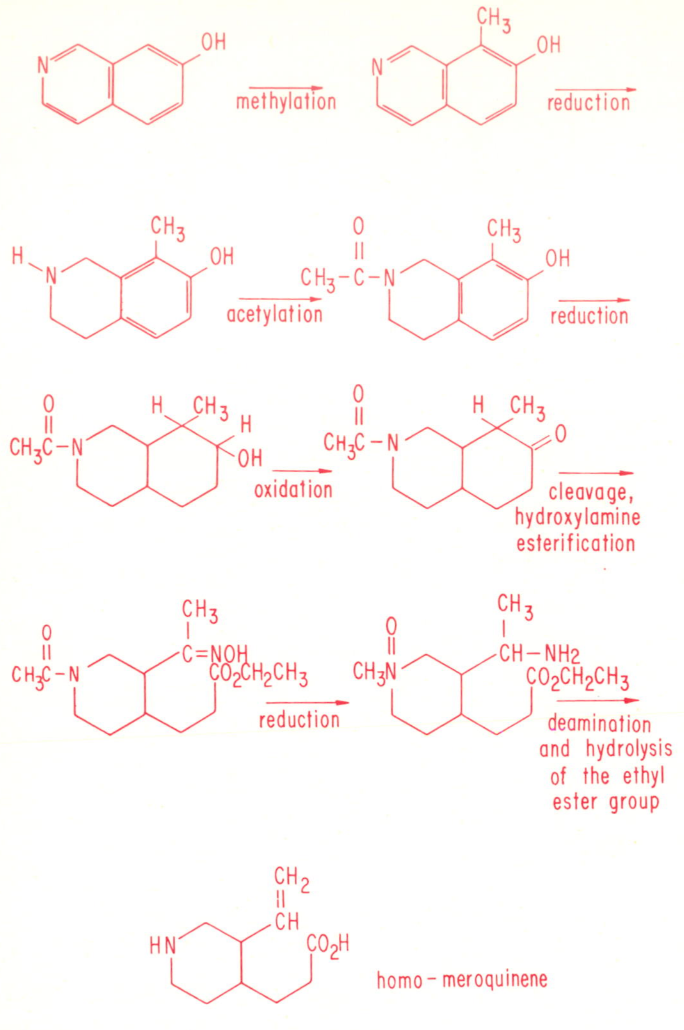

homo – meroquinene

Conversion of homo-meroquinene into quinotoxine, as had been done in the case of the dihydro derivative, completed the total synthesis of quinine, thus fulfilling a boyhood ambition of R. B. Woodward, now a Nobel prize winner and acknowledged master of organic synthesis.

As part of the continuing effort to develop new easily synthesizable anti-malarial molecules, it was naturally of interest to find out which parts of the quinine molecule are responsible (or essential) for its antimalarial activity. Rather surprisingly, perhaps, considering the flutter to synthesize quinine during the early 1940s, reduction of the vinyl group to produce dihydroquinine does not significantly alter quinine's antimalarial activity.

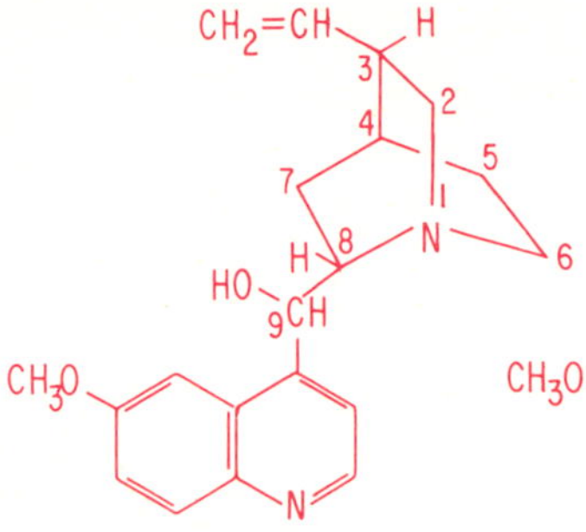

quinine 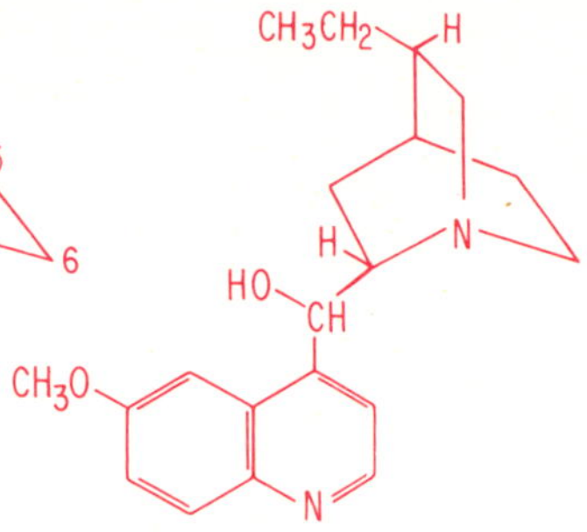   dihydroquinine

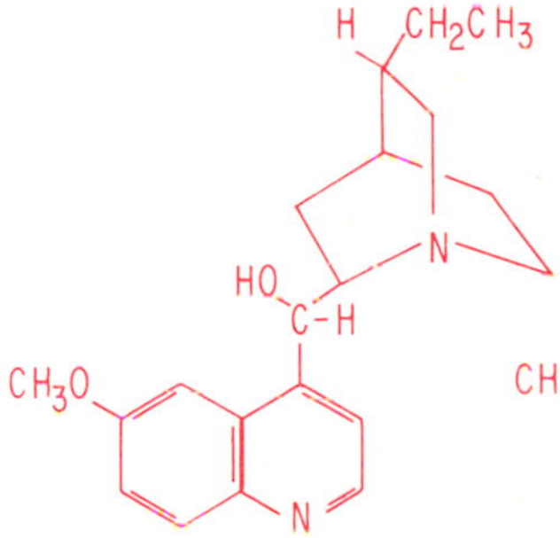

epi – dihydroquinine 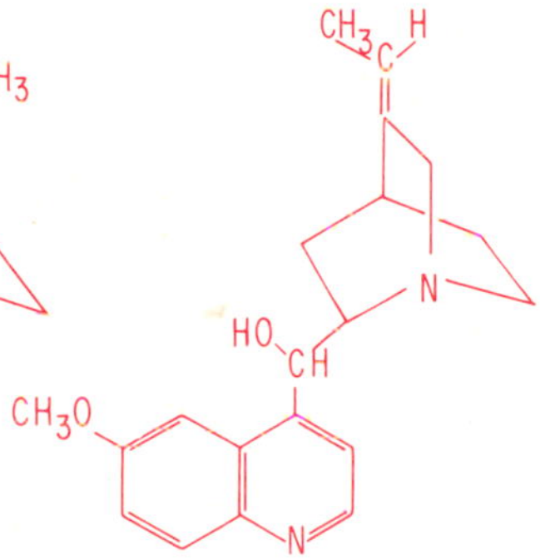   isoquinine

Even when the configuration of carbon atom-3 is inverted to form epi-dihydro-quinine, the molecule retains its potency. Furthermore, migration of the double bond, producing isoquinine, has little effect on the antimalarial properties. However, if the configuration at carbon atom-9 (the carbinol group) is inverted to form epi-quinine, the molecule does lose its potency. Also essential to antimalarial activity is the bond between carbon atom-8 and the quinuclidine nitrogen atom. Breaking this bond converts quinine into an inactive piperidine derivative.

Synthetic Antimalarial Drugs

The search for an easily synthesizable quinine substitute began in Germany during the First World War. Curiously, the First World War situation was the reverse of the Second World War situation in that quinine sources during World War I were in the hands of the Allies.

German scientists began their quest with the methylene blue molecule, which was known since 1891 to have some antimalarial properties.

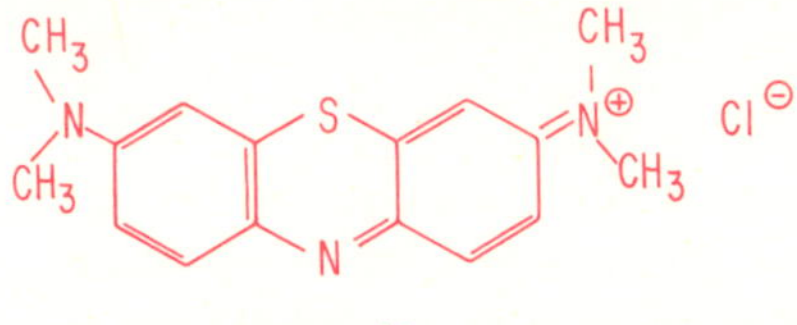

methylene blue

After a variety of structural modifications of the various portions of the methylene blue molecule, there emerged the first synthetic antimalarial, called pamaquine. Pamaquine contains the 6-methoxyquinoline structure of quinine with a dialkylaminoalkylamino side chain at position 8.

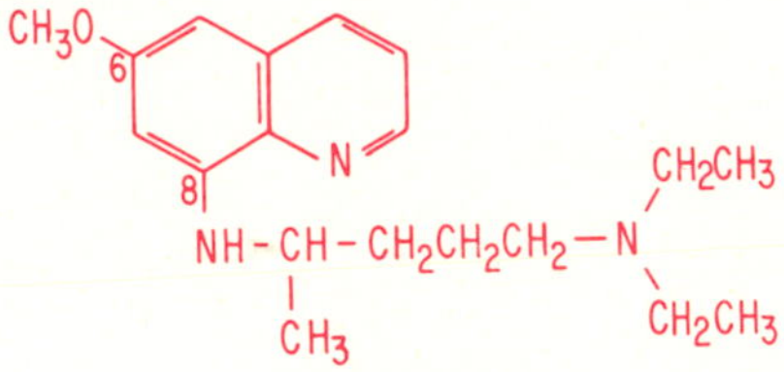

pamaquine

Further experimentation, however, resulted in a less toxic molecule, more effective in acute cases of malaria. This material, named Atebrin (also called quinacrine or mepacrine), contains the same side chain that pamaquine does, but contains an additional aromatic ring substituted with a chlorine atom. Atebrin was used extensively by Allied troops during the Second World War.

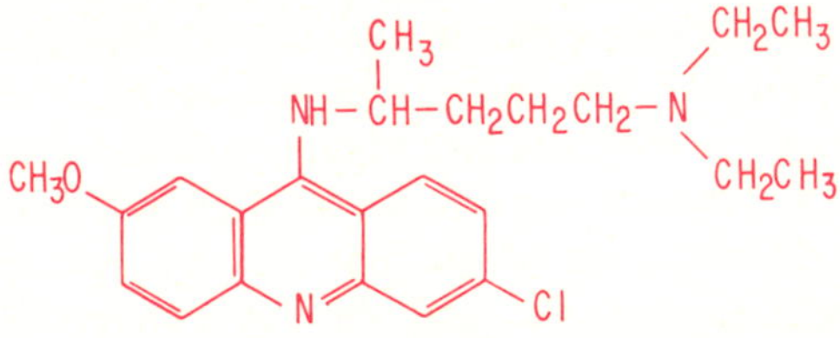

Atebrin

A simplified version of Atebrin called chloroquine also became an important antimalarial during the war. But neither chloroquine nor Atebrine nor even quinine itself could prevent relapses of the vivax type of malaria, and so the

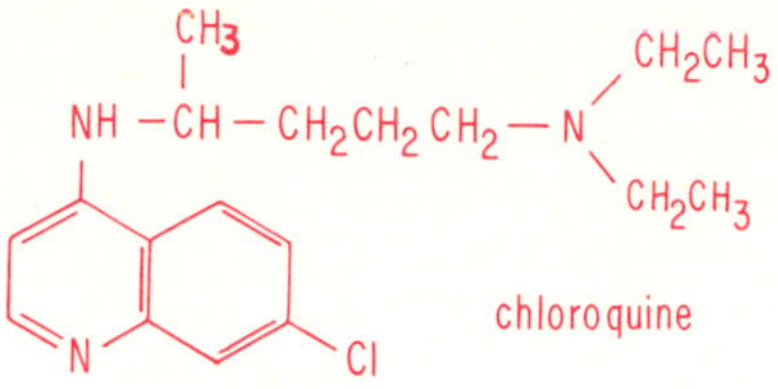

chloroquine

8-aminoquinoline structure of pamaquine, the first synthetic antimalarial, was reinvestigated. By 1945, when the war was about over, structural variations of the side chain resulted in pentaquine, and then in primaquine, the safest, most powerful of the 8-aminoquinolines.

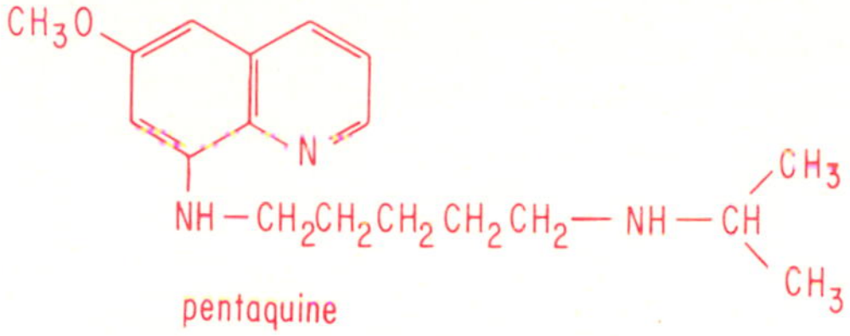

pentaquine

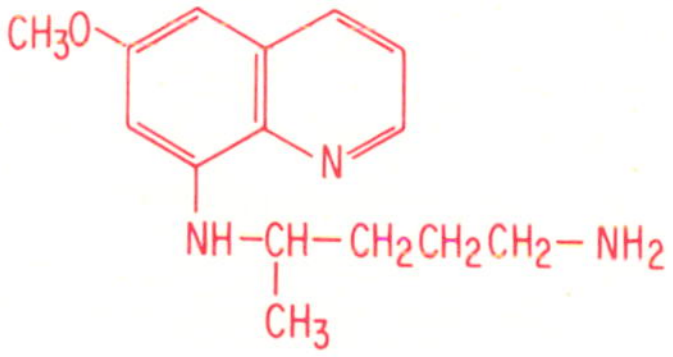

primaquine

After all those years, the only difference between pamaquine and primaquine is the two ethyl groups on the end of the side chain. This is obviously an important difference, however, since the absence of these alkyl groups provides the molecule with a hydrogen-bond-donor-NH_2 group at chain end.

A completely different kind of antimalarial molecule was developed in England as a result of wartime research. An example is chloroguanide. Unfortunately, use of this type of antimalarial appeared to give rise to chloroguanide

resistant strains of malaria. Some of these are apparently resistant to other kinds of synthetic antimalarials as well, necessitating recourse to the natural antimalarial quinine. The search for new synthetic antimalarial drugs is continuing with renewed vigor because of the disconcerting and unexpected appearance of resistant strains of this age-old disease.*

Sulfanilamide

The modern era of chemotherapy, in which organic molecules synthesized in the laboratory are used to cure disease, began with an explosion of research activity throughout the world following reports in 1936 that the long-known sulfanilamide molecule has almost miraculous powers to fight a wide variety of infections often lethal to man. First synthesized in 1908, sulfanilamide was being produced inexpensively in tonnage quantities for use as an intermediate in the dye industry prior to its emergence as the first modern "wonder drug".

Considering its effectiveness, the sulfanilamide molecule is surprisingly simple, as aspirin is. It is merely aniline (phenylamine) substituted in the *para* position by a sulfonamide group.

aniline
(phenylamine)

sulfanilamide
(*p* − aminobenzenesulfonamide)

---

* We say that an organism is resistant to a chemical when that given chemical has been used to select for those very few organisms which are not harmed by the chemical. Those few organisms then breed and the population soon has a large number which are resistant. Mutant forms of an organism (those with a different DNA code for a given gene or genes) usually arise by chance, but they can also be caused by exposure of the sex cells of the parent to radiation or certain chemicals, like nitrogen mustard. The first step in natural selection therefore, is the appearance of an alternate form of a given trait. The environment can then select for or against that trait. In this case, a synthetic-antimalarial-resistant *Plasmodium* was selectively favored.

In contrast with quinine, which is really a specific drug against malaria, sulfanilamide and its derivatives are active against a broad spectrum of bacterial infections. Even compared with penicillin, which is widely useful but limited in activity to Gram-positive bacteria (bacteria which bind the Gram stain), sulfanilamide can destroy both Gram-positive and Gram-negative bacteria. Areas of application include malaria, tuberculosis, leprosy, meningitis, pneumonia, scarlet fever, plague, respiratory infections, and infections of the intestinal and urinary tracts.

In spite of the simplicity of the molecule, some 10,000 structural modifications have been prepared during the blaze of research activity following the discovery of its anti-microbial powers. Only about thirty derivatives have gained clinical acceptance.

How did the discovery happen to be made? Investigations into selective staining of bacteria by dyes showed that certain dyes have the ability to kill microbes *in vitro*, that is, in a bacterial culture outside a living host. This led to the hope that the same anti-microbial action might occur *in vivo*, that is, within an infected living host. Comparisons of *in vitro* and *in vivo* testing, however, showed poor correlation. Thus, the standard practice of screening compounds for anti-bacterial activity only by *in vitro* tests began to give way to direct *in vivo* testing, usually in infected mice.

German dye chemists in 1935 knew that the introduction of a sulfonamide group ($-SO_2NH_2$) into a dye enhanced the wool-fastness of the dye. With this in mind, they put a sulfonamide group into an azo dye to produce prontosil.

azo dye

prontosil

In vivo tests of prontosil showed anti-bacterial activity, although *in vitro* tests showed no activity. The then unknown key to this seeming inconsistency lies in the fact that animal cells metabolize prontosil and release sulfanilamide. Prontosil itself is inactive against bacteria. It is fortunate, therefore, that *in vivo* testing was becoming part of routine screening methods.

The German chemists, however, placed no great significance on the discrepancies between *in vitro* and *in vivo* testing. But French chemists at the Pasteur Institute did, and so began a series of structural modifications of prontosil. They found that a wide variety of azo dyes containing the sulfanilamide portion of prontosil showed little variation in *in vivo* anti-streptococcal activity. But if they altered the sulfanilamide portion, for example by replacing the sulfonamide group ($-SO_2NH_2$) by a carboxamide group ($-CONH_2$) or by a cyanide group ($-CN$), the antibacterial activity disappeared.

This indicated that the sulfanilamide portion of the dye molecule contained the antibacterial activity, and they proposed that this activity was due to metabolic cleavage of the azo group ($-N{=}N-$). Following this hypothesis, they found that animals do convert prontosil into sulfanilamide and that sulfanilamide itself is just as active as is the parent dye.

How does this little molecule act like a "magic bullet" to kill bacteria so cleanly without doing any apparent damage to animal cells? Although in most cases, details of the mechanism of drug action are not known, the case of sulfanilamide is an exception.

Sulfanilamide kills bacteria by interfering with the synthesis of folic acid. Synthesis of folic acid is an essential life process for bacteria. In contrast, animal cells do not synthesize folic acid, but do require folic acid molecules in the diet. This is why sulfanilamide is toxic to bacteria but not to man.

folic acid

pteridine derivative | *p*-aminobenzoic acid | glutamic acid

The folic acid molecule may be visualized as containing three component molecules—a pteridine derivative, a *p*-aminobenzoic acid molecule, and glutamic acid (one of the common amino acids).

Sulfanilamide hinders the biosynthesis of folic acid by competing against *p*-aminobenzoic acid for incorporation into the folic acid molecule. The structure of sulfanilamide is evidently similar enough to the structure of *p*-aminobenzoic acid to allow the sulfanilamide molecule to "fool" the enzymes responsible for linking the three portions of the molecule together.

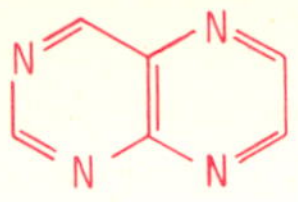

pteridine 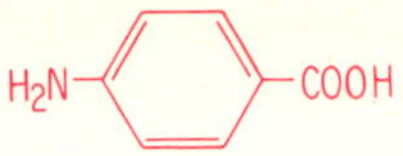   para-aminobenzoic acid 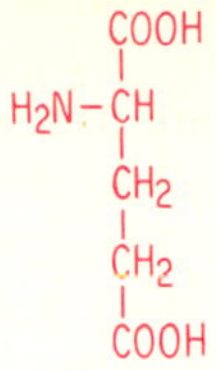   glutamic acid

Thus, sulfanilamide does actually become incorporated into a "mock" folic acid molecule, which presumably cannot perform the vital functions of "true" folic acid within the bacterium. This is the secret to the anti-bacterial activity of sulfanilamide—the first of the modern wonder drugs.

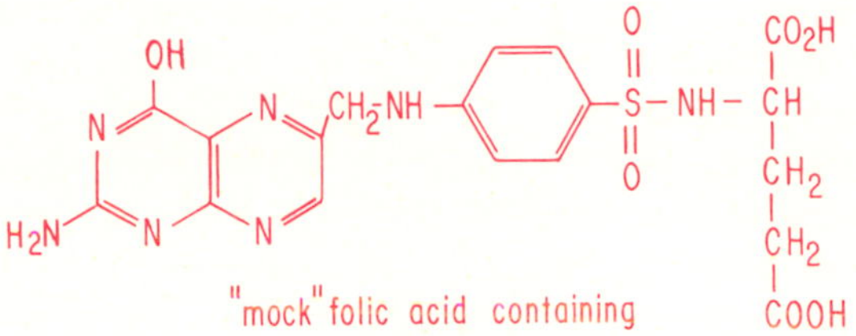

"mock" folic acid containing sulfanilamide

Penicillin and Other Antibiotics

The next tremendous development in the use of organic molecules in the fight against disease was the discovery of penicillin. To qualify as an antibiotic in the generally accepted sense of the term, a molecule must be produced by a micro-organism and must be capable of hindering the growth of another micro-organism (or killing it).

The anti-microbial activity of penicillin was first observed in 1929 by the British bacteriologist Alexander Fleming. The circumstances of the discovery resulted from a fortunate accident. Cultures of staphylococcus germs growing on the common nutrient material known as agar happened to become contaminated by a green mold. Fleming noticed that growth of the green mold resulted in destruction of the staphylococcus germs in the vicinity of the mold. He then found that broth in which the mold had been grown was not harmful to white blood cells, and he suggested that the active principle therein might be useful as an antiseptic. The mold itself was subsequently identified as *Penicillium notatum*, but the possibility of its exudate being used to treat general infections was not immediately investigated.

Then, a few years before the beginning of World War II, H. W. Florey and
E. B. Chain at Oxford decided to look into the matter. Their efforts led in 1940
to the isolation of an impure sodium salt of penicillin. Having this, they were
able to show a remarkable anti-staphylococcal activity in infected mice. Early
the following year, penicillin was used for the first time on a human being—to
treat an infection in an Oxford policeman—and the antibiotic was on its way
to fame.

Because of the war, penicillin became a strategic material, and investigations
in England and in the United States were conducted under tight security
wraps. Primary effort aimed at fermentative production of the antibiotic in
large amounts. However, on the molecular level, crash programs attempted to
discover the structure of the molecule in the hope that it might be more easily
obtained in the pure state synthetically. Although not a particularly large nor
complex molecule, penicillin led structural organic chemists a merry chase
during the war years. It was not until 1945, and then only with the aid of x-ray
crystallography (see Chapter XI), that the correct structure was finally nailed
down. Much of the difficulty lay in the fact that pure samples were difficult to
obtain (and to maintain) because of an extreme sensitivity of the molecule to
all but the mildest experimental conditions. This annoying lability turned out
to be primarily due to the unexpected, and hard to believe, presence of a
four-membered cyclic amide function in the molecule. This β-lactam structure
had never before been observed in a natural product.

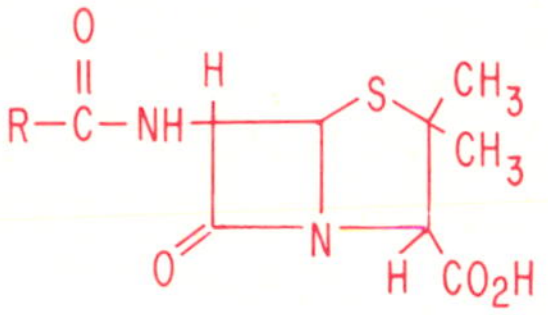

penicillin

where

R = ⬡–CH_2–

in penicillin G

On the bright side, perhaps, elucidation of the structure indicated that, even
if the structure had become known earlier, artificial synthesis would not have
been a practical source of the antibiotic.

The different kinds of *Penicillium* molds produce different kinds of penicillin
molecules, but these molecules differ only in the side chain designated R in the
structure above. A widely used penicillin, called penicillin G, contains a benzyl

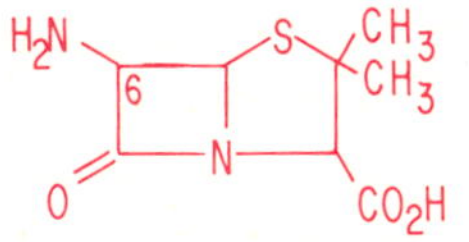

6 - aminopenicillanic acid

side chain (C_6H_5—CH_2—), and is often referred to as benzylpenicillin. The heart of the molecule is really the bicyclic portion known as 6-aminopenicillanic acid.

This molecule may be considered as having been derived from two naturally occurring amino acids—cysteine and valine.

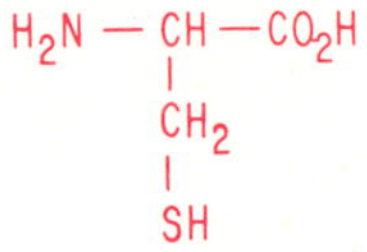

cysteine 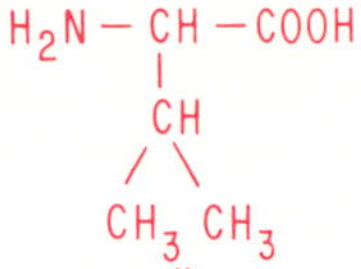   valine

The relationship between these two amino acids and 6-aminopenicillanic acid can be seen more easily by arranging the molecules in a more penicillin-like orientation.

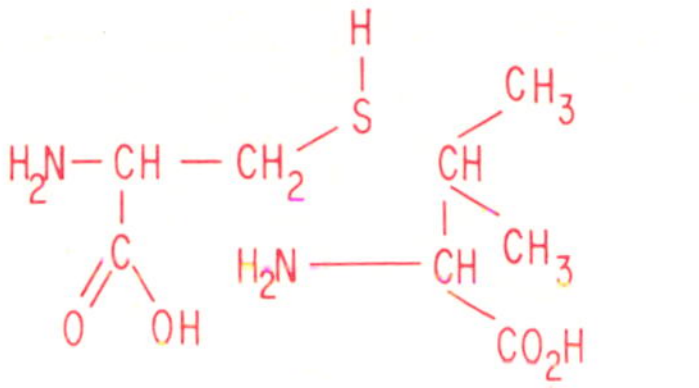

cysteine and valine arranged as in penicillin

Chemical degradation of penicillin, studied extensively during the work on the structure determination, does not split the molecule into these two amino acids, but it does produce related structures. First, the most sensitive part of the molecule—the β-lactam ring—opens to produce a dicarboxylic acid called penicilloic acid.

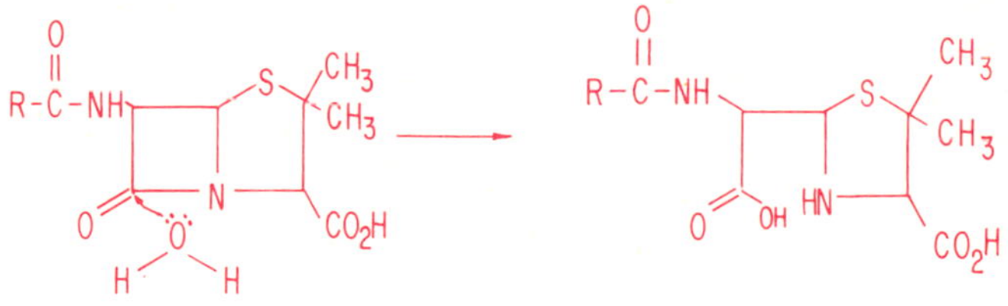

mild hydrolysis of penicillin   penicilloic acid

Some bacteria, particularly among the Gram-negative types, contain an enzyme called penicillinase which does just this. Since penicilloic acid is not active antibiotically, these bacteria are immune to treatment by penicillin.

Treatment of penicilloic acid with mercuric chloride splits off valine-thiol (called penicillamine).

penicilloic acid

valine – thiol
(penicillamine)

When the structures of penicillamine and of the other fragment were finally worked out, three possible structures were proposed for penicillin, based on the three most likely recombination products which could be visualized for the two fragments. Only one of these contained a β-lactam ring. With the information available at the time, however, no final decision could be made as to which was the true structure of the antiobiotic. X-ray crystallography soon showed the β-lactam formulation to be the correct one.

Over the years since then, much work has been done to try to find out how penicillin works—how it kills bacteria. As with sulfanilamide, many details are now known, and again as with sulfanilamide, the mode of operation takes advantage of a difference between a bacterial cell and an animal cell.

A bacterium requires a cell wall which, among other functions, protects the bacterium against changes in environmental osmotic pressures. Animal cells do not have cell walls; they merely have cell membranes. Penicillin prevents the production of new cell wall necessary for bacterial growth and reproduction. A faulty cell wall (or no cell wall) could cause a bacterium to rupture and spill its vital contents into the surroundings. It could also allow the natural defense molecules of the body, antibodies, to penetrate the bacterial membrane, and destroy the invader from within.

Penicillin is particularly, and almost exclusively, effective against Gram-positive organisms. At one stage in the construction of Gram-positive cell walls, an enzyme called transpeptidase causes the formation of a peptide bond between a carboxyl group of a D-alanine* and the amino group of a glycine. This is a cross-linking reaction between two short peptide chains. During the reaction, a molecule of D-alanine is lost from the end of the chains. This terminal D-alanine (with the free carboxyl group) probably binds to the active site of the transpeptidase enzyme.

---

* We have seen earlier that L-amino acids are the normal building blocks of proteins. In certain bacteria, however, D-amino acids frequently occur.

D—alanine

~~~ NH — CH — CH$_3$ CH$_3$
            | |
            C — NH ——— CH   D— alanine (bound to enzyme)
         ⁄⁄   \
        O)   CO$_2$H
           |
          ··
          NH$_2$
           | transpeptidase
glycine   CH$_2$
           |
           C = O
           ⸽

                          D- alanine
               ~~~ NH — CH—CH$_3$
                        |
                   O = C
                        | CH$_3$
                        NH   ⁄
           glycine | +   H$_2$N — CH   D- alanine
                        CH$_2$ \
                        | CO$_2$H
                        C = O
                        ⸽

Concerning the mode of action of penicillin, the important thing is that the alanine with the free carboxyl group which binds to the enzyme transpeptidase has the "unnatural" D-configuration. This matches the configuration of the carbon atom bearing the free carboxyl group in penicillin. This similarity in geometries and other evidence has led to the belief that penicillin interferes with cell-wall construction by binding to transpeptidase and inactivating this enzyme needed for the peptide-chain cross-linking reaction.

peptide—chain end   penicillin

~~~NH — CH — CH$_3$ CH$_3$ ~~~NH— CH —CH   S   C⟨CH$_3$
             | | | | ⟨CH$_3$
            C — NH — C   C — N ——— C
         ⁄⁄   \CO$_2$H   ⁄⁄   \CO$_2$H
        O   H   O   H

carbon atom bearing free carboxyl group
in each case has "unnatural" D-configuration

The cell walls of Gram-negative bacteria do not become stained when treated with the Gram dye. This indicates that the structure of the Gram-negative cell wall is different from the structure of the Gram-positive wall. A different cell-wall structure presumably requires different enzymes for its construction. Therefore, the absence of the Gram-positive enzyme transpeptidase in Gram-negative cell-wall construction would account for the general lack of effectiveness of penicillin against Gram-negative bacteria.

Streptomycin

With the demonstration of the antibiotic activity of penicillin in England, scientists in the United States began looking for other antibiotics from other sources. An especially fruitful source has been a group of soil microbes known as the *streptomycetes* (or *actinomycetes*). In 1940, S. A. Waksman at Rutgers University reported the isolation of an antibiotic named actinomycin. He followed this in 1942 with streptothricin, and in 1943 with streptomycin. At first streptomycin held tremendous promise as an antibiotic because it is active against both Gram-positive and Gram-negative bacteria. It soon became apparent, however, that bacteria often develop resistance to streptomycin during treatment. For this reason, its initial widespread use in treating Gram-negative infections has given place to more recently discovered antibiotics. However, its particular effectiveness against tuberculosis has sustained its usefulness. It is also sometimes used at low dosage in combination with penicillin or sulfanilamide. The discovery of streptomycin, following so closely on the heels of the discovery of the clinical value of penicillin, provided an explosive thrust to the search for new antibiotics, particularly in the pharmaceutical industry. This has resulted in the modern era of antibiotics, most of which were discovered in cultures of various types of *streptomycetes*. Thousands of new antibiotics have been discovered during the three ensuing decades, but relatively few are routinely used to any great extent. The structures of these are remarkably diverse.

The streptomycin molecule is carbohydrate derived, and therefore contains a large number of oxygen atoms. It has three principal component parts—streptamine, streptose, and glucosamine.

streptamine   streptose   glucosamine

These three components are linked to each other through the same kind of glycosidic linkages seen in the polysaccharides (starch, cellulose, and glycogen). Also, in streptomycin, the three amine functions are substituted. The two amine the molecule—the β -lactam ring—opens to produce a dicarboxylic acid called penicilloic acid.

functions in streptamine appear as guanidine groups (N_2H—C(=NH)—NH—)
and the one in glucosamine is methylated.

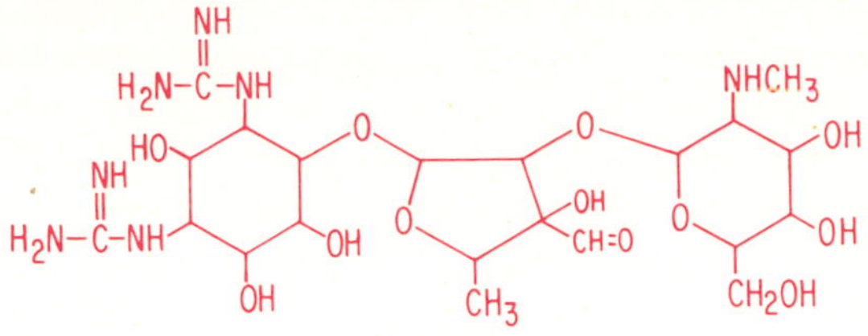

streptomycin

The therapeutic popularity of streptomycin waned mainly because of the
appearance of a new family of antibiotics known as the tetracyclines. These
molecules are also produced by *streptomycetes,* but their structures are
nothing like the streptomycin structure. The tetracyclines are all derivatives of
the tetracyclic fused-ring system known as naphthacene.

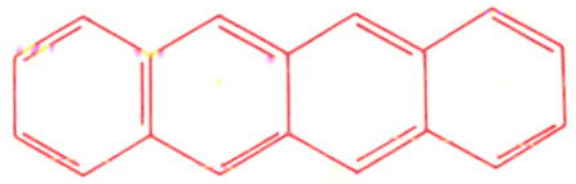

naphthacene

The first member of the family to be discovered was isolated from a culture
of *Streptomyces aureofaciens* in 1947 and named aureomycin. Actually, it is
7-chlorotetracycline, one of the few natural organic molecules which contain
a chlorine atom.

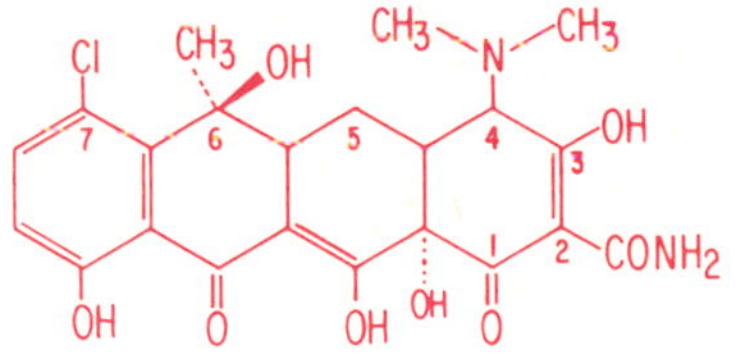

7-chlortetracycline
(aureomycin)

In 1957, mutant strains of *Streptomyces aureofaciens* were found to be
producing a tetracycline molecule missing the 6-methyl group. This molecule

also contains a chlorine atom in the 7-position, which makes it 6-demethyl-7-chlorotetracycline, or more simply declomycin. The 7-chloro substituent is not necessary for antibiotic activity, however. Its removal produces tetracycline itself, a clinically valuable member of the family.

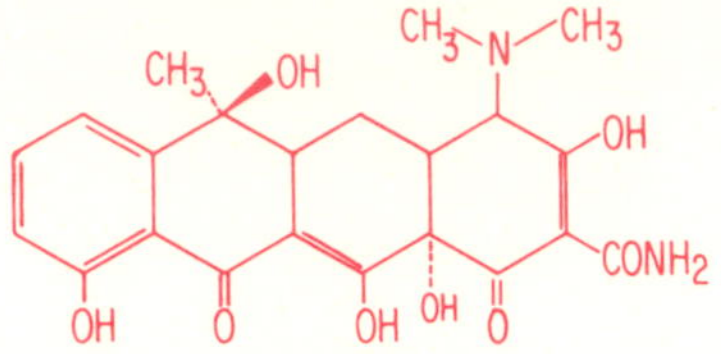

tetracycline

Structural modifications of the tetracyclines have shown that the minimum degree of substitution in which reasonable antibiotic activity is retained occurs with the removal of the 6-hydroxyl group from 6-demethyltetracycline to produce 6-demethyl-6-deoxytetracycline.

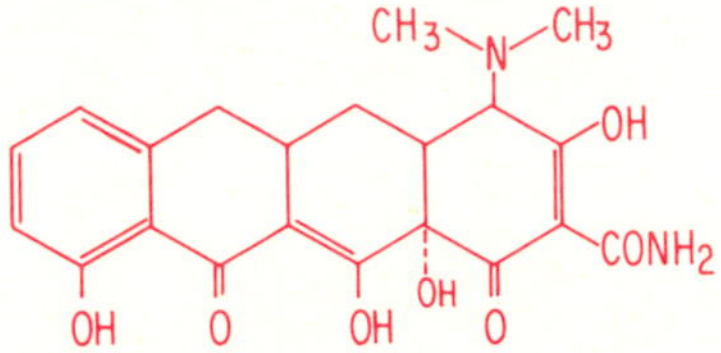

6-demethyl-6-deoxytetracycline

The tetracyclines are called broad-spectrum antibiotics because they are active not only against Gram-positive and Gram-negative bacteria but also against spirochetes, rickettsiae, and even some large viruses. In fact, they have a broader range of activity than any other antibiotic has. Except for the penicillins, they are used more than any other antibiotic. The total synthesis of 6-demethyl-6-deoxytetracycline was accomplished by R. B. Woodward in 1962, but clinical quantities of the tetracyclines are obtained more easily and more economically by bacterial fermentation.

Another broad-spectrum antibiotic, also produced by *streptomycetes,* has a much simpler structure and is produced synthetically in commercial quantities. Coincidentally, this molecule is another rare chlorine-containing natural molecule. Called chloramphenicol, this antibiotic is not used as extensively as it once was because of the danger of serious side effects. As in the case of aureomycin, the rather unexpected observation is made that the chlorine atoms are not essential for antibacterial activity.

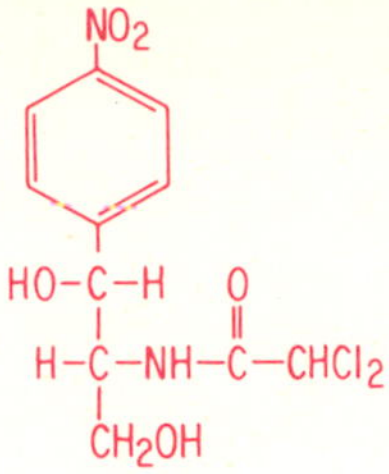

chloramphenicol

Chloramphenicol can be made from *p*-nitrophenyl methyl ketone.

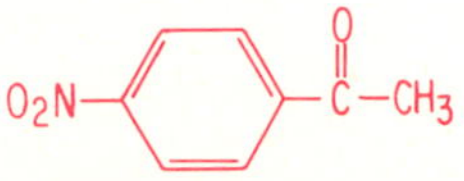

p-nitrophenyl methyl ketone

Bromination produces the bromomethyl derivative, which then reacts with ammonia to give a primary amine.

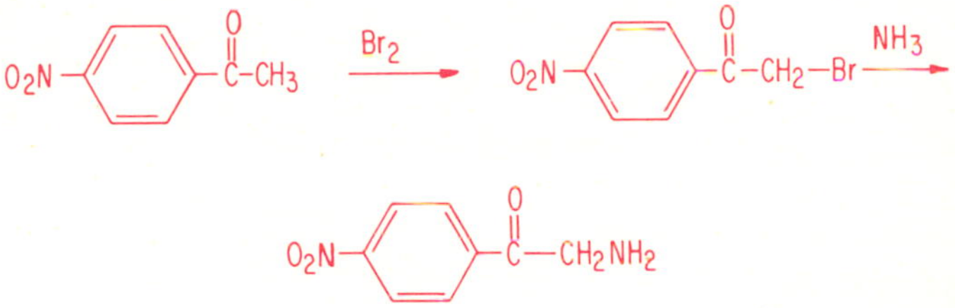

The acetyl derivative of the amine then reacts with formaldehyde to produce a primary alcohol.

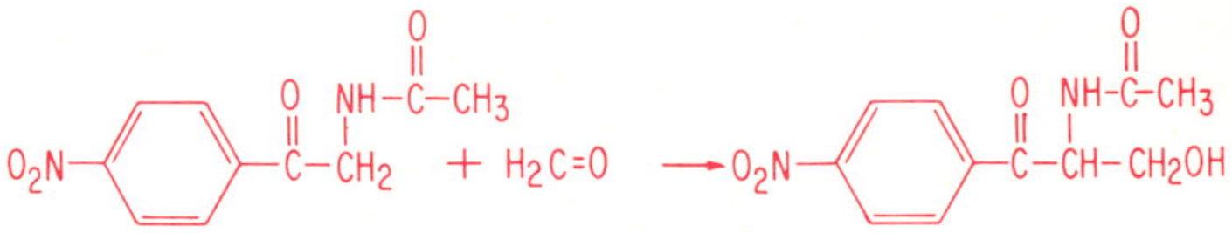

This completes the carbon skeleton of the antibiotic. The only remaining conversions needed are reduction of the ketone to an alcohol and replacement of the acetyl group with a dichloroacetyl group.

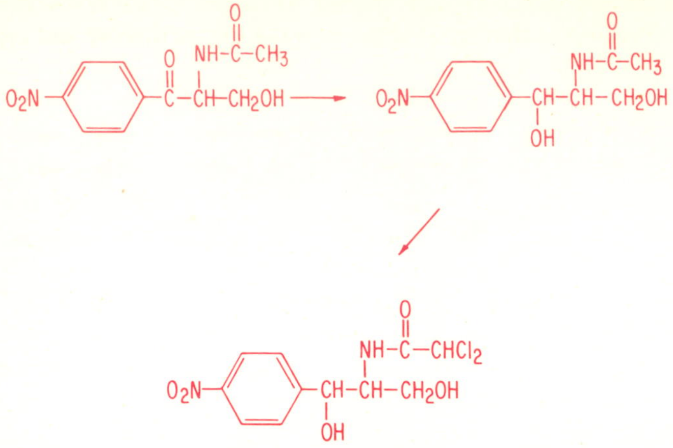

chloramphenicol

The chloramphenicol molecule contains two asymmetric carbon atoms. Since each of these asymmetric centers can exist in two configurations, four stereoisomers are possible.

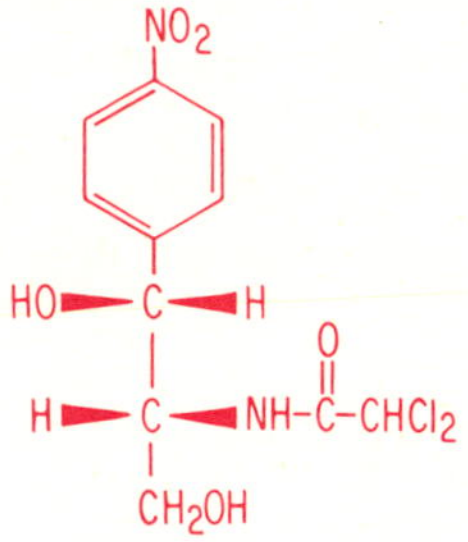

I

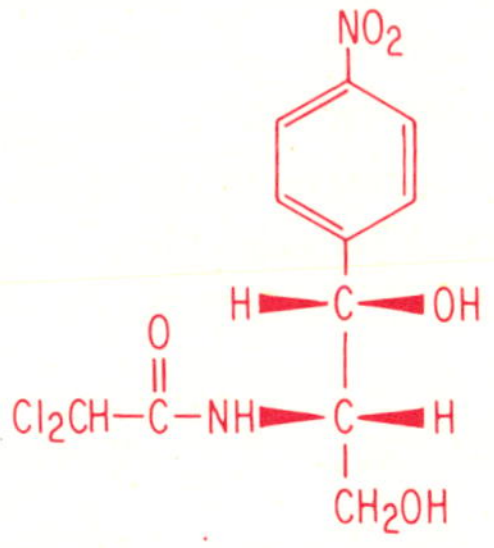

II

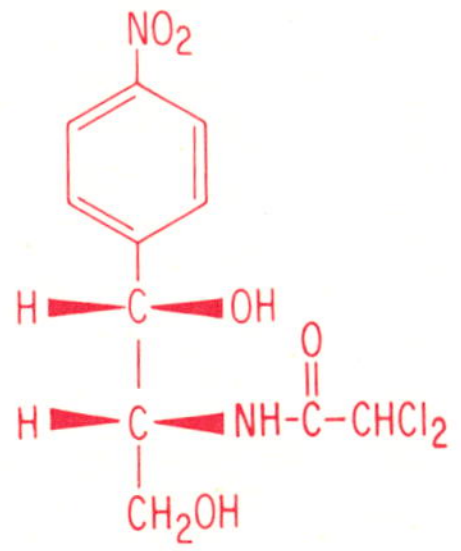

III

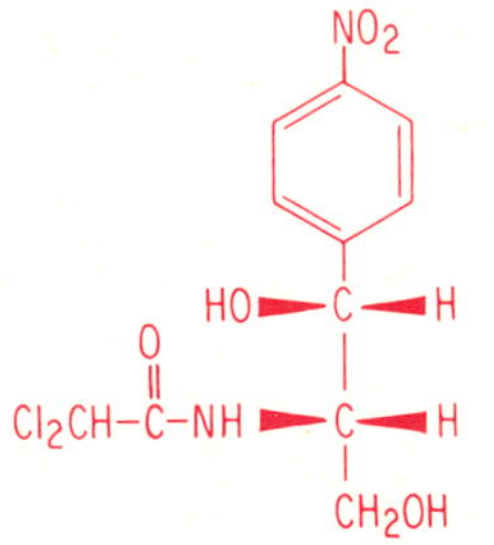

IV

The stereoisomers I–IV are named D-threo, L-threo, D-erythro, and L-erythro because of their resemblance to the sugars threose and erythrose.

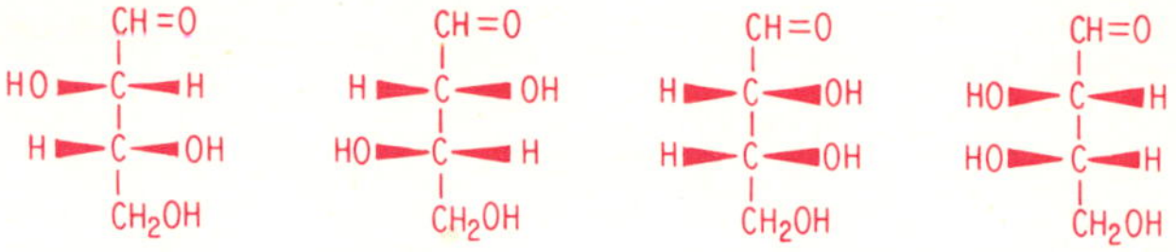

D-threose   L-threose   D-erythrose   L-erythrose

One of the interesting facts to come out of the synthetic work, which of course produces all four of the possible stereoisomers of chloramphenicol, is that only one of the four is active antibacterially. The active isomer has the D-threo configuration. Chloramphenicol is believed to owe its antibiotic potency to its ability to inhibit protein synthesis.

SUMMARY

This chapter was concerned with molecules isolated by the chemist which are used to control and/or eliminate specific diseases. Quinine and other anti-malarials, sulfanilamide, penicillin, streptomycin, chloramphenicol and tetracyclines were discussed. We observed the skill necessary to isolate, characterize and synthesize these molecules of might. It should also be clear that medicinal chemists must be alert to the development of resistant strains of microorganisms and must be prepared to synthesize new powerful antibiotics to control these virulent mutants.

SOURCE MATERIALS AND SUGGESTED READING

Burger, A., ed., 1970. *Medicinal Chemistry*, Vols. 1 and 2, second edition, John Wiley and Sons, New York.

Krieg, M., 1964. *Green Medicine*, Rand McNally, Chicago.

CHAPTER VII

THE STEROID FAMILY
Hardening of the Arteries to Family Planning

LIVING THINGS synthesize a large number of molecules, called steroids, which perform many specific biological functions, and yet which all have the same fundamental carbon skeleton. Some of these, for example the sex hormones, are synthesized in only minute quantities and are very powerful physiologically. In contrast, the steroid cholesterol is synthesized in abundance, and yet nature's purpose in doing this is unknown.

As with morphine, the fundamental carbon skeleton of the steroids can be considered a derivative of phenanthrene. The skeleton contains four hydrocarbon rings and is called the cyclopentano-perhydrophenanthrene ring system.

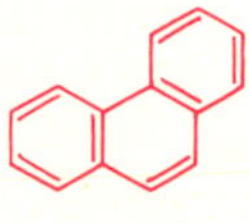

phenanthrene    cyclopentano-perhydrophenanthrene

The prefix perhydro means that enough hydrogen atoms have been added to phenanthrene to saturate the molecule. Cyclopentano means that the structure contains a five-membered ring.

The steroid family points out an interesting diversity between life forms. Steroids synthesized by animals are different from steroids synthesized by plants. Furthermore, plant steroids are not metabolized by animals. If an animal wants an external supply of usable steroids, it must eat another animal, not a plant. This is in sharp contrast to the striking unity of life seen earlier in such things as the universality of the anaerobic metabolism of glucose.

As an example of the almost unbelievable specificity of some metabolic enzymes, compare the structure of the animal steroid cholesterol with the structure of the plant steroid sitosterol. (The conventional enumeration of the carbons in cholesterol is indicated in the formula on p. 127.)

cholesterol
(an animal steroid)

sitosterol
(a plant steroid)

The only difference between these two molecules is that sitosterol contains an "extra" ethyl group (—CH_2CH_3) out near the end of the side chain. Even the stereochemical configurations of all eight asymmetric carbon atoms in each molecule are identical. And yet sitosterol is not metabolized by the higher animals, but cholesterol is.

From another point of view, consider the number of stereoisomers possible for a structure such as cholesterol. With eight asymmetric centers, there are 2^8 or 256 possible stereoisomers (see appendix). And yet the enzymes responsible for synthesizing cholesterol are so stereospecific that only one of these stereoisomers exists in nature.

Cholesterol has long been recognized as the principal constituent of gall stones, but in fact cholesterol is present in all the tissues of the body either as the free alcohol or in ester form. It is present in particularly high proportions in nerve tissue, in the spinal cord, and in the brain, where it accounts for about one-sixth of the dry weight. It is sometimes found lining the walls of arteries, where it "hardens" the arteries and diminishes the inside diameter of the blood vessel, which impairs circulation and contributes to high blood pressure. High levels of cholesterol in the blood seem to correlate with the prevalence of this pathological condition called arteriosclerosis, but the primary cause has yet to be discovered.

Despite its having been known as a pure material since 1812, the structure of cholesterol was not known completely until 1955. One of the difficulties in the structural determination, besides the intricate ring system and the centers of asymmetry, was that so much of the molecule is saturated hydrocarbon. The inertness of this skeleton makes it difficult to degrade or break down the molecule into recognizable fragments, which has traditionally been an essential tool in structure determinations. The necessity of resorting to severe degradation conditions produces a more random set of degradation reactions at a large number of positions in the molecule, often producing an assortment of small fragments difficult to separate and identify. But, fortunately, cholesterol is plentiful so that a single experimental blunder would not have wasted the world's supply, as can happen with natural products difficult to obtain.

An important breakthrough in the structural study came when cholesterol was shown to contain the same ring system that the bile acid, cholic acid, contains. The value of this discovery was that cholic acid contains two more hydroxyl groups than cholesterol does, and so cholic acid can be oxidatively degraded more easily and more systematically. The similarity of the ring systems of cholesterol and cholic acid was demonstrated by converting both molecules to cholanic acid.

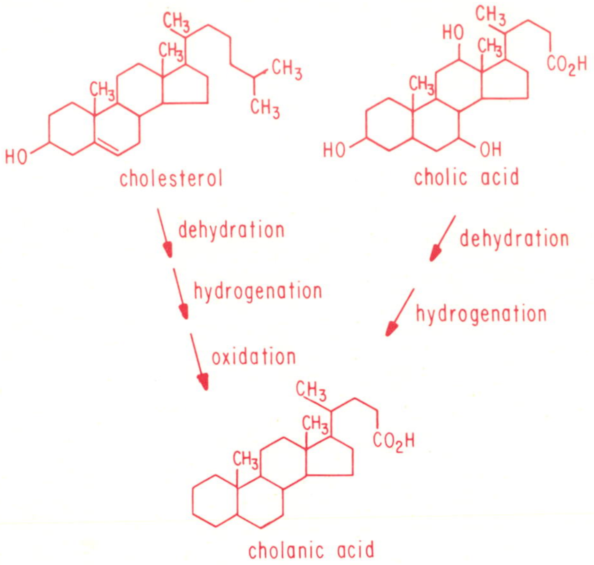

Dehydration of cholesterol removes the hydroxyl group and puts a second double bond into the molecule.

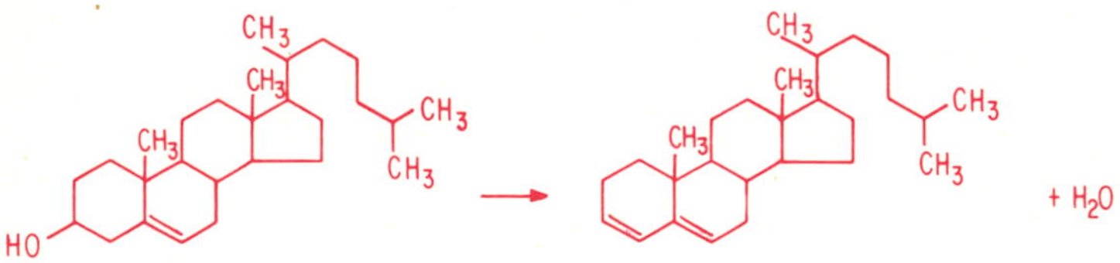

Addition of hydrogen to the diene creates a new asymmetric center at position 5. This produces two stereoisomers called cholestane and coprostane.

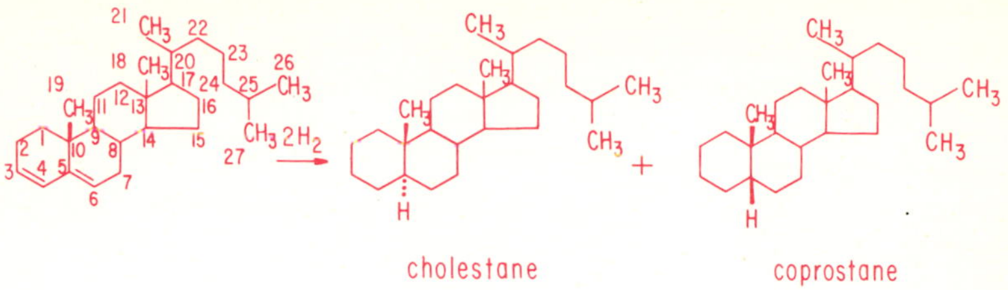

cholestane   coprostone

Only one of these stereoisomers—coprostane—has the same configuration at position 5 that cholic acid does, and so only coprostane gives cholanic acid on oxidation.

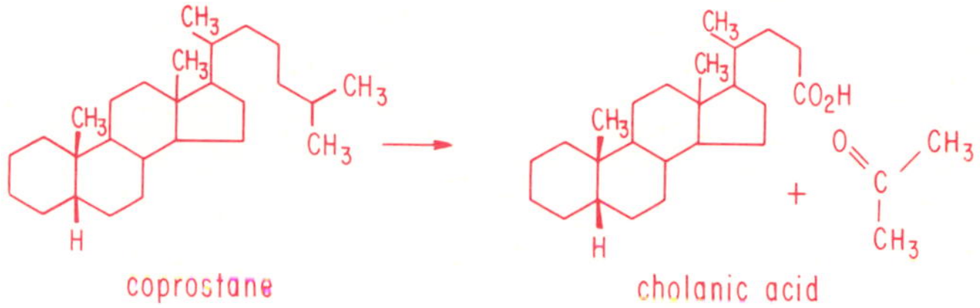

coprostane   cholanic acid

Dehydration of cholic acid followed by hydrogenation can produce cholanic acid without forming any new asymmetric centers.

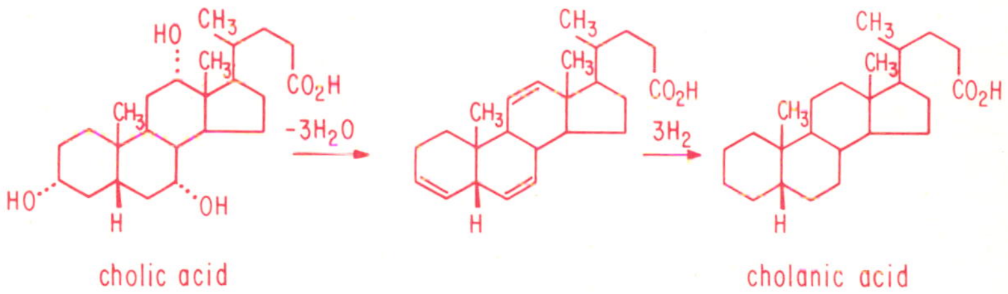

cholic acid   cholanic acid

The fact that cholesterol contains four rings can be deduced from the molecular formula, $C_{27}H_{46}O$, since the formation of a ring has the same effect on a molecular formula that the formation of a double bond has, that is, it diminishes the number of hydrogen atoms by two. For example, the molecular formula of the hexanes is C_6H_{14}, and the molecular formula for cyclohexane and for the hexenes is C_6H_{12}. Chain branching does not affect the molecular formula. A double bond can be distinguished from a ring because a ring does not react with hydrogen gas as a double bond does. The number of double bonds in a molecule can be determined by measuring the amount of hydrogen required to saturate the molecule. One mole of cholesterol takes up only one mole of

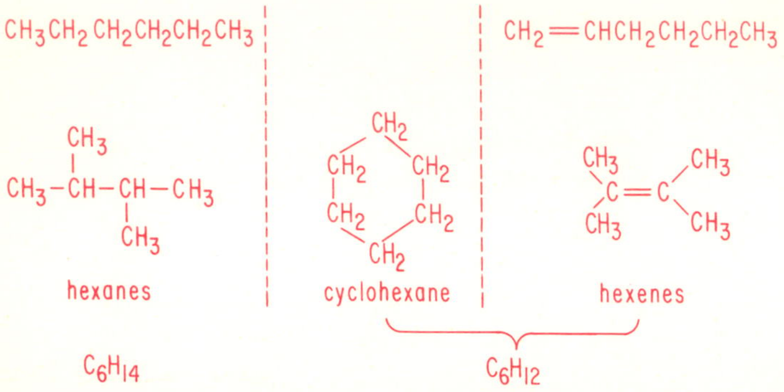

hydrogen, so it contains only one double bond. If it were an open-chain alcohol containing one double bond, for example allyl alcohol CH_2=CH—CH_2OH, its molecular formula would fit the generalized form $C_nH_{2n}O$, as allyl alcohol C_3H_6O does. Instead, cholesterol $C_{27}H_{46}O$ fits the form $C_nH_{2n-8}O$. It is eight hydrogen atoms short of an open-chain alcohol with one double bond, and so it must contain four rings.

An indication of ring size is obtained by oxidizing the ring open to a dicarboxylic acid, followed by pyrolysis (heating). A five-membered ring usually produces an anhydride during this treatment, and a six-membered ring usually ends up as a cyclopentanone.

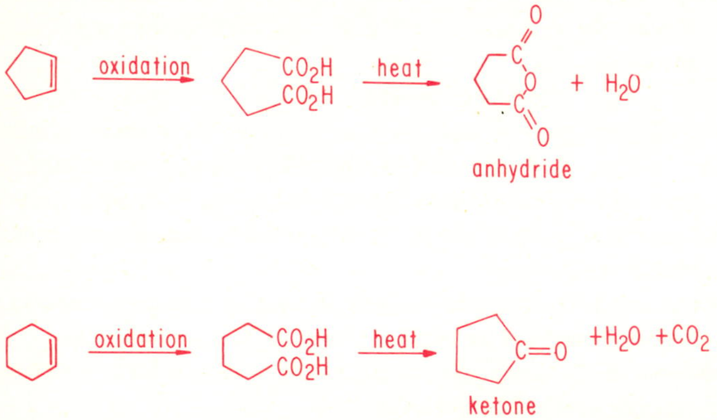

This was done in the cholesterol/cholic acid structural study and gave the correct answer for three of the rings and the wrong answer for one of the rings. On the basis of this evidence, cholesterol was believed to contain two five-membered rings and two six-membered rings. This prompted a proposal in 1928 of a structure for cholesterol, which was wrong, but which won a Nobel prize for the proposers.

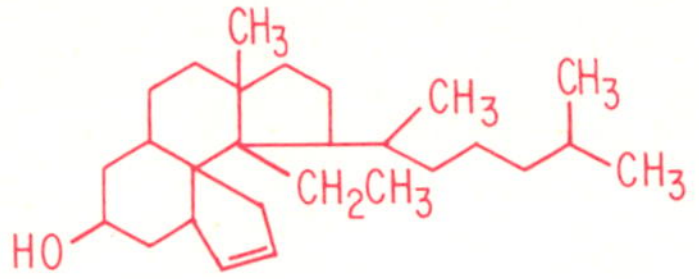

Incorrect structure proposed
in 1928 for cholesterol

Shortly thereafter, fortunately, it was discovered that selenium dehydrogenation of cholesterol produced chrysene and methylcyclopentanophenanthrene.

chrysene   methycyclopentano-
                  phenanthrene

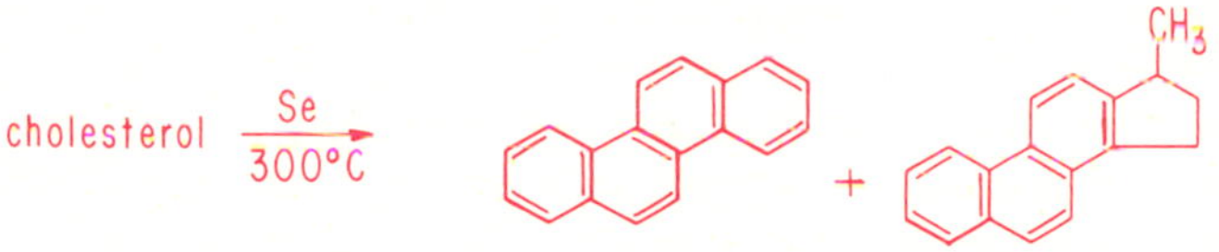

This led to the postulation of the correct ring system for cholesterol in 1932. The remainder of the structure, principally the stereoisomer aspects, required 23 more years of work.

Now comes the intriguing question of how cholesterol is synthesized in the body. The question became even more intriguing when it became known that cholesterol can be synthesized biologically using only acetate as a source of carbon atoms. How does this happen? The answer in detail is quite complicated, but the overall view is simple since it is reminiscent of the formation of natural rubber from isoprene.

First, three acetates combine to form an isoprene unit and carbon dioxide. Then six isoprene units combine to form a long open-chain compound called squalene. Then squalene simply cyclizes to produce a steroid called lanosterol—a precursor of cholesterol.

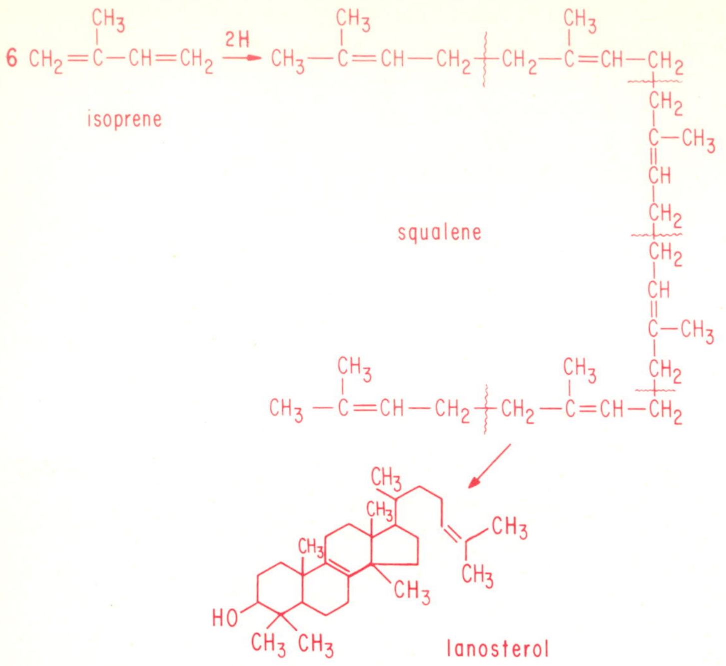

$$6\ CH_2=\underset{\underset{isoprene}{|}}{\overset{\overset{CH_3}{|}}{C}}-CH=CH_2 \xrightarrow{2H} CH_3-\overset{\overset{CH_3}{|}}{C}=CH-CH_2\!-\!\!\mid\!\!CH_2-\overset{\overset{CH_3}{|}}{C}=CH-CH_2$$

squalene

$$CH_3-\overset{\overset{CH_3}{|}}{C}=CH-CH_2\!-\!\!\mid\!\!CH_2-\overset{\overset{CH_3}{|}}{C}=CH-CH_2$$

lanosterol

As seen earlier in the citric-acid cycle, p. 37, acetate or acetic acid is activated by reacting with coenzyme A (RSH) to form the thiol-ester called acetyl-coenzyme A.

$$\underset{acetic\ acid}{CH_3-\overset{\overset{O}{\|}}{C}-OH} + \underset{coenzyme\ A}{H-S-R} \xrightarrow{ATP} \underset{acetyl\ coenzyme\ A}{CH_3-\overset{\overset{O}{\|}}{C}-S-R}$$

This reaction and those that follow for the biosynthesis of lanosterol are enzyme–catalyzed. For each reaction, there is a different, very specific enzyme (see Chapter III).

Loss of a slightly acidic proton (adjacent to the carbonyl group) provides an anion which can attack another molecule of acetyl-coenzyme A.

$$CH_3-\overset{\overset{O}{\|}}{C}-S-R \longrightarrow \overset{\ominus}{C}H_2-\overset{\overset{O}{\|}}{C}-S-R + H^{\oplus}$$

$$CH_3-\overset{\overset{O}{\|}}{C}-S-R + \overset{\ominus}{C}H_2-\overset{\overset{O}{\|}}{C}-S-R \longrightarrow CH_3-\overset{\overset{O}{\|}}{C}-CH_2-\overset{\overset{O}{\|}}{C}-S-R + \overset{\ominus}{S}-R$$

acetoacetyl coenzyme A

The next step in the formation of the biologically active isoprene unit from acetate is another nucleophilic attack by another anion of acetyl-coenzyme A.

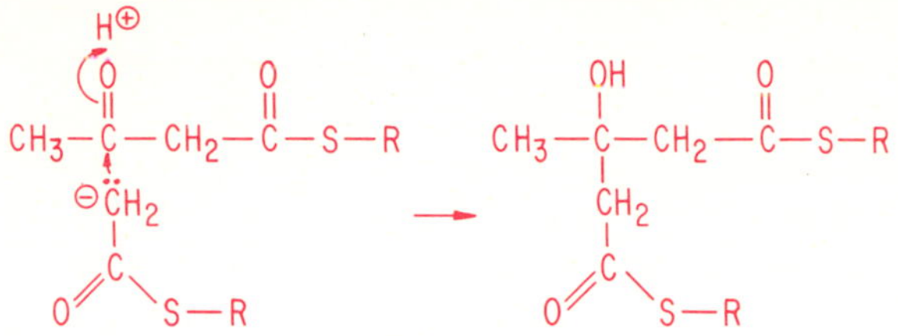

$$CH_3-\overset{\overset{\displaystyle H^{\oplus}}{\underset{\displaystyle O}{\parallel}}}{C}-CH_2-\overset{O}{\overset{\parallel}{C}}-S-R \qquad CH_3-\overset{\overset{\displaystyle OH}{\mid}}{\underset{\displaystyle CH_2}{C}}-CH_2-\overset{O}{\overset{\parallel}{C}}-S-R$$

Hydrolysis at one carboxyl group and reduction at the other produces mevalonic acid.

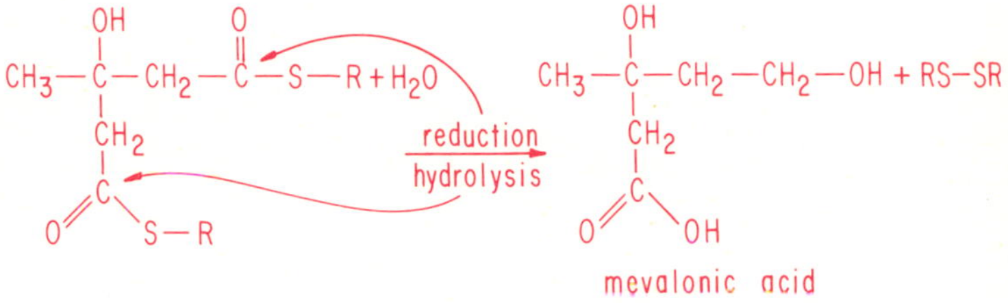

$$CH_3-\overset{OH}{\underset{CH_2}{C}}-CH_2-\overset{O}{\overset{\parallel}{C}}-S-R+H_2O \qquad \underset{\text{hydrolysis}}{\overset{\text{reduction}}{\longrightarrow}} \qquad CH_3-\overset{OH}{\underset{CH_2}{C}}-CH_2-CH_2-OH + RS-SR$$

mevalonic acid

Mevalonic acid undergoes dehydration and decarboxylation to provide the five-carbon skeleton of the isoprene unit. (Compare with formula for isoprene, p. 132.)

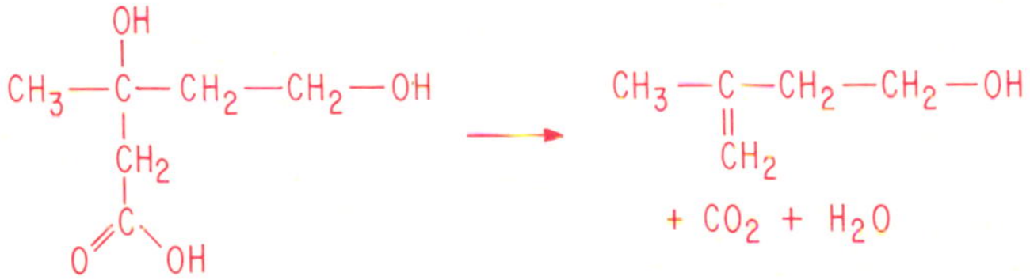

$$CH_3-\overset{OH}{\underset{CH_2}{C}}-CH_2-CH_2-OH \qquad \longrightarrow \qquad CH_3-\overset{}{\underset{CH_2}{C}}-CH_2-CH_2-OH$$

$$+ CO_2 + H_2O$$

This isopentenyl alcohol is activated by ATP to the pyrophosphate ester which acts as the fundamental physiological isoprene unit in biosynthesis.

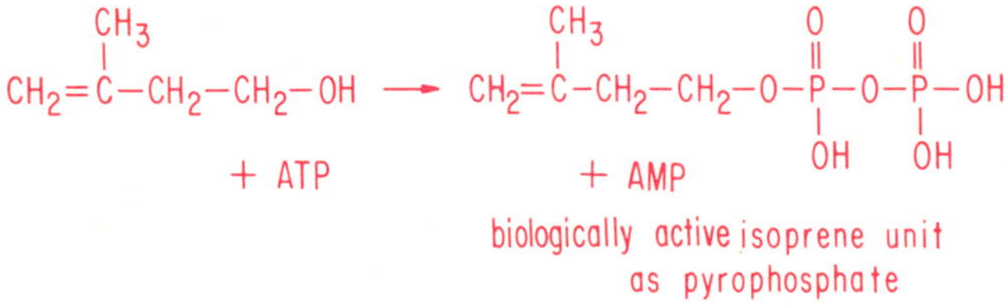

$$CH_2=\overset{CH_3}{\overset{\mid}{C}}-CH_2-CH_2-OH \longrightarrow CH_2=\overset{CH_3}{\overset{\mid}{C}}-CH_2-CH_2-O-\overset{O}{\overset{\parallel}{\underset{OH}{P}}}-O-\overset{O}{\overset{\parallel}{\underset{OH}{P}}}-OH$$

$$+ \text{ATP} \qquad\qquad + \text{AMP}$$

biologically active isoprene unit
as pyrophosphate

Biosynthesis of Squalene

The formation of squalene by combination of six isoprene units (p. 132) begins with isomerization of the double bond from the less stable terminal position to the more stable internal position. This is followed by loss of pyrophosphate anion to produce an allylic cation similar to the cation seen in the polymerization of isoprene in the laboratory.

$$CH_2=\overset{\overset{\displaystyle CH_3}{|}}{C}-CH_2-CH_2-O-\overset{\overset{\displaystyle O}{\|}}{\underset{\underset{\displaystyle OH}{|}}{P}}-O-\overset{\overset{\displaystyle O}{\|}}{\underset{\underset{\displaystyle OH}{|}}{P}}-OH \longrightarrow$$

$$CH_3-\overset{\overset{\displaystyle CH_3}{|}}{C}=CH-CH_2-O-\overset{\overset{\displaystyle O}{\|}}{\underset{\underset{\displaystyle OH}{|}}{P}}-O-\overset{\overset{\displaystyle O}{\|}}{\underset{\underset{\displaystyle OH}{|}}{P}}-OH \longrightarrow$$

$$CH_3-\overset{\overset{\displaystyle CH_3}{|}}{C}=CH-\overset{\oplus}{CH_2} \quad + \quad \overset{\ominus}{O}-\overset{\overset{\displaystyle O}{\|}}{\underset{\underset{\displaystyle OH}{|}}{P}}-O-\overset{\overset{\displaystyle O}{\|}}{\underset{\underset{\displaystyle OH}{|}}{P}}-OH$$

allylic cation   pyrophosphate

The allylic cation combines with an unrearranged isoprene pyrophosphate unit to produce a new cation.

$$CH_3-\overset{\overset{\displaystyle CH_3}{|}}{C}=CH-\overset{\oplus}{CH_2} \quad + \quad CH_2=\overset{\overset{\displaystyle CH_3}{|}}{C}-CH_2-CH_2-O-\overset{\overset{\displaystyle O}{\|}}{\underset{\underset{\displaystyle OH}{|}}{P}}-O-\overset{\overset{\displaystyle O}{\|}}{\underset{\underset{\displaystyle OH}{|}}{P}}-OH$$

$$\downarrow$$

$$CH_3-\overset{\overset{\displaystyle CH_3}{|}}{C}=CH-CH_2-CH_2-\overset{\overset{\displaystyle CH_3}{|}}{\underset{\underset{\displaystyle \oplus}{|}}{C}}-CH_2-CH_2-O-\overset{\overset{\displaystyle O}{\|}}{\underset{\underset{\displaystyle OH}{|}}{P}}-O-\overset{\overset{\displaystyle O}{\|}}{\underset{\underset{\displaystyle OH}{|}}{P}}-OH$$

This cation, to continue the head-to-tail polymerization, must lose a proton and pyrophosphate to form a new allylic cation.

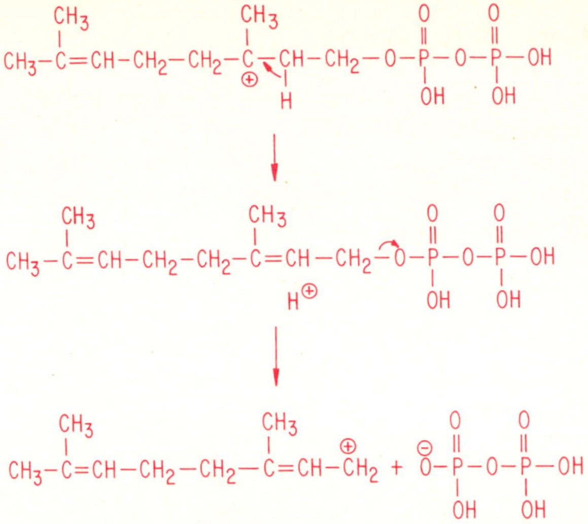

$$CH_3-\underset{\oplus}{C}=CH-CH_2-CH_2-\underset{\oplus}{\overset{CH_3}{C}}-\underset{H}{\overset{|}{C}H}-CH_2-O-\underset{OH}{\overset{O}{\overset{\|}{P}}}-O-\underset{OH}{\overset{O}{\overset{\|}{P}}}-OH$$

$$CH_3-\overset{CH_3}{\overset{|}{C}}=CH-CH_2-CH_2-\overset{CH_3}{\overset{|}{C}}=CH-CH_2-O-\underset{OH}{\overset{O}{\overset{\|}{P}}}-O-\underset{OH}{\overset{O}{\overset{\|}{P}}}-OH$$
$$H^{\oplus}$$

$$CH_3-\overset{CH_3}{\overset{|}{C}}=CH-CH_2-CH_2-\overset{CH_3}{\overset{|}{C}}=CH-\overset{\oplus}{C}H_2 \ + \ \overset{\ominus}{O}-\underset{OH}{\overset{O}{\overset{\|}{P}}}-O-\underset{OH}{\overset{O}{\overset{\|}{P}}}-OH$$

Combination with one more isoprene pyrophosphate unit produces a three-unit molecule called farnesyl pyrophosphate. In contrast to the natural-rubber *cis*-polyisoprene structure, the double bonds in farnesyl pyrophosphate are *trans*.

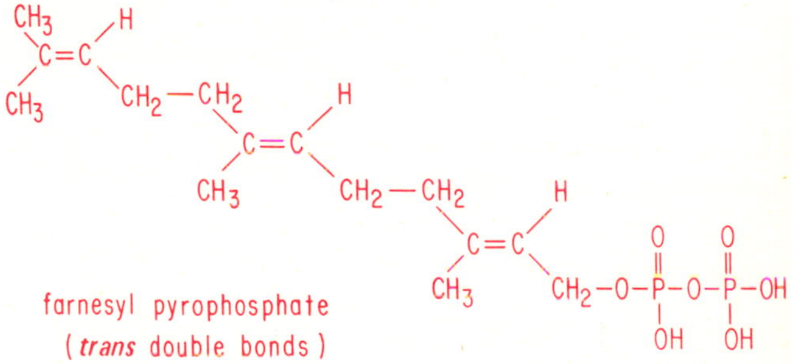

farnesyl pyrophosphate
(*trans* double bonds)

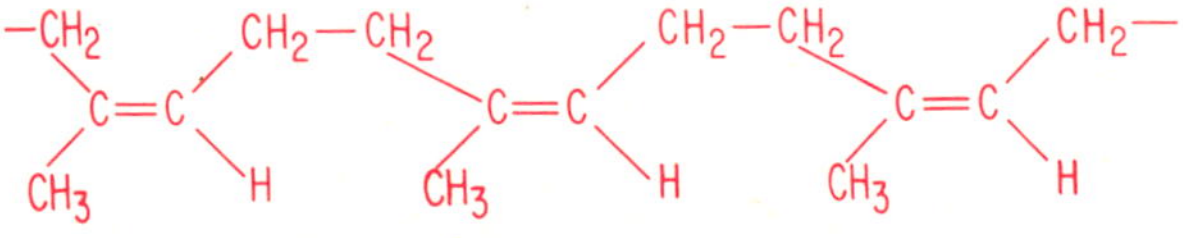

natural rubber structure
(*cis* double bonds)

Head-to-head coupling between two farnesyl pyrophosphate molecules produces squalene.

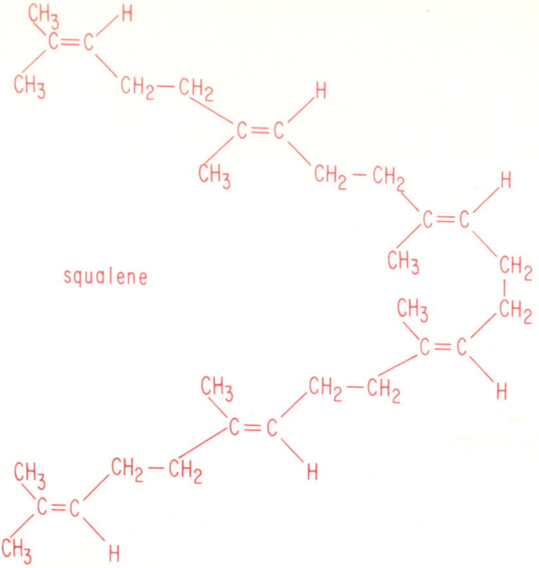

squalene

The structural relationship between squalene and the steroid ring system
becomes apparent in rewriting the structure. The utility of the *trans* double
bonds is also evident in this diagram.

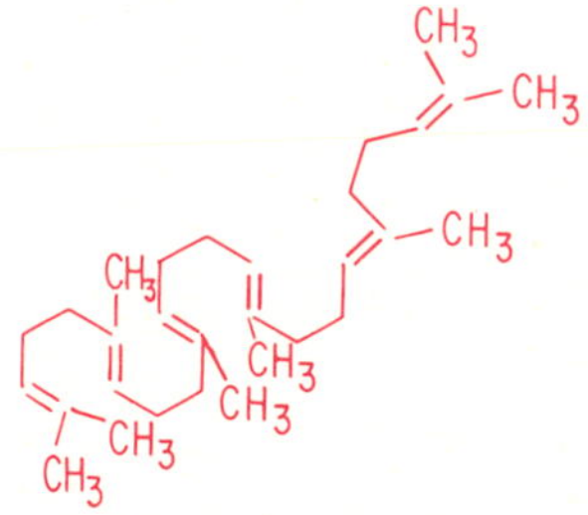

An electrophilic oxidizing agent, represented as HO^+, could initiate the ring
closures needed to convert squalene to a steroid.*

---

* The oxidation of one of the two terminal double bonds of squalene has recently been
carried out in the laboratory. This yields an epoxide shown below. The acid catalyzed ring
opening of the epoxide will lead to the ring closures shown on p. 137 (top) and a steroid
results.

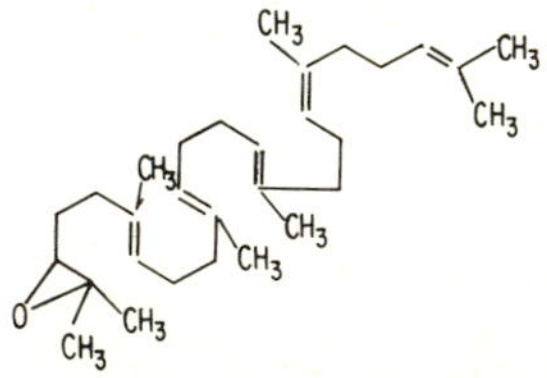

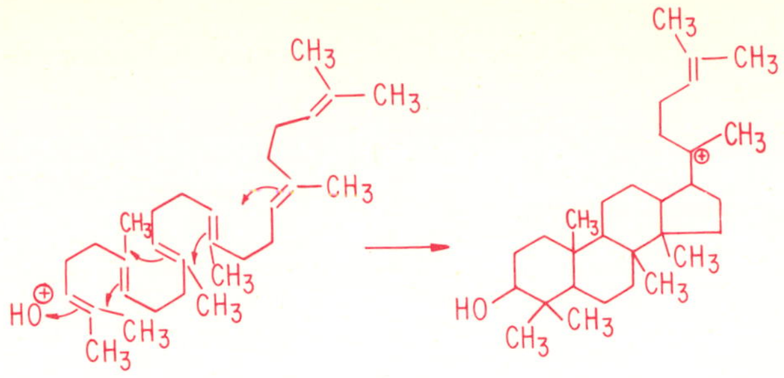

From this steroidal carbonium ion, the lanosterol carbon skeleton is just two hydride shifts and two methyl shifts away.

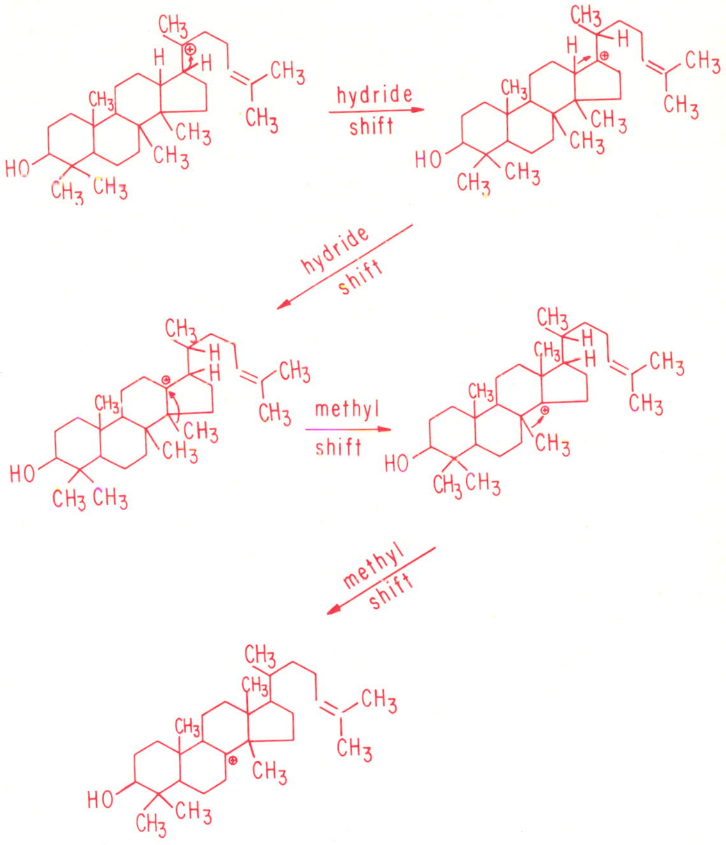

lanosterol skeleton

Loss of a proton completes the conversion of squalene to lanosterol.

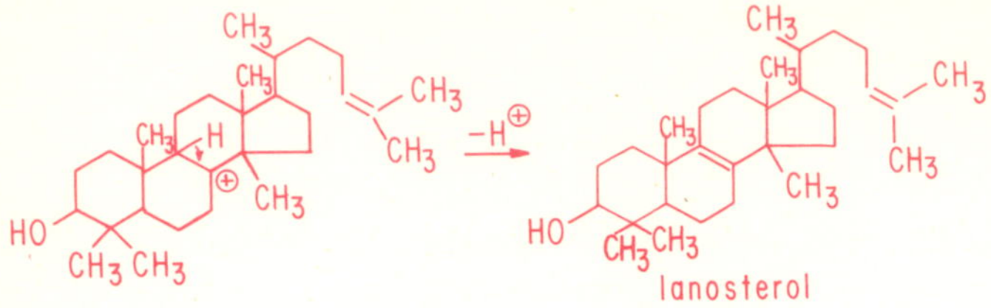

Of course, a series of reactions like this, which begins with acetic acid and produces a steroid such as lanosterol, is quite different from the ordinary random molecular-collision type of reactions encountered in non-biological systems. One would wait forever and in vain for a cyclopentanoperhydrophenanthrene molecule. The above reactions are highly specific, enzyme-catalyzed and directed toward a set of particular goals—a set of single stereoisomers selected from the thousands which are possible in complex systems such as these. This is the marvel of natural organic synthesis.

Evidence that the reactions pictured above in skeleton form do actually occur as shown comes mainly from experiments using isotope-labelling techniques. The general method involves feeding a molecule containing a radioactive isotope in a known position into a biological system, and then isolating the product molecule believed to be derived from the original feed molecule and degrading the product molecule systematically to locate the position of the radioactive atom. A Nobel prize was awarded in 1964 to Dr. Konrad Bloch and his associates for carrying out this immensely important work.

For example, if acetic acid containing carbon-13 in the carboxyl position is fed into a system capable of synthesizing mevalonic acid (see pp. 132–133), carbon-13 appears in all three oxygenated positions in mevalonic acid.

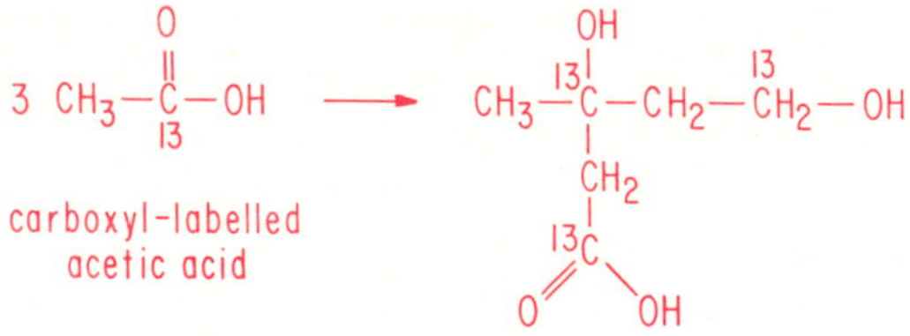

If the methyl group of acetic acid contains carbon-13, then the non-oxygenated positions of mevalonic acid become labelled with the stable heavy carbon isotope.

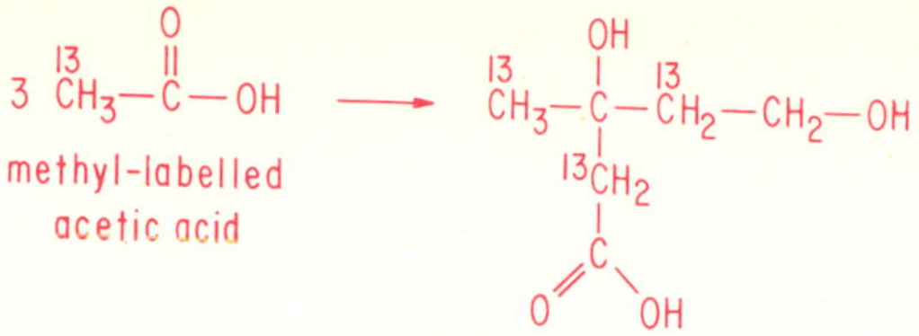

$$3 \ \overset{13}{C}H_3 - \overset{O}{\underset{\|}{C}} - OH \longrightarrow \overset{13}{C}H_3 - \overset{OH}{\underset{\underset{O}{\overset{|}{\underset{\|}{C}}}}{\underset{|}{\overset{|}{\underset{13}{C}H_2}}}} - \overset{13}{C}H_2 - CH_2 - CH_2 - OH$$

methyl-labelled
acetic acid

Studies like these have also shown that two methyl shifts must occur during the biosynthesis of lanosterol because the methyl group at the 8-position of the precursor steroid ends up at the 14-position in lanosterol, and the methyl group at the 14-position in the precursor ends up at the 13-position in lanosterol. In the following reactions radioactive carbon, C^{14}, is indicated by C^*.

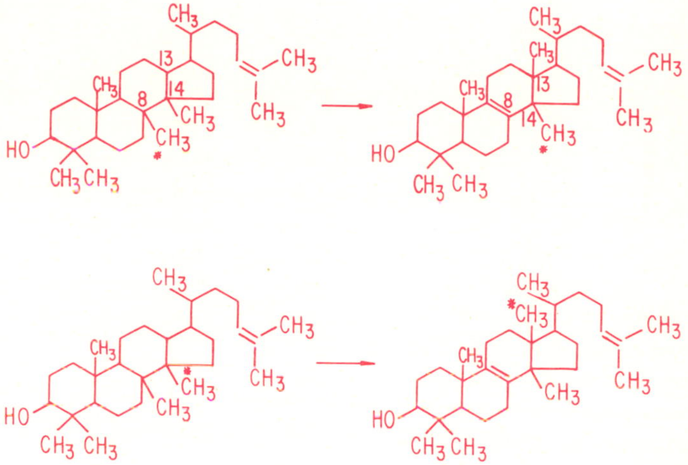

Finally, the validity of the overall scheme is shown by the fact that labelled squalene, prepared from methyl-labelled acetic acid and added to homogenized liver tissue (a source of necessary enzymes), produces lanosterol and cholesterol labelled in all the proper positions.

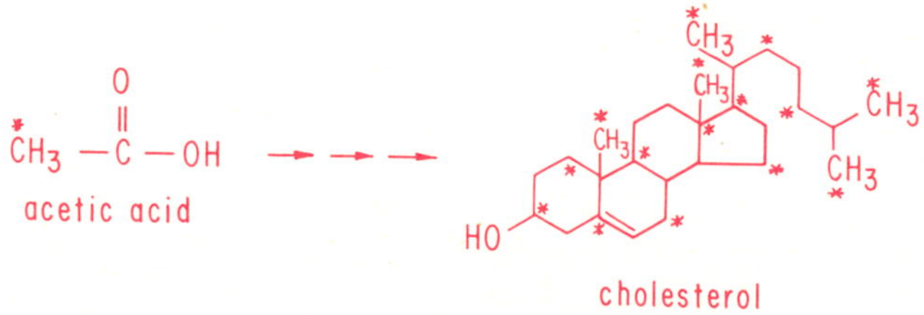

$$\overset{*}{C}H_3 - \overset{O}{\underset{\|}{C}} - OH \longrightarrow \dashrightarrow \dashrightarrow \dashrightarrow$$

acetic acid

cholesterol

The biosynthesis of cholesterol by animal tissue is a prolific process. An average-sized man has a total cholesterol content of about half a pound, the purpose of which is not known. It has been demonstrated that cholesterol is metabolized by the body into bile acids and steroid hormones, but these molecules are required in very small quantities compared to the amount of cholesterol in the body.

Vitamin D

One use cholesterol has found outside the body, however, is as a precursor for synthetic vitamin D_3. Entry into the vitamin D field marked the medical debut of steroids.

As work on determining the basic steroidal skeletal structure was nearing completion in 1930, it was already suspected that the active ingredient in fish-liver oil capable of curing rickets was a steroid. (Rickets is a vitamin-deficiency involving abnormal calcium and phosphate metabolism, which produces structural changes in bones and teeth.) It was soon discovered that food exposed to sunlight also counteracted rickets, and that even simple exposure of the patient to sunlight had a beneficial effect. The question then arose as to whether ultraviolet radiation might convert a simple steroid such as cholesterol into a substance with anti-rickets activity similar to the unknown factor in fish-liver oil, called vitamin D.

Investigation soon showed that some samples of cholesterol on irradiation became slightly active against rickets, but this proved to be due to the presence of small amounts of another steroid in some of the cholesterol samples. This other steroid was very like ergosterol, a steroid found in yeast. Indeed, irradiation of ergosterol itself produced material with high potency against rickets. The active substance was shortly isolated and named vitamin D_2.

ergosterol   radiation →   vitamin D_2

At first, it seemed reasonable to assume that the active substance obtained by irradiating ergosterol was identical to the active substance in fish-liver oil, but it soon became apparent that other molecules also showed vitamin D activity. Thus, the ergosterol derivative was designated D_2.

Another very active material was obtained from cholesterol by introducing another double bond into the molecule (in conjugation with the double bond already there) to produce the ergosterol ring system and then irradiating the diene. The active product, obviously different from vitamin D_2, was named vitamin D_3.

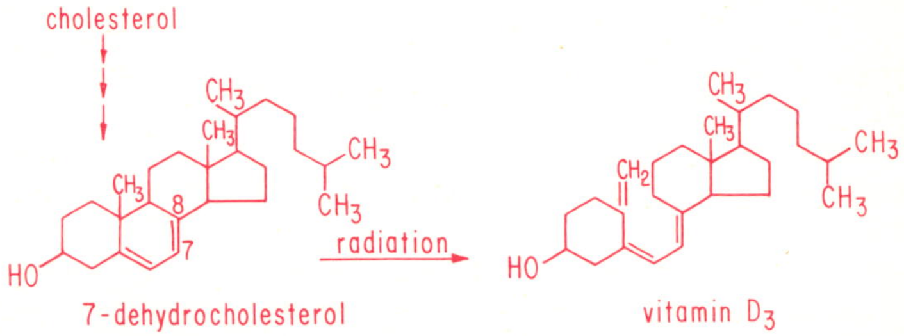

In general, irradiation reactions are very complex and usually produce many different structural changes. The present example is no exception. However, in the direct pathway to the single active product, vitamin D_3, the first thing that happens is probably rupture of the 9,10-covalent bond to produce a diradical.

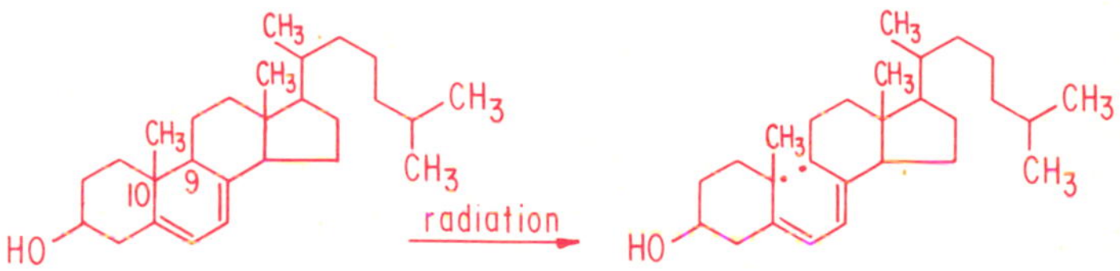

The radical at position 9 must then abstract a hydrogen atom from the nearby methyl group.

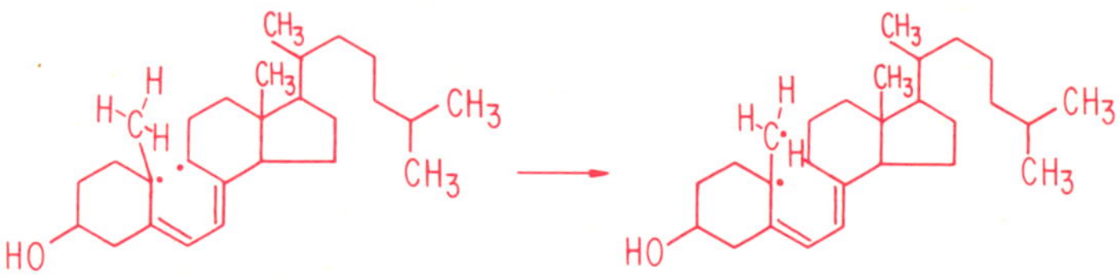

The resulting diradical is vicinal (adjacent atoms) and quickly collapses to form the third double bond of vitamin D_3.

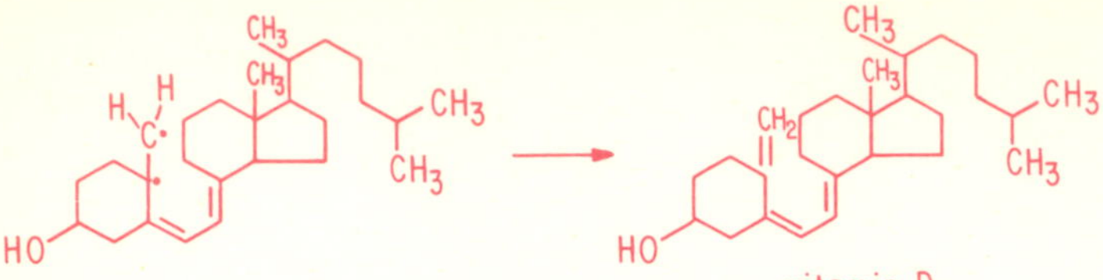

vitamin D$_3$

When the naturally occurring vitamin D was finally isolated in 1936, it turned out to be identical to vitamin D$_3$. However, since ergosterol is more readily available than dehydrocholesterol, vitamin D$_2$ is usually used as a food supplement or additive in human nutrition. Vitamin D$_3$ is much more effective in preventing rickets in chickens, though, and so it is used for this purpose instead of the ergosterol derivative. It should be noted that all these steroids of the vitamin D family are not actually the functioning vitamins but are the precursors of the biologically active form of the vitamin.

Sex Hormones and the Pill

Prominent among the body's many chemical regulators, called hormones, are the two steroidal sex hormones estradiol and testosterone. Estradiol is the female hormone and testosterone is the male hormone.

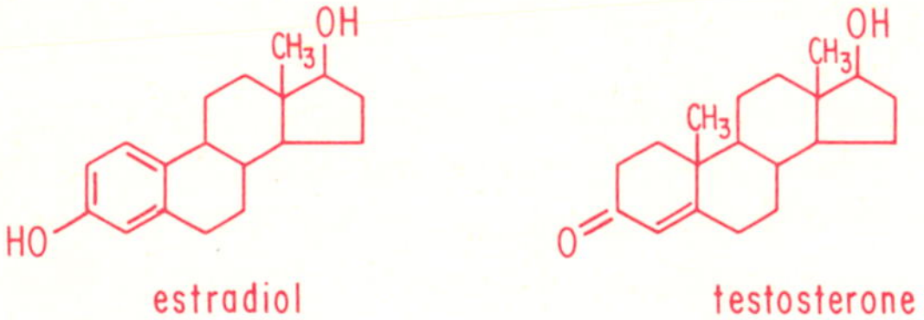

estradiol   testosterone

These hormones are responsible for the development of the various organs required for reproduction, and are also required for the maintenance of the normal functions of these organs. They also stimulate the development of secondary sex characteristics. The physiological functions of these two steroids are very different, one from the other, and yet their structures differ only by a molecule of methane. If a methyl group and a hydrogen atom were properly added to one of the double bonds of estradiol, the product would be the enol form of testosterone.

The sex hormones are very powerful physiologically and are normally present only in minute quantities in the body. They are capable of producing

estradiol   enol form of   testosterone

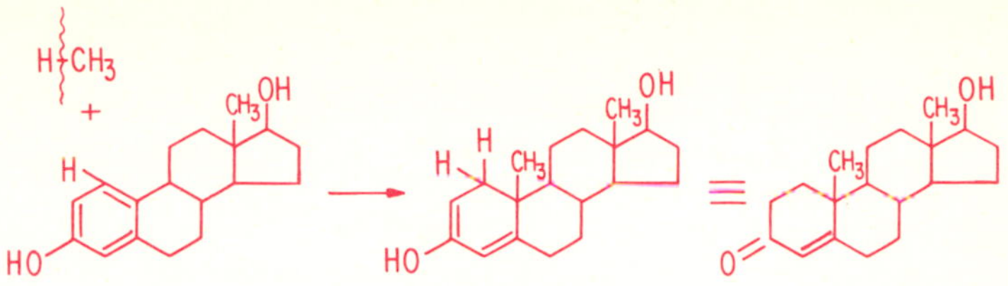

responses even in castrated and immature animals. For example, a capon (castrated rooster) injected with testosterone rapidly grows a comb, and an immature female mouse treated with estradiol goes into heat. As little as one-tenth of a microgram (not enough to see) of estrone induces this response in spayed female mice. This quantity has therefore been defined as the international mouse unit. A microgram is one-millionth of a gram, and estrone is the ketonic oxidation product of estradiol. Estradiol is even more potent than estrone.

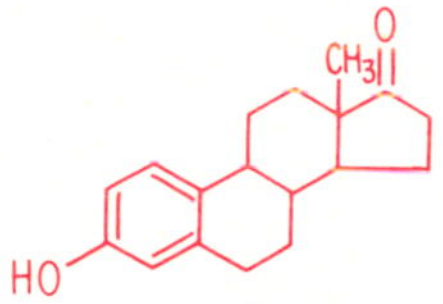

estrone

The normal sequence of events in mammalian reproduction begins in the master gland—the pituitary—a small gland located under the brain. The pituitary is considered the master gland because it secretes hormones which regulate the hormonal secretions of other glands. To start the reproductive cycle in females, the pituitary releases a hormone called follicle stimulating hormone. This hormone is a protein which stimulates the ovaries to develop follicular tissue. The follicular tissue produces the egg and also secretes estradiol. Estradiol does two important things. It inhibits further secretion of follicle stimulating hormone, and it stimulates the growth of uterine tissue. The first of these acts is necessary to prevent simultaneous production of a large number of eggs. The second function begins the preparation of a site where the mature, fertilized egg can implant itself after release from the ovary. When the mature egg is released by the ovary, the follicular tissue of the ovary is converted into a so-called *corpus luteum* which secretes another steroid hormone, called progesterone.

progesterone
(the pregnancy hormone)

Progesterone is also known as the pregnancy hormone, since it is secreted continuously throughout pregnancy to prevent the ovary from producing any more eggs. It does this by preventing the pituitary from secreting follicle stimulating hormone. In this sense, progesterone behaves like estradiol. It does in another way, too. It causes the proliferation of the growing uterine tissue at the site of egg implantation, which would be needed for the nutritive support of a fertilized egg.

Surprisingly, the structure of progesterone resembles that of the male hormone testosterone more than it resembles the structure of the female hormone estradiol. Progesterone and testosterone differ only in the side chain (hydroxyl versus acetyl group), while estradiol, in addition to the above difference, is missing the methyl group at C-10, and contains an aromatic ring. This is remarkable in view of the similarity of physiological response to progesterone and to estradiol, principally inhibition of secretion of follicle stimulating hormone. It indicates that they act in different ways to produce the same result.

This result—the prevention of egg production—is tantamount to temporary sterility in the female. The discovery some 20 years ago that this state can be induced and maintained artificially has given rise to the popular new method for preventing pregnancy known as oral contraception. The method itself is very simple. It involves merely setting up and supporting a hormonal state of pseudo-pregnancy by swallowing highly potent artificial derivatives of estradiol and progesterone. The artificial derivatives are preferred over the natural hormones because they are much more powerful when administered orally. In particular, oral potency is greatly enhanced by putting an acetylenic group into the molecule at the side-chain position (C-17).

synthetic estrogen
(17-ethynylestradiol)

natural hormone
(estradiol)

The synthetics are called estrogens or progestins according to which natural hormone they resemble. One popular synthetic estrogen is simply the 17-ethynyl (acetylenic) derivative of estradiol. Some of the widely used synthetic progestins, however, resemble the male sex hormone testosterone more than they do progesterone. One such is called norethindrone.

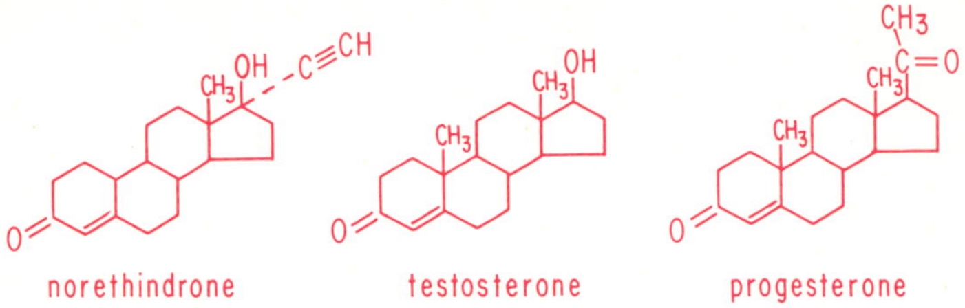

norethindrone   testosterone   progesterone

Norethindrone contains the sex-hormone side chain (hydroxyl) rather than the progesterone side chain (acetyl). It also lacks a methyl group at C-10, which makes it more like the female sex hormone. Removal of this methyl group reportedly reduces male-hormone-like responses.

The synthetic estrogens and progestins are very powerful physiologically and are very effective in preventing conception by interfering with the natural hormonal activity of the body. As might be expected from the fact that they are not "natural" molecules, they have been reported to cause undesirable effects. The World Health Organization in 1965 warned of apparent high pregnancy rates on discontinuance of use, a risk of diabetes, an increased coagulability of blood, and liver disturbance. However, the Food and Drug Administration in 1966 described their use as "not unsafe."

The Prostaglandins

As part of our discussion of steroids and related molecules, we should now turn to a relatively new and fascinating class of molecules, namely, the prostaglandins. The term was devised by von Euler who extracted lipid-soluble acids from human semen.

The class of compounds called prostaglandins are named as derivatives of prostanoic acid.

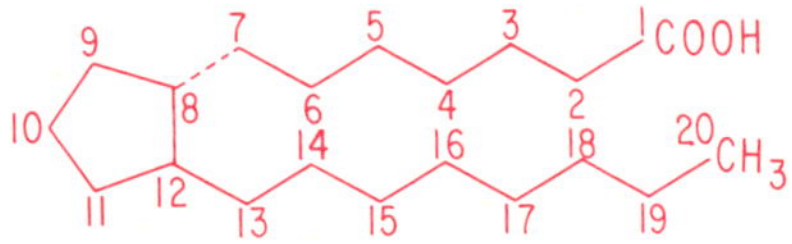

Substituents on the cyclopentane ring are designated α if they are on the same side of the plane of the ring as carbon 7; they are designated β if they are on the opposite side of the plane of the ring from carbon 7.

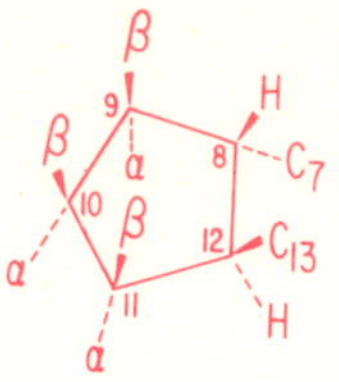

A schematic representation of the α and β positions is shown above. When substitution occurs in the side chains, the stereochemistry is designated by the Cahn-Ingold-Prelog nomenclature system (see appendix).

The discovery of the prostaglandins involved observations on the effect of seminal extracts on uterine tissue, of adrenal glands on circulation and of renal extracts on blood pressure. Over many years scientists noted special biological effects of fresh extracts of many organs including the prostate gland. In the 1930s von Euler found an active principle in the vesicular gland of sheep. With the help of Dr. Hugo Theorell and his newly developed electrophoresis apparatus,* von Euler discovered and "baptized" the prostaglandins. The task of purifying and identifying these molecules took nearly another 30 years. Von Euler's students ultimately succeeded in elucidating the structural features of many of the prostaglandins. Primary among these people were Bergstrom and Sjövall. They isolated and characterized the so-called prostaglandins E and F from the prostate gland of rams and showed that reduction of prostaglandin E with sodium borohydride gave two products,

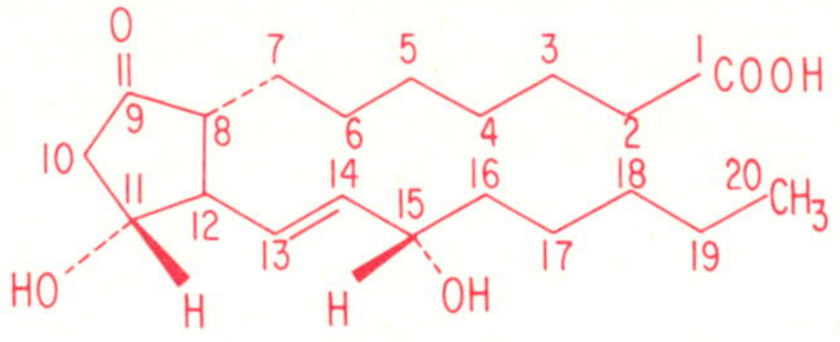

11 α, 15 (S) dihydroxy–9–oxo–13–transprostenoic acid
(prostaglandin E)

---

* Electrophoresis is a technique whereby a mixture of compounds is placed in the center of an inert support sheet which is wetted by an ion containing aqueous neutral solution. Electrodes are attached at each end and a current is passed through the support sheet. Neutral molecules remain in place while cationic molecules move toward the cathode and anions move toward the anode. Since the prostaglandins are carboxylic acids, Theorell and von Euler noted that the active principle moved toward the anode.

one of which was identical to the naturally occurring prostaglandin F. Structural elucidation involved absorption spectroscopy, degradation and identification of fragments, synthesis and x-ray crystallography. From these results, the Swedish scientists were able to assign the above structure to prostaglandin E.

It is clear how prostaglandin F is structurally related to prostaglandin E when one recalls that sodium borohydride selectively reduces aldehydes and ketones to alcohols under mild conditions.

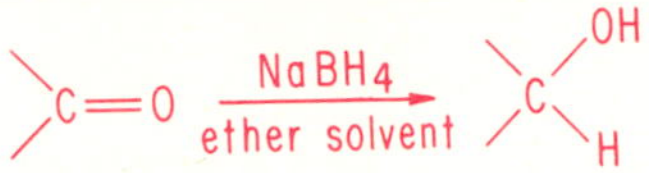

When the ketonic group at carbon 9 of prostaglandin E is reduced to an alcohol, two optical isomers can result; one with the hydroxyl group carbon 9 in the α position and the other with the hydroxyl in the β position. The α isomer was shown to be identical with the naturally occurring prostaglandin F.

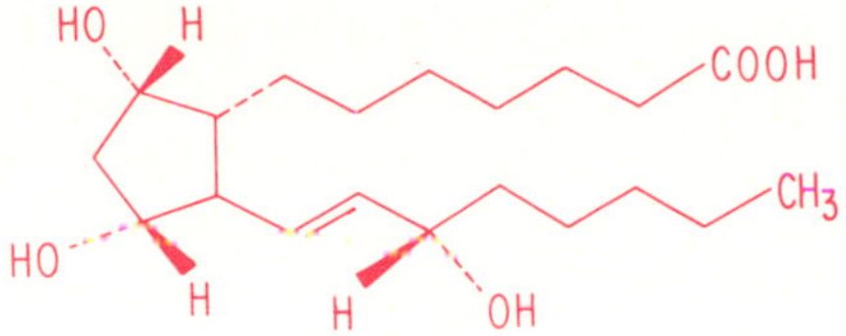

9 α,11α, 15(S) —13—*trans*—prostenoic acid
(prostaglandin F)

About a dozen naturally occurring prostaglandins and their derivatives have been identified in recent years. Many have been synthesized. The first successful synthesis was carried out by Samuelson and Stollberg who allowed a β-ketodiester to react with an α-bromoketone in the presence of strongly basic conditions.

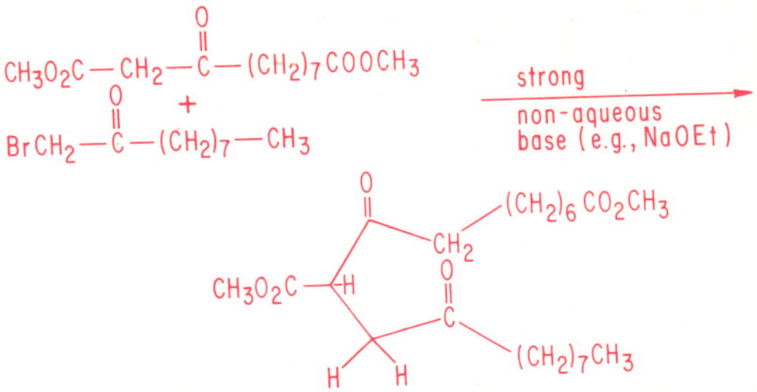

This reaction is a classical synthetic route to form carbon bonds. It involves the removal of an active hydrogen from a β-ketoester by a strong base.

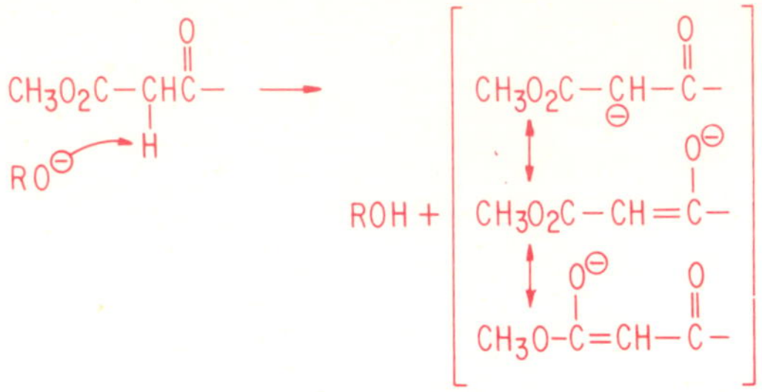

$$CH_3O_2C-\underset{RO^{\ominus} \nearrow H}{\overset{\overset{O}{\parallel}}{CHC-}} \longrightarrow \quad ROH + \left[\begin{array}{c} CH_3O_2C-\underset{\ominus}{CH}-\overset{\overset{O}{\parallel}}{C-} \\ \updownarrow \quad O^{\ominus} \\ CH_3O_2C-CH=\overset{|}{C}- \\ \updownarrow \quad O^{\ominus} \\ CH_3O-\overset{|}{C}=CH-\overset{\overset{O}{\parallel}}{C}- \end{array} \right]$$

The resulting carbanion is resonance stabilized by delocalization of the electrons as shown. Simplistically we can view the first form as representative of the reacting delocalized carbanion. It can react with the bromoethyl ketone in a typical S_N2 displacement reaction (see appendix, p. 317).

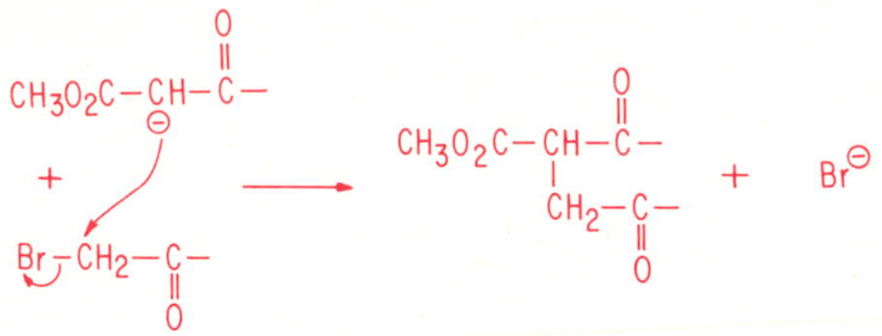

$$CH_3O_2C-\underset{\ominus}{CH}-\overset{\overset{O}{\parallel}}{C}- \quad + \quad Br-CH_2-\underset{\overset{\parallel}{O}}{C}- \longrightarrow CH_3O_2C-\underset{CH_2-\underset{\overset{\parallel}{O}}{C}-}{\overset{|}{CH}}-\overset{\overset{O}{\parallel}}{C}- \quad + \quad Br^{\ominus}$$

In this manner the first intermediate was synthesized. Another base-catalyzed carbanion formation was allowed to proceed under more drastic conditions.

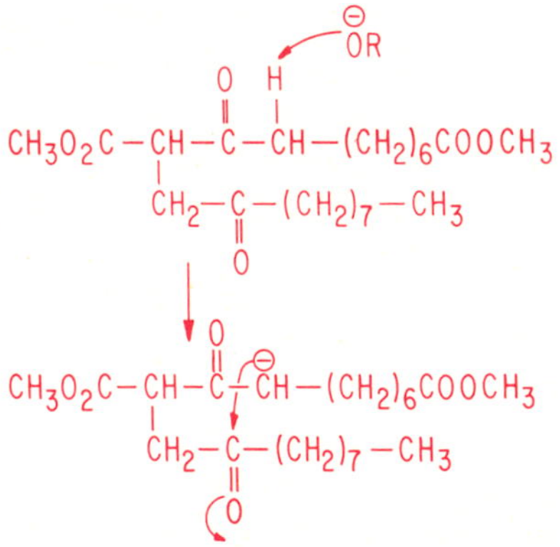

$$CH_3O_2C-\underset{CH_2-\underset{\overset{\parallel}{O}}{C}-(CH_2)_7-CH_3}{\overset{|}{CH}}-\overset{\overset{O}{\parallel}}{C}-\underset{}{\overset{\overset{\ominus}{\overset{OR}{\nearrow}}}{\overset{|}{CH}}}-(CH_2)_6COOCH_3$$

$$\downarrow$$

$$CH_3O_2C-\underset{CH_2-\underset{\overset{\parallel}{O}}{C}-(CH_2)_7-CH_3}{\overset{|}{CH}}-\overset{\overset{O}{\parallel}}{C}-\overset{\ominus}{CH}-(CH_2)_6COOCH_3$$

Ring closure occurs by attack of the carbanion on the ketone (four carbons away). This reaction requires more drastic conditions because the carbanion is not nearly as stable as the one arising from the β-ketone ester.

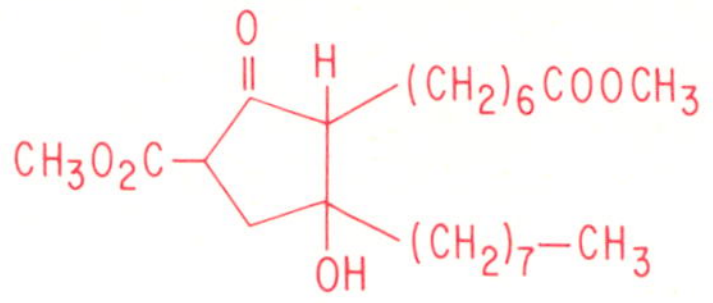

The ester groups are hydrolyzed and the alcohol group is removed by an elimination reaction under basic conditions.

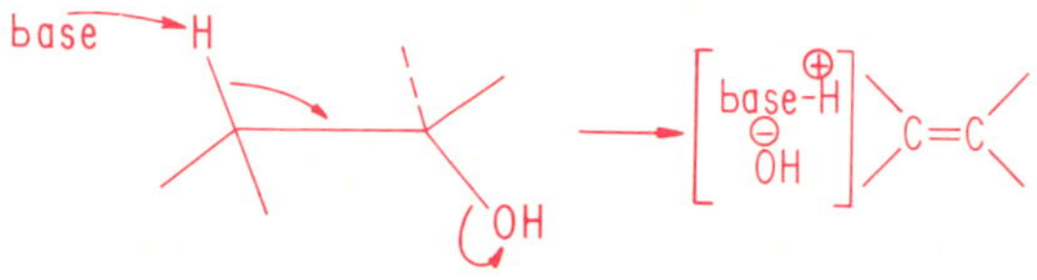

Acidification and heat removes the carboxyl from the cyclopentane ring yielding a C_{20} prostanoic acid derivative.

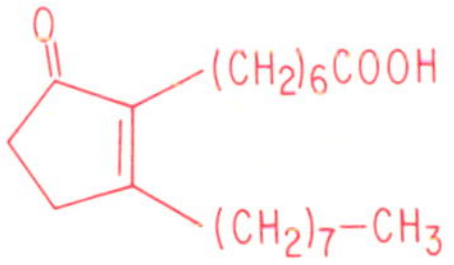

Syntheses of biologically active prostaglandins have been reported. Elegant and complex organic reactions have been employed which are beyond the scope of this discussion. The synthetic scheme outlined above should suffice to illustrate the basic principles necessary for total synthesis of prostaglandins.

We alluded to some of the biodynamic effects of these C_{20} unsaturated acids. At present it appears that prostaglandins are involved in reproduction. They have been shown to be readily absorbed through the human vaginal walls to facilitate fertilization. Prostaglandins of the E type inhibit reproductive smooth muscle contraction while those of the F type stimulate contraction.

Cardiovascular effects have been documented. Increase of heart rate and decrease of blood pressure are typical responses to infusion of prostaglandin E into healthy human beings. It has been shown that this vasodepressor effect is unaffected by the adrenalin system. Prostaglandin F derivatives can act as pressor agents in rats, dogs and chickens which may result from increased cardiac output because of constriction of the heart blood vessels.

The prostaglandins are also involved in the release of free fatty acids from adipose tissues. Prostaglandin E inhibits the accumulation of cyclic AMP* which is believed to be an intermediate hormone in the activation of lipase of fatty tissue.

Two English research teams have recently published results implicating the prostaglandins in aspirin action. They showed that aspirin and related drugs inhibit or prevent the biosynthesis of prostaglandins. Since the prostaglandins are found in almost all body tissue, they have been implicated in fever production and inflammations. Clearly, prevention of prostaglandin formation can then lower both fever and swelling. No explanation was offered to explain aspirin's pain-killing properties. Prostaglandins have not been shown to be involved in analgesic activities. The English workers' results do go a long way in explaining the aspirin mystery. They even speculate that some failures of intrauterine contraceptive devices may be attributable to aspirin. The intrauterine device (IUD) usually prevents implantation of the human embryo into the uterine lining. Implantation occurs at the blastocyst stage of human development, about eight days after fertilization. The blastocyst stage follows fertilization and many sets of cell division, and is characterized by the differentiation of two cell types—those of the inner cell mass and those in the enveloping layer. Chemically speaking, the IUD may prevent implantation by causing an accumulation of uterine prostaglandin, which then stimulates uterine muscle contraction. Since aspirin blocks prostaglandin synthesis, it is possible that some IUD failures are caused by frequent consumption of aspirin.

It is clear that we have only begun to understand the workings of these C_{20} fatty acids. Many researchers are studying the biosynthesis and metabolism of the prostaglandins. These compounds appear to play a role in such widely divergent areas as deposition of lipids (arteriosclerosis), blood pressure nerve transmission, cyclic AMP accumulation, reproduction (smooth muscle contraction), etc. These data show the biological effects of prostaglandins; but scientists have yet to discover their metabolic functions.

SUMMARY

After reading this chapter, it should be clear that steroids and related compounds have many biological functions. Vitamins, sex hormones, and arteriosclerosis represent typical areas of steroid involvement. The "pill" in its

---

* The ubiquitous character of cyclic adenosine monophosphate and its importance as a metabolic mediator was early recognized by Professor Earl Sutherland. We noted in Chapter III that he was awarded the Nobel Prize in Medicine for his discovery of the importance of cyclic AMP.

present form is nothing more than chemist-created female steroid sex hormones.

We examined the biosynthesis of steroids and related substances. The biogenesis of these compounds was shown to be based on highly selective enzyme-catalyzed reactions of isoprene and acetate.

Lastly, we covered a new class of compounds, the prostaglandins. Their significance and functions are just now being elucidated. They have been implicated in release of fatty acids from adipose tissue, mechanism of aspirin action, reproduction and many other areas.

SOURCE MATERIALS AND SUGGESTED READING

Ferreira, S. H., S. Moncada, and J. R. Vane, 1971. "Indomethacin and aspirin abolish prostaglandin release from the spleen." *Nature New Biol.*, **231** (25), p. 237.

Fieser, L. F., and M. Fieser, 1959. *The Steroids*, Reinhold Publishing Corp., New York.

Hanson, J. R., 1968. *Introduction to Steroid Chemistry*, Pergamon Press, New York.

Pincus, G., 1970. "Control of Conception by Hormonal Steroids," p. 46, in *Chemicals and Life*, K. E. Maxwell, ed., Dickenson Publishing Co., Belmont, California.

Pincus, G., T. Nakao, and J. F. Tait, ed., 1966. *Hormones*, Academic Press, New York.

Ramwell, P. W., J. E. Shaw, G. B. Clarke, M. F. Grostic, D. G. Kaiser, and J. E. Pike, 1968. "Prostaglandins," *Prog. Chem. Fats Lipids*, **9**, Part 2, Polyunsaturated Acids, p. 231.

Smith, J. G., and A. L. Willis, 1971. "Aspirin selectively inhibits prostaglandin production in human platelets," *Nature New Biol.*, **231** (25), p. 235.

Vane, J. R., 1971. "Inhibition of prostaglandin synthesis as a mechanism of action for aspirin-like drugs," *Nature New Biol.*, **231** (25), p. 232.

CHAPTER VIII

MOLECULES OF MYSTICISM
The Mind Changers

MAN'S UNENDING search for life's meaning, for deep psychological satisfactions, and for insights into immortality have led him to collect, catalog and even concoct substances that allow him to transcend normal physical limitations and routines. Throughout recorded history (and before) men have eaten, imbibed, sniffed and injected extracts of plant or animal origin to achieve altered consciousness. Herodotus (484–424 B.C.) noted that the Scythians steamed hemp seeds on extremely hot stones. Breathing the vapors made them shout for joy and otherwise exhibit feelings of euphoria.

Societies in every part of the world developed rituals and religious foundations for use of certain extracts since the mental state created by these drugs was likened to paradise. North American indians (and before them the Aztecs, Incas and the Mayas) have long used peyote, mushrooms and ololiuqui. The use of hashish (or other forms of cannabis) is worldwide. It developed extensively in the Arabic lands. The word hashish is derived from the Arabic hashishin or hemp-eaters. In the eleventh century, a group of Hashishin practised murder of enemies as a sacred religious obligation. Prior to assigned assassinations, they would take hashish in order to feel the ecstasy of heaven. No wonder that by a play on words the sect became known as the "assassins"!

Asians, Africans and Polynesians have incorporated substantial use of mood-altering materials in their cultures. Although some scholars have claimed that hallucinogenic agents were used extensively in biblical times, it remained for the Judeo-Christian culture to develop strong taboos on "drugs for escape from reality." The one exception was the use of alcohol. Even in that, limited indulgence was recommended. In many Protestant groups, alcohol, caffeine and nicotine were and are proscribed. It is clear that the intent of classical Western ethics is against mind-altering drugs. It has been said that while people of other cultures were weakened by drugs that altered the mind, Europeans were diligently developing the social structures and technology that altered the world.

Now, the separations and purity of cultures have essentially been destroyed by the selfsame science and technology. Chemistry has played its part in this transformation. Most illusions and magical aspects of age-old extracts have been exposed. We are now able to isolate and identify the active agents in many of the naturally occurring "mystical" extracts. We are also able to synthesize numerous analogs with varied and intense pharmacological properties. It is common knowledge today that poppies grown in Asia can appear as heroin in Chicago; that coca leaves grown in Central America are extracted to yield cocaine that may be sniffed in Paris, and that amphetamines and other synthetics are easily shipped all over the world. What are the structures of these "molecules of mysticism"? In this chapter, we hope to provide some answers to these and other questions.

Before proceeding to examine specific molecular systems, let us consider the fascinating story of Soma, the "divine mushroom". Many people thought it to be fictitious, a creation of the mind of Aldous Huxley. Yet a Vedic culture did exist 3000 years ago and from it we have a volume of 1028 hymns dedicated to Soma. What type of plant was it? Its identity remains obscure. In 1963, an amateur ethnobotanist, R. G. Wasson, undertook a study of the problem. Wasson, a retired banker, and his wife have spent years examining mushrooms throughout the world. Through their study of mushrooms, they made substantial contributions to our understanding of the migrations and of the linguistic patterns among tribes in ancient Asia. They also uncovered the long-hidden mushroom cult of Mexico (see p. 174).

After extensive scholarly and scientific investigation, Wasson concluded that Soma actually is *Amanita muscaria*, a brilliant red mushroom with white spots common to all of northern Eurasia (Figure 8.1). He based his assignment among other things on folklore and Vedic writing, the Rg Veda.

Amanita muscaria has long been known as a "fly agaric". In other words, the resinous material from this mushroom is used to attract flies so that they may be killed. The mushroom may be crushed and its juices spread on a surface; alternatively, it may be cut longitudinally to expose the juices. In both cases, flies are attracted, suck the juices and fall into a stupor. It takes hours for them to recover. In the meantime they are, of course, easy prey for their enemies.

The Wassons found that the Koryak, the Ostyaks and other Siberians established complex procedures for consuming mushrooms. For example, the women chewed the *Amanita muscaria* and rolled the cuds into elongated rods. They then provided the masticated specimens for the men to swallow whole. Early the Aryans (Indo Europeans) discovered that the intoxicating principles were excreted essentially unaltered in their urine. This gave rise to the religious rite in which celebrants drank their own or another's urine in order to maintain the intensity of the intoxication. Urine cults in many forms abounded in Asia in protohistorical periods and even to recent times. In the fourth century, St. Augustine, a Manichaean before his conversion to Christianity, railed against this Iranian-based religion because of its hedonistic values which included enormous consumption of the red mushroom (*Amanita muscaria*). Later the selfsame Manichaeans ruled China. In the twelfth century, a Chinese official wrote that the Manichaeans considered urine as ritual water and used it for their ablutions.

It is Wasson's contention that the identification of *Amanita muscaria* with killing flies came from Oriental and Christian moralists in order to hide the true character of *Amanita muscaria*. Incidentally, John Allegro developed this theme extensively in the *Sacred Mushroom and the Cross*.

Wasson presents a detailed analysis of the Rg Veda to support his thesis that Soma and *Amanita muscaria* are one and the same. He compares the hymns to Soma appearing in the Rg Veda to the references to Huoma from the Avesta, the bible of the Zoroastrians. He considers the substitutes for the lost Soma in the post-Vedic times, the other suggestions as to its identity and the relationship of the Chinese *Ling Chih* to Soma. Chinese people believe that good fortune comes to those who collect and eat the *Ling Chih*, that old people achieve even greater longevity and sick people become remarkably well. As an aside, it has recently been shown by some Japanese scientists that a modern version of *Ling Chih*, namely lentin, reduces cholesterol levels in the blood. Perhaps the ancient Chinese understood the practical aspects of diet and disease at least as well, if not better than we do in our highly "scientific" society.

It remained for Wasson and his associates to experiment with *Amanita muscaria*. He wrote:

In 1965 and again in 1966 we tried out the fly-agarics repeatedly on ourselves. The results were disappointing. We ate them raw, on empty stomachs. We drank the juice, on empty stomachs. We mixed the juice with milk, and drank the mixture, always on empty stomachs. We felt nauseated and some of us threw up. We felt disposed to sleep, and fell into a deep slumber from which shouts could not rouse us, lying like logs, not snoring, dead to the outside world. When in this state, I once had vivid dreams, but nothing like what happened when I took the *Psilocybe* mushrooms in Mexico, where I did not sleep at all. In our experiments at Sugadaira there was one occasion that differed from the others, one that could be called successful. Rokuya Imazeki took his mushrooms with *mizo shiru*, the delectable soup that the Japanese usually serve with breakfast, and he toasted his mushroom caps on a fork before an open fire. When he rose from the sleep that came from the mushrooms, he was in full elation. For three hours he could not help but speak; he was a compulsive speaker. The purport of his remarks was that this was nothing like the alcoholic state; it was infinitely better, beyond all comparison. We did not know at the time why, on this single occasion, our friend Imazeki was affected in this way.

Wasson does not present an appealing picture. Even the one apparently successful encounter is vaguely described. We are left with the clear impression that euphoria is only one ingredient of the fly agaric experience. To the chemist, it is not possible to forget that *Amanita muscaria* contains a deadly poison, muscarin. Ingestion of "Soma" is certainly not free of dangers.

muscarin

Let us now proceed to discuss four different classes of molecules that affect mental processes—most fit into the classification of hallucinogenic or psychotomimetic drugs.* Our list includes:

1) Lysergic acid derivatives (LSD, alkaloids from ololiuqui, etc.)
2) Phenethylamine compounds (mescaline, amphetamines, etc.)
3) Tryptamine-based molecules (psylocin, bufotenine, etc.)
4) Cannabinoids from *Cannabis sativa* (hashish, marihuana, etc.)

This is by no means a complete list. It will suffice for us to illustrate the underlying chemical and biological principles that govern the effects of these mind changers.

1. Lysergic Acid Derivatives

A "devil's curse", "St. Anthony's fire",† the "diabolical seed", were typical descriptions of ergotism, a disease long known in Europe when rye (or even wheat) was infected by the fungus, *Claviceps purpurea*. Two types of ergotism were clearly observed, the gangrenous and the convulsive. With the gangrenous infection, people experienced tingling fingers and toes, followed by vomiting and diarrhea. In a few days, they experienced painful rotting of toes and fingers.

---

* Hallucinogens and psychotomimetics have been used to apply to molecules that alter thought, perception and mood. Short-term side effects such as addiction, nervous system deterioration and memory impairment are considered minimal for this group of drugs. It should be recalled that we considered narcotic drugs in the chapter dealing with pain killers. This separation is not accidental since we believe their mode of action and side effects are such that they fall outside of the definition of hallucinogens and psychotomimetics.

† The term actually arose from the fact that many sufferers of ergotism ran extremely high fevers. They had the sensation of being on fire. From the eighth century, the disease was called *ignis sacer* (the holy fire). St. Anthony was an early Christian monastic (second-third century) whose bones were enshrined in Constantinople. In the eleventh century, they were brought to Vienna during a coincidental but particularly virulent siege of ergotism. After St. Anthony's bones were situated in Vienna, the disease began to abate. From that time, afflicted people have prayed to St. Anthony and the disease became known as *St. Anthony's fire*.

Dry gangrene of entire limbs ensued. In many cases, arms and/or legs fell off. Death came slowly in agonizing stages. If the disease were of the convulsive types, the symptoms began as noted above. Instead of gangrene, intense convulsive muscle spasms of the limbs developed followed by epileptic-type seizures. Ultimately, the patients became delerious, fell into a coma, and slowly died.

Epidemics of both types occurred before and during the Middle Ages in such widely separated places as the Ukraine and the southern regions of France, typical grain-growing areas of Europe. Imagine the view of the ancient or medieval man when entire villages were overwhelmed by the disease. Not only were people overcome, but livestock were even more susceptible. When people died in the thousands, horses and cattle died in the tens of thousands.

Julius Caesar's legions were obsessed with fear of poisons and food spoilage. Yet their ranks were ravaged by ergotism arising from spoiled grain during one of the campaigns in southern Gaul. In the year A.D. 994, it was estimated that 50,000 people died in an epidemic in France alone. European history is replete with recurrent outbreaks. After the severe epidemics of 1770 and 1777, legislation was introduced in France and other countries. As late as 1926, an extensive ergot infection of rye occurred in Russia leading to 11,000 cases of ergotism. Two years later, several hundred Jewish refugees in England from Eastern Europe came down with ergotism after consuming coarse rye bread from spoiled grain grown in England.

The medieval man must certainly have considered the infected villages to be in league with the devil. Fortunately, the fungal infections were easy to see since the kernels of the grain were replaced by a protruding brown-violet horn-shaped mass (Figure 8.2). Farmers early recognized that the disease could be controlled by prohibiting use of grain with substantial ergot growths. As

ergotamine

FIGURE 8.2 Ergot on Rye. The fungus *Claviceps purpurae* forms horny nodules rich in ergot alkaloids. Ingestion of infected grain can lead to the often fatal disease, ergotism. [From the book by Pamela M. North, 1967. *Poisonous Plants and Fungi in Colour*, Blandford Press Ltd., London]

noted above, laws were passed in many countries. These proscribed use of grain more than three ergot nodules per thousand kernels of grain. Recently, it has been suggested that even this number may be toxic to livestock.

What are the active ingredients of this pernicious fungus? How do they achieve such hideous effects? Earlier in this century, structures of an alkaloid family responsible for ergotism were elucidated. All are amides of lysergic acid. The amine portions are constructed from cyclic tripeptides. A prime example of this group is ergotamine.*

This compound can be hydrolyzed to form lysergic acid and various hydrolysis products of the cyclic tripeptide including: ammonia, pyruvic acid, proline and phenylalanine.

Fragmentation of Ergotamine

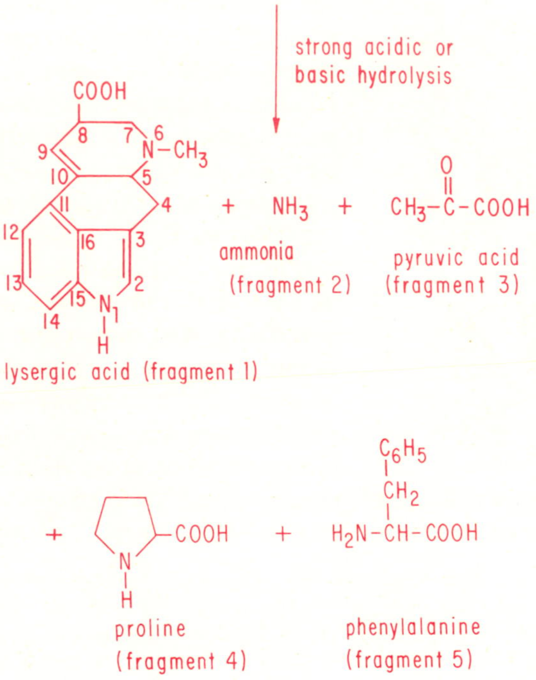

strong acidic or
basic hydrolysis

lysergic acid (fragment 1)

$+ NH_3 +$

O
‖
$CH_3-C-COOH$

ammonia
(fragment 2)

pyruvic acid
(fragment 3)

proline
(fragment 4)

C_6H_5
|
CH_2
|
$H_2N-CH-COOH$

phenylalanine
(fragment 5)

In lysergic acid, carbons 5 and 8 are asymmetric; hence there are four optical isomers. All four optical isomers of lysergic acid are psychotomimetically inactive. However, derivatives of lysergic acid are biologically active; the D-isomers particularly so.

---

* This compound is useful in treatment of migraine headaches, and has been used to reduce intense labor pains during childbirth.

Spoiled grain is not the only natural source of lysergic acid derived psychotomimetic agents. Ololiuqui, an hallucinogen-producing vine of the Aztecs, was
described by Francisco Hernandez, personal physician to the King of Spain.
He worked in Mexico from 1570–1575. To the trained botanist, the picture of

De OLILIVHQVI, seu planta orbicularium foliorum. Cap. XIV.

OLILIVHQVI, quam *Coaxihuitl*,
seu herbam Serpentis alij vocant,
volubilis herba est, folia viridia ferens, te-
nuia, cordis figura. caules teretes, virides, te-
nuesq;. flores albos, & longiusculos. semen
rotundum simile Coriandro, vnde nomen.
radices fibris similes. calida quarto ordine
planta est. luem Gallicam curat. dolores è
frigore ortos sedat. flatum, ac præter natu-
ram tumores discutit. puluis resina mixtus
pellit frigus. luxatis aut fractis ossibus, &
lumbis fœminarum laxis, aucto robore mi-
rum auxiliatur in modum. Seminis etiam
est vsus in medicina, quod tritum, ac deuoratum, illitumq; capiti, & fronti, cum lacte
& *Chilli*, fertur morbis oculorum mederi.
deuoratum verò, venerem excitat. Acri est
sapore, & temperie, veluti & planta eius,
impensè calida. Indorum sacrifici cum videri volebant versari cum Superis, ac respó-
sa accipere ab eis, ea vescebátur planta, vt de-
siperent, milleq; phantasmata, & dæmonũ obuersátium effigies circumspecta-
rent. qua in re Solano maniaco Dioscoridis similis fortasse alicui videri possit.

FIGURE 8.3   Text and illustration from Hernandez' study of plants and animals
of the new world. In this passage Dr. Hernandez describes the Mexican morning
glory, ololiuqui (*Rivea corymbosa*). [From Francisco Hernandez' *Rerum medicarum
Novae Hispaniae thesaurus, seu plantarum, animalium, mineralium mexicanorum
historia* (*Rome*, 1651)]

the plant, shown in Figure 8.3, clearly shows that ololiuqui (*Rivea corymbosa*)
is a morning glory, a member of the Convolvulaceae family.

The early Spanish chroniclers were ecclesiastics. They viewed the use of
these plants for divination as satanic. Very quickly Christian persecution
drove the users of the seeds underground.

It was not until the 1960s that the active psychoactive ingredients were shown to be of the lysergic acid type. A whole group of related alkaloids were characterized and identified as ingredients of the morning glory seed. A partial list of some of these compounds includes ergine, isoergine, ergometrine, chanoclavin, elymoclavin, and lysergol.

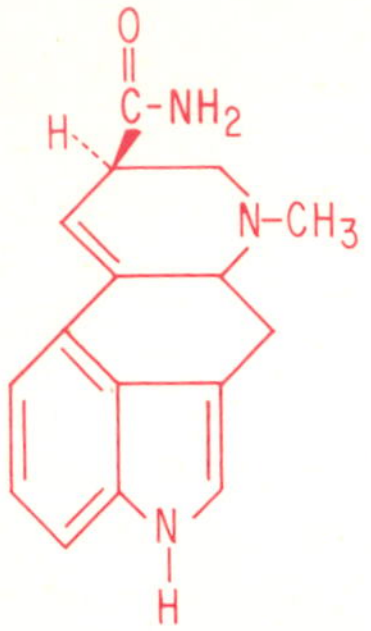

ergine (lysergic acid amide)

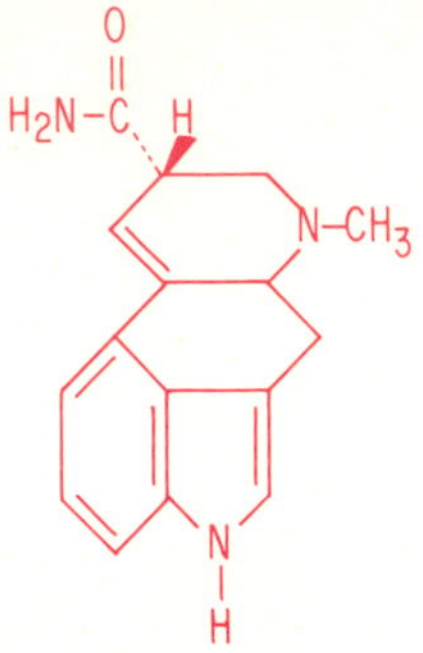

isoergine (isolysergic acid amide)

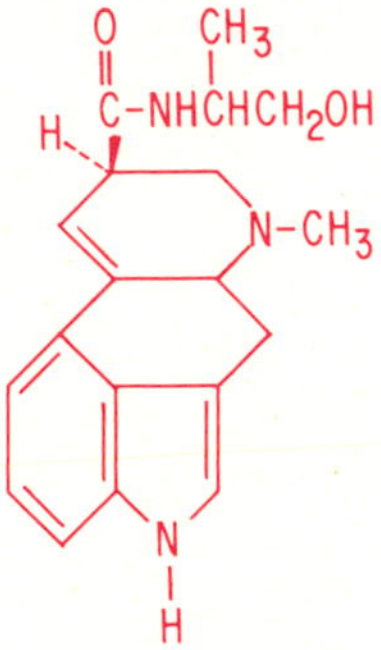

ergometrine

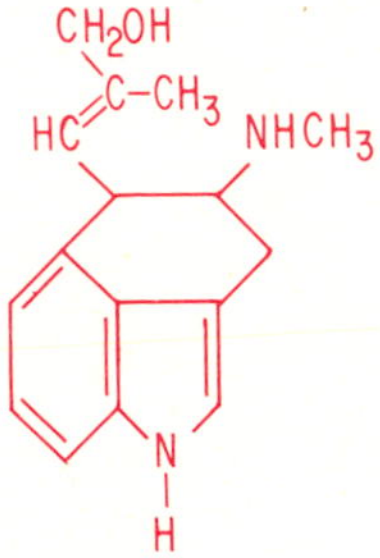

chanoclavin

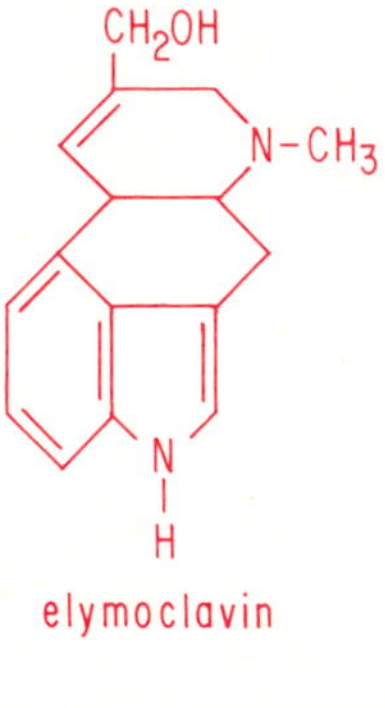

elymoclavin

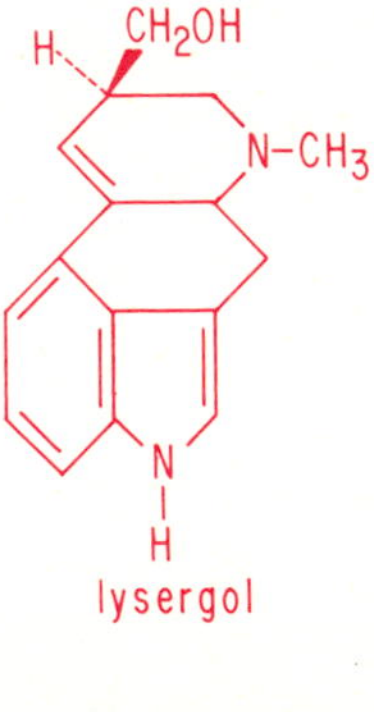

lysergol

It remained for the chemist to outdo nature in producing a lysergic acid derivative with enormous physiological and psychodynamic activity. In 1943, Dr. Albert Hofmann of the Sandoz Laboratories at Basel, Switzerland, was engaged in making various amides of lysergic acid in the hopes of obtaining an improved analeptic agent related to nikethamide, the structural formula for which is shown below.

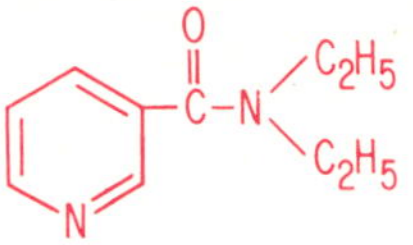

nikethamide

In the course of this work, he synthesized lysergic acid diethylamide (LSD) from lysergic acid and diethylamine. As shown in the reaction sequence below, a standard preparation of an acid chloride is undertaken. This material can then be allowed to react with excess diethylamine in an aqueous-organic solvent. The neutral product represents the desired lysergic acid, diethylamide (LSD).

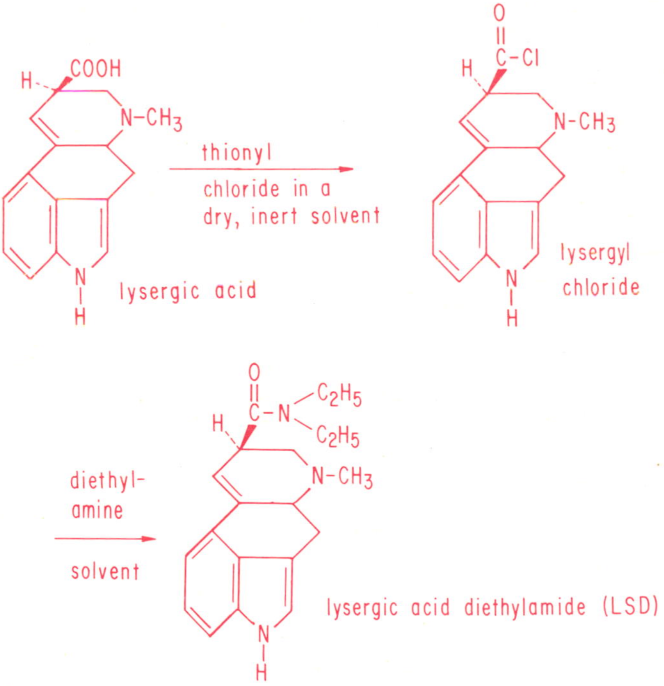

While working with the new materials, Dr. Hofmann was stricken with severe hallucinations which he knew arose from accidental inhalation of LSD. Hofmann described the psychological effects of the drug as follows:

On a Friday afternoon, April 16, 1943, while working in the laboratory, I was seized by a peculiar sensation of vertigo and restlessness. Objects, as well as the shape of my associates in the laboratories, appeared to undergo optical changes. I was unable to concentrate on my work. In a dreamlike state, I left for home, where an irresistible urge to lie down and sleep overcame me. Light was so intense as to be unpleasant. I drew the curtains and immediately fell into a peculiar state of "drunkenness", characterized by an exaggerated imagination. With my eyes closed, fantastic pictures of extraordinary plasticity and intensive color seemed to surge towards me. After two hours, this state gradually subsided and I was able to eat dinner with a good appetite.

And so, in an anticlimactic fashion, Dr. Hofmann concluded his description of the first LSD "trip" ever. From his own words it does not appear that he achieved the state of nirvana or divine happiness. Yet there is no question that he discovered a powerful drug with unpredictable social and medical implications.

Almost immediately following the initial synthesis of LSD, medical researchers began their studies with psychotomimetic drugs. Within a short time, the use of LSD in psychiatry came under heavy attack. Experts disagreed violently with each other; the cases for and against the use of LSD spilled over into the popular press. The time was correct, the drugs available and the consequences sufficiently vague so as to be ignored. The clear and deep experiences with LSD and other drugs were obvious. And thus Western society, for the first time, faced a new and complex pattern of social behavior; that of a significant number of people seeking altered consciousness through psychotomimetics and narcotics.

Numerous experiments have now been reported on the pharmacology, mechanism of action and toxicity of lysergic acid based compounds. LSD is quickly absorbed into the bloodstream. Tests with various tissue preparations *in vitro* and whole animals *in vivo* have demonstrated that in the body the molecule is altered. It has been shown that LSD is a substantial analgesic in addition to its hallucinogenic properties. Many patients, however, objected so strongly to the psychodynamic experience that they refused additional LSD treatment even though they were completely devoid of pain under the influence of the drug.

To date, the chemical reactions underlying the psychological experiences induced by hallucinogens are not known. There appears to be good evidence that LSD interferes with the transmission of impulses in the nervous system. LSD may be antagonistic to the biosynthesis and release of serotonin, adrenalin and related compounds. Adrenalin* and serotonin have long been implicated in brain function. High levels of these and other related amines can produce anesthesia.

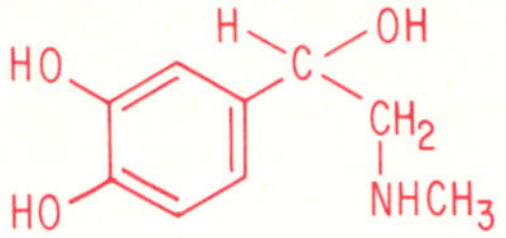

adrenalin (epinephrine) 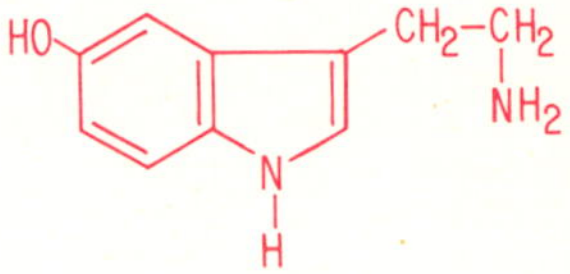   serotonin

Adrenalin is convertible *in vitro* and *in vivo* into indoles some of which are highly active psychotomimetics such as adrenochrome.

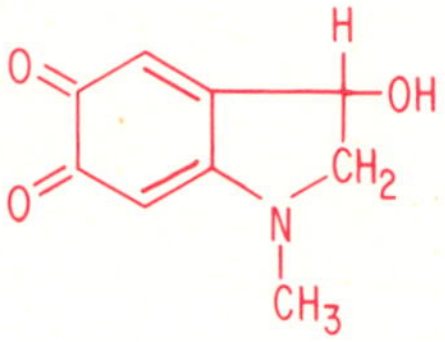

adrenochrome

This conversion occurs by the following steps.

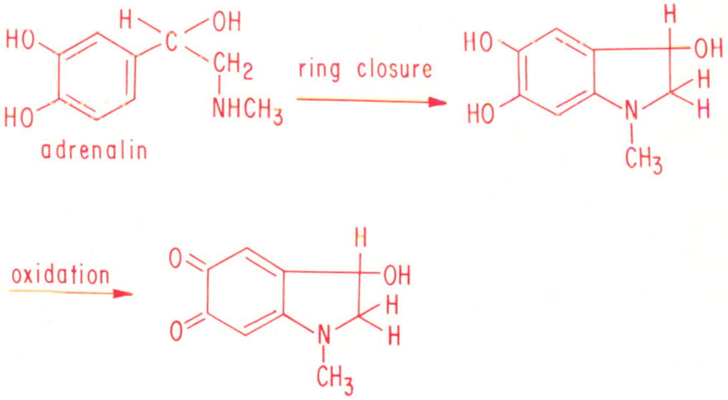

adrenalin   ring closure

oxidation

adrenochrome

It is known that adrenochrome interferes with nerve actions. With all of this information, it is reasonable to speculate that LSD competes with adrenalin

* Adrenalin is also called epinephrine.

and serotonin as amine components of nerve action. Evidence has also been presented to show a substantial antagonism between LSD and histamine.

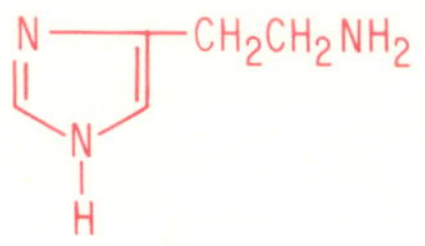

$$\begin{array}{c} N{-}\!\!\!-\!\!\!-CH_2CH_2NH_2 \\ \| \\ N \\ | \\ H \end{array}$$

histamine

LSD appears to be a powerful inhibitor of acetylcholinesterase, an enzyme involved in the hydrolysis of acetylcholine. This molecule acts as a chemical transmitter of nerve stimuli. The scheme which follows outlines the conversion of acetylcholine into choline and acetate and the resynthesis of acetylcholine.

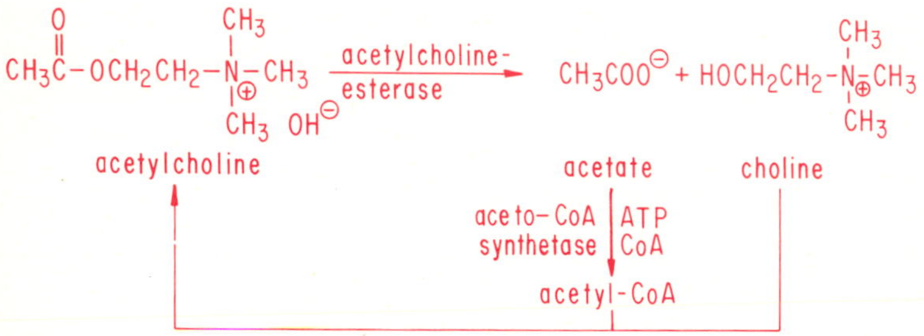

$$CH_3\overset{O}{\overset{\|}{C}}{-}OCH_2CH_2{-}\overset{CH_3}{\underset{\underset{CH_3}{|}}{\overset{|}{\underset{\oplus}{N}}}}{-}CH_3 \quad OH^{\ominus} \xrightarrow{\text{acetylcholine-esterase}} CH_3COO^{\ominus} + HOCH_2CH_2{-}\overset{CH_3}{\underset{\underset{CH_3}{|}}{\overset{|}{\underset{\oplus}{N}}}}{-}CH_3$$

acetylcholine   acetate   choline

aceto–CoA | ATP
synthetase | CoA

acetyl-CoA

choline acetylase

By blocking acetylcholinesterase activity, LSD disrupts the parasympathetic nervous system and elevated acetylcholine levels are encountered. This effect combined with the above-mentioned interference with metabolism of adrenalin, serotonin and histamine should suffice to indicate that the overall biochemistry of LSD is enormously complex.

Some toxicology results are available on LSD and related lysergic acid derivatives. For man, the LD-50* for LSD is about 0.2 mg/kg of body weight or about 15 mg for the average adult male. When LSD is used in the treatment of schizophrenia of alcoholism, psychiatrists have employed doses between 25

---

* LD-50 refers to the dosage at which 50% of the subjects die. Literally it stands for the lethal dose to kill 50% of the test animals.

and 1500 μg (1.5 mg). The latter value is certainly close to the upper limit of safety.*

Toxic reactions from continued use of morning glory seeds have been encountered. Though the active agents as noted above are not LSD, they certainly are related alkaloids. At this time, it is fair to say that the ergot alkaloids, natural and synthetic, involve poorly understood biochemical and psychological effects. The boundaries are not clear for these compounds with regard to their toxicity and psychotomimetic activities.

2.   Phenethylamine Compounds

Mescaline, the most active agent of a Mexican cactus, became the object of interest of such important literary-scientist types as Havelock Ellis, William James and Heinrich Kluver. Aldous Huxley, scion of an illustrious scientific family and a great writer in his own right, became one of its most enthusiastic users and boosters. The cactus was called *peyotyl* by the Aztecs from which the Spanish took the name peyote. A peyote cult existed in Mexico for several centuries and spread to many North American Indian tribes. By 1880, a fully developed religion existed in the southwestern regions of the United States that combined Christian beliefs with use of the divine cactus. This religion became the most widespread among the American Indians. It incorporated in 1918 as the Native American with a theological tenet that the peyote cactus is divinely endowed to shape men's lives. From the sixteenth century when Bernadino de Sahagun first described use of peyote by the Aztecs to the time of the Native American Church, governmental and traditional Christian authorities have tried to eradicate peyote cults. Today in the United States only *bona fide* members of the Native American Church are legally allowed the use of peyote in their rites.

Heffter isolated mescaline from the cactus plant in 1896 but it was not until 1918 that Spaeth elucidated the structure and synthesized the molecule. It is clearly of the phenethylamine class of compounds.

$$-CH_2-CH_2-NH_2 \qquad phenethylamine$$

---

* We are not here discussing the so-called "good" or "bad" trips arising from LSD but rather the poisonous nature of the chemical itself. With small doses, LSD cannot be considered a poison. With large doses, toxic reactions are to be expected.

A formalistic picture of the biosynthetic pathway for mescaline is shown below.

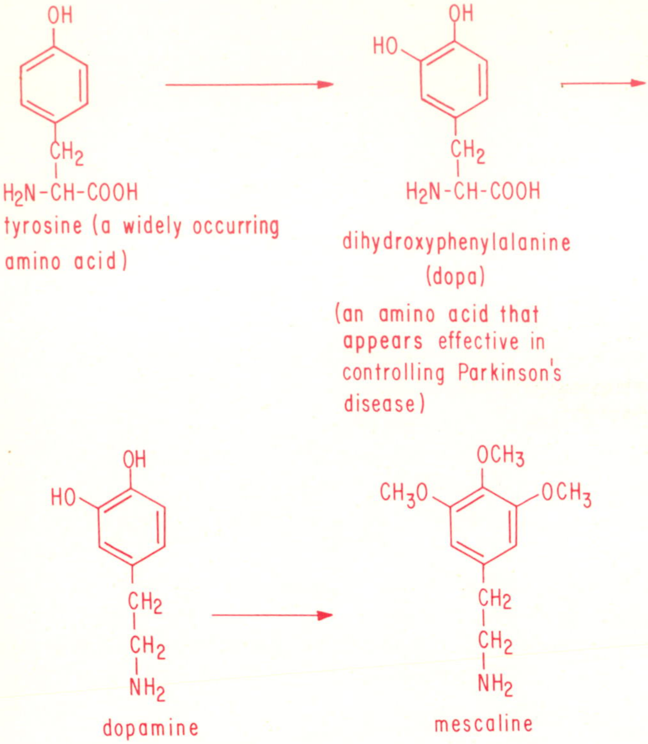
OH
CH_2
H_2N-CH-COOH
tyrosine (a widely occurring amino acid)
OH
HO
CH_2
H_2N-CH-COOH
dihydroxyphenylalanine (dopa)
(an amino acid that appears effective in controlling Parkinson's disease)
OH
HO
CH_2
CH_2
NH_2
dopamine
OCH_3
CH_3O
OCH_3
CH_2
CH_2
NH_2
mescaline

Mescaline has been synthesized in the laboratory by several independent routes. One of the approaches involves gallic acid as a starting material.

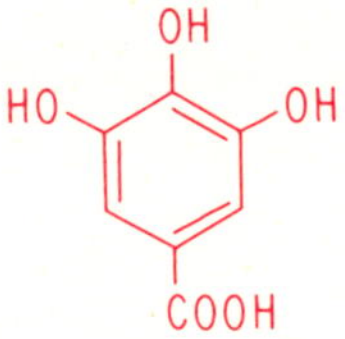
OH
HO
OH
COOH
gallic acid

Gallic acid is treated with excess methyl iodide and heat to convert it to the trimethoxy derivative.

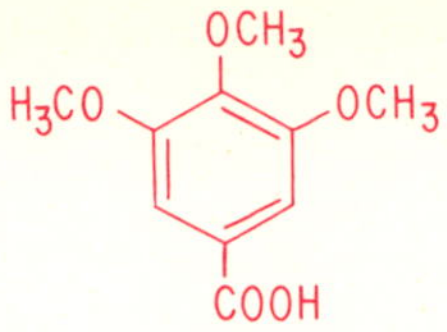

OCH$_3$

H$_3$CO — | — OCH$_3$

COOH

This acid is esterified by reaction with methanol and hydrogen chloride.

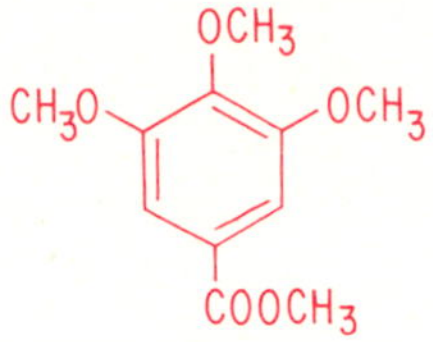

OCH$_3$

CH$_3$O — | — OCH$_3$

COOCH$_3$

The ester is allowed to react with lithium aluminum hydride to obtain the suitably substituted benzylalcohol.

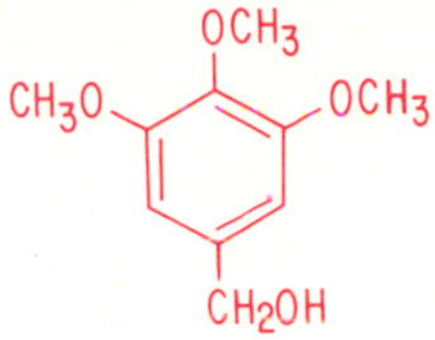

OCH$_3$

CH$_3$O — | — OCH$_3$

CH$_2$OH

This in turn can be converted to the benzyl chloride by reaction with phosphorus trichloride in an organic inert solvent.

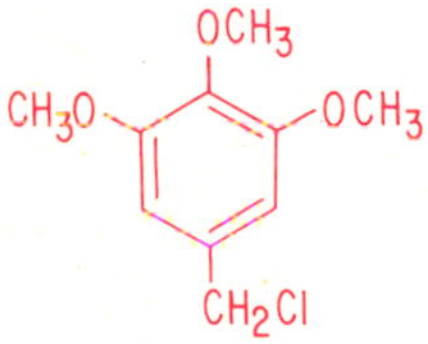

OCH$_3$

CH$_3$O — | — OCH$_3$

CH$_2$Cl

The benzyl chloride is then allowed to react with sodium cyanide with heat in a nucleophilic substitution by cyanide ion of chloride.

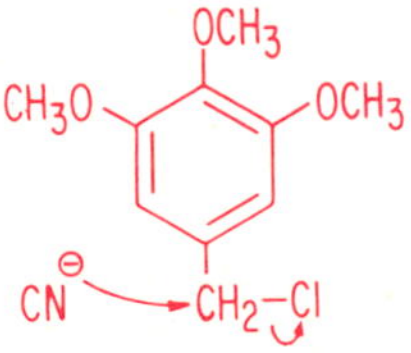

OCH$_3$

CH$_3$O — | — OCH$_3$

CN$^{\ominus}$ → CH$_2$—Cl

This reaction yields the trimethoxybenzyl cyanide.

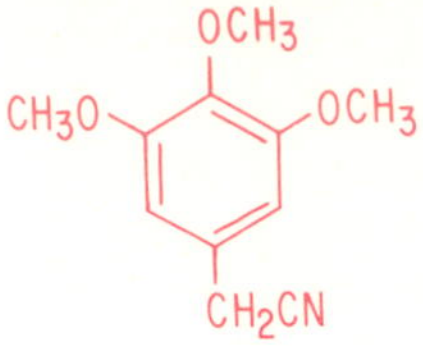

This in turn can be reduced to mescaline by lithium aluminum hydride. (The hydride reduction involves adding hydrogen to the triple bond of the nitrile group.)

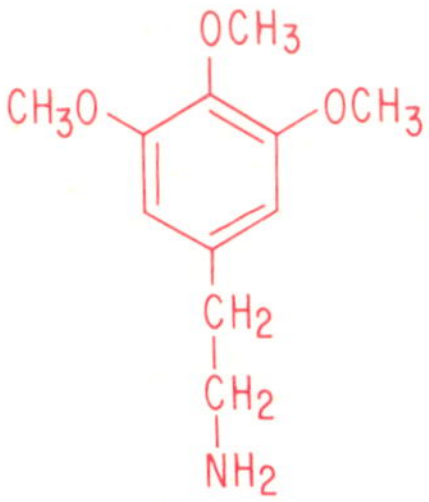

mescaline

It was the elucidation of the structure of adrenalin that aroused great interest in amines of similar structure. A comparison of the structure of mescaline above with that of adrenalin, p. 163, indicates that they are closely related.

We noted earlier that adrenalin is involved in transmission of nerve impulses. It is attractive to speculate that mescaline alters nerve function and competes with adrenalin and serotonin. Mescaline appears to possess many of the same pharmacological properties as LSD. Normally, this drug is administered in doses of 5 mg per kg of body weight or 350 mg for the physiologically average adult male (70 kg). (The typical total dose for LSD for the same adult male is approximately 0.5 mg.) Toxicity tests with white mice, rats, guinea pigs and frogs showed the LD-50 for mescaline is about 320 mg/kg of body weight. The animals experienced convulsions, respiratory and finally cardiac arrest. A well-known psychiatrist, experienced with the use of psychoactive drugs in treating mentally ill patients, took 750 mg of mescaline. He underwent an extreme psychodynamic experience far in excess of our definition of a psycho-tomimetic reaction. It lasted 16 hours. For the next two months, the poor man was haunted with the delusion that a man in one of his paintings was about to attack him. Removing the picture did not help. He was forced to sleep elsewhere. After two months, the severe anxieties were less intense, finally disappeared completely, and at long last he was able to return to a normal life.

Smaller doses unquestionably lead to hallucinations which have been described by the literary figures noted above. In 1928, Kluver reported the most complete study of the visual changes induced by mescaline. Initially, subjects encountered changes in brightness of color. These were then followed by geometric forms and color combinations. Kluver believed these stages to be common to all mescaline experiences. Symmetrical forms were seen in brilliant colors. Among the typical patterns were Gothic domes, helices, prisms, spider webs, oriental rugs and wallpaper designs. All shone with peculiar and intense hues.

We need only to turn to Aldous Huxley to obtain a more literary and emotional description of the effects of mescaline. In his essay "Doors of Perception" he wrote:

At ordinary times the eye concerns itself with such problems as Where?—How far?—How situated in relation to what? In the mescaline experience the implied questions to which the eye responds are of another order. Place and distance cease to be of much interest. The mind does its perceiving in terms of intensity of existence, profundity of significance, relationships within a pattern.

In another section of the essay, Huxley wrote:

Though the intellect remains unimpaired and though perception is enormously improved, the will suffers a profound change for the worse. The mescaline taker sees no reason for doing anything in particular and finds most of the causes for which, at ordinary times, he was prepared to act and suffer, profoundly uninteresting. He can't be bothered with them for the good reason that he has better things to think about.

Lastly, under the influence of mescaline Huxley mused, "A rose is a rose is a rose. But these chair legs were chair legs were St. Michael and all angels."

Both LSD and mescaline have been employed by psychiatrists and psychologists in treatment of such diseases as schizophrenia and alcoholism. Although neither drug proved to be a panacea, some beneficial results have been clearly demonstrated. Studies have appeared which clearly indicate that severe side reactions occur with both drugs. Both drugs involve similar and complex psychological responses. Subjects become temporarily psychotic.

We can now consider the input of the chemists who early became aware of the special activities of naturally occurring phenethylamine-like molecules. Amphetamines were synthesized and were found to be central nervous system stimulants.

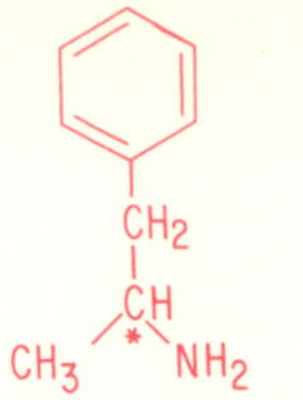

amphetamine

* asymmetric center

This molecule contains one asymmetric center as shown above and therefore can exist in two optical antipode forms (Fischer convention).

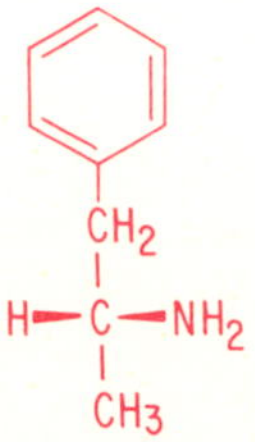

D-amphetamine

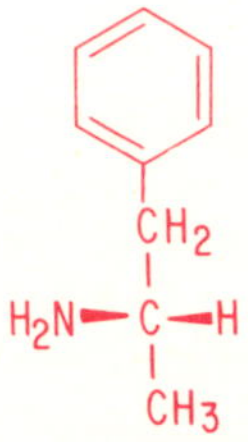

L-amphetamine

The D-amphetamine as its sulfate salt is more active than the corresponding L-compound. Typical oral doses for adults are in the 5–15 mg range. The LD-50 levels for mice and rabbits are in the range of 50 mg/kg of body weight.

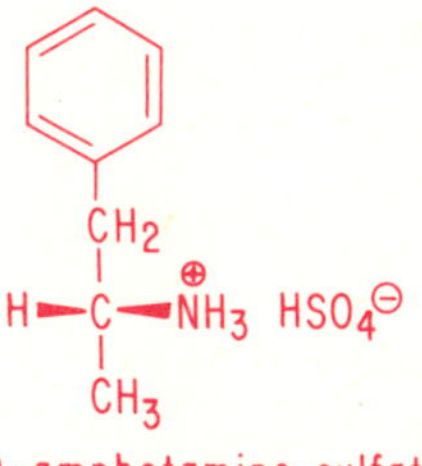

D-amphetamine sulfate

Psychoses are a common major side reaction from amphetamines. Typical clinical analysis of subjects under the effect of amphetamines include paranoia with delusions and auditory hallucinations. In addition, many subjects given 50 mg of D-amphetamine become violent. A typical case involved a subject who attacked the doctors and nurses and hallucinated cockroaches and worms crawling on the wall. Instances of true narcotic addiction have been reported.

A closely related drug to amphetamine is methamphetamine. Both are so-called "pep pills" in that they enhance the adrenalin response. Methamphetamine has been termed "speed" by some users and the popular press because of the rapid and intense psychological effects resulting from its use. Tests have shown it to be active at dose levels typical for amphetamine. Chemically, methamphetamine is simply N-methyl amphetamine.

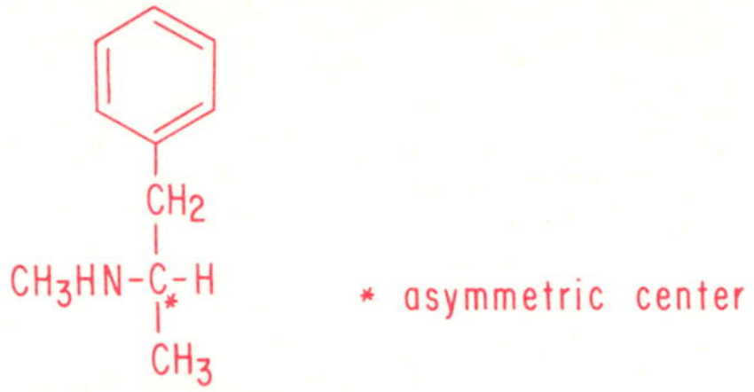
CH2
CH3HN-C-H
CH3
* asymmetric center

methamphetamine

Like amphetamine, this compound contains one asymmetric center and therefore can exist in two optically active forms. The optically active D-form is the more active. It can be synthesized from ephedrin, a naturally occurring methylamino alcohol isolated from the ephedra plant.

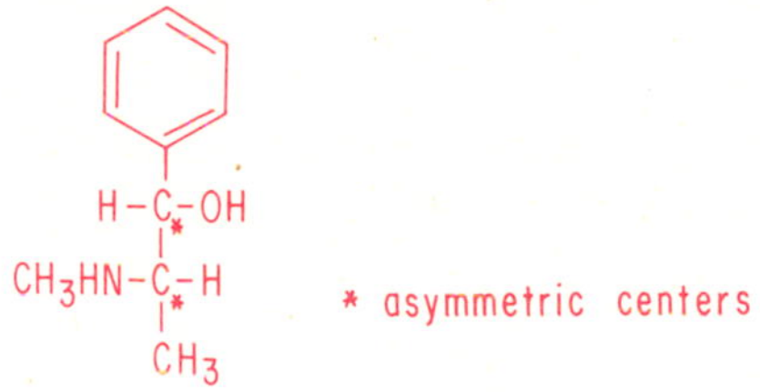
H-C-OH
CH3HN-C-H
CH3
* asymmetric centers

ephedrin

This compound contains two asymmetric centers adjacent to each other. Thus there are four optical isomers. The natural material exists with the N-methylamino group in the proper configuration for maximum physiological activity. Since methamphetamine has only one asymmetric center, the synthetic scheme involves the conversion of the secondary alcohol function of ephedrin

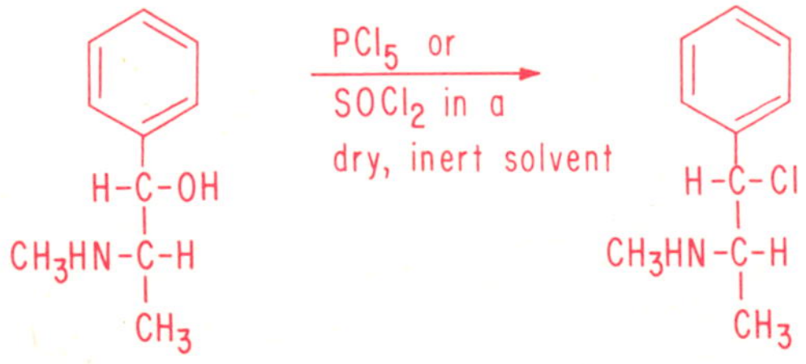
H-C-OH
CH3HN-C-H
CH3
PCl5 or
SOCl2 in a
dry, inert solvent
H-C-Cl
CH3HN-C-H
CH3

into a methylene group. In the first step of this transformation, ephedrin is allowed to react with either phosphorus pentachloride or thionyl chloride.

This displacement reaction is facile because it is carried out at a benzyl carbon site. Formally, the benzyl carbonium ion is somewhat stabilized by the neighboring phenyl and methylamino group. Thus displacements can be accomplished with relative ease.

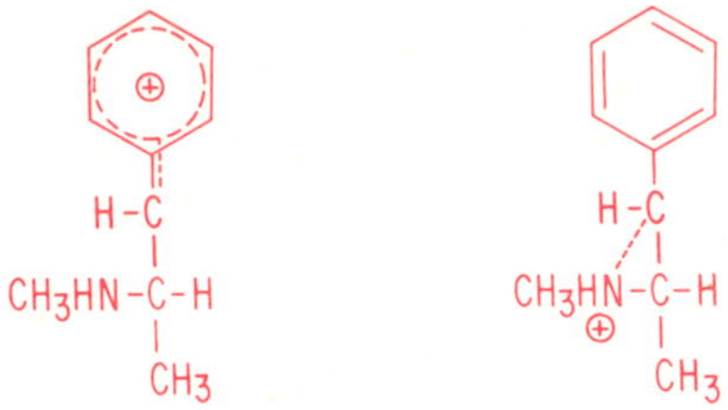

Stabilizing factors for the ephedrin-derived carbonium ion

The chloro compound is allowed to react with hydrogen in the presence of a noble metal catalyst.

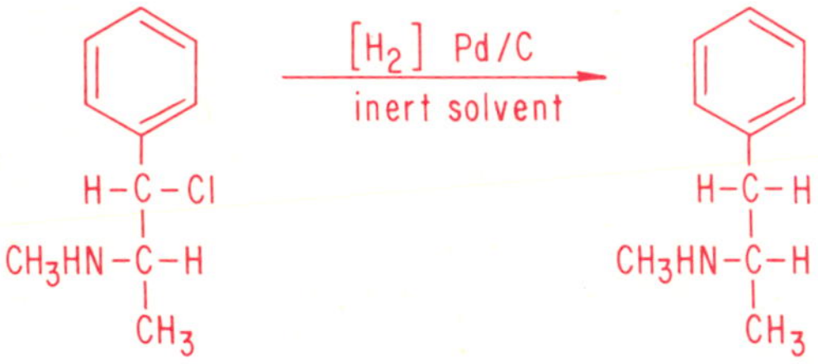

As noted above, reactions at the benzyl carbon are relatively facile. In addition to carbonium ion reactions, benzyl carbon–halogen bond cleavages (reductions) are enhanced by stabilization forces for the intermediates of the reactions.

3.   Tryptamine Based Molecules

The biologically important amino acid tryptophan can be shown to be a precursor in the biogenesis and synthesis of various indole alkaloids such as serotonin, psilocin (a major active principle in Mexican mushrooms), bufotenine (the prime hallucinogen of Haiti's ancient snuff).

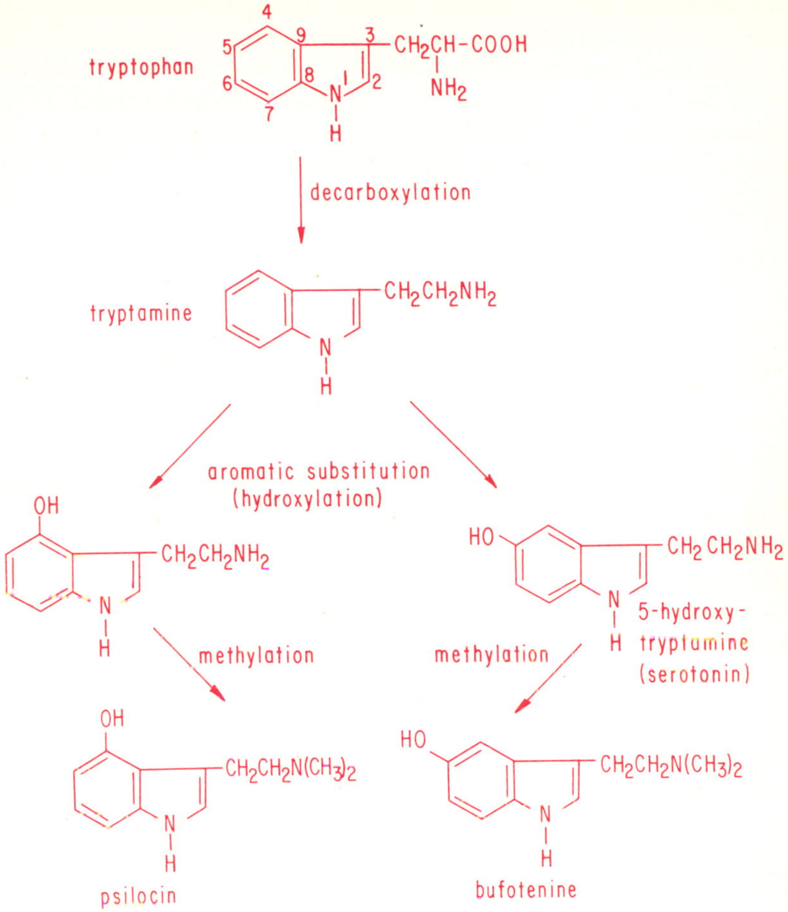

tryptophan

decarboxylation

tryptamine

aromatic substitution
(hydroxylation)

methylation

methylation

5-hydroxy-
tryptamine
(serotonin)

psilocin

bufotenine

The amino acid tyrosine can also function as a source for indole molecules through phenethylamine derivatives as shown earlier.

Psilocybin is a phosphorylated 4-hydroxydimethyltryptamine.

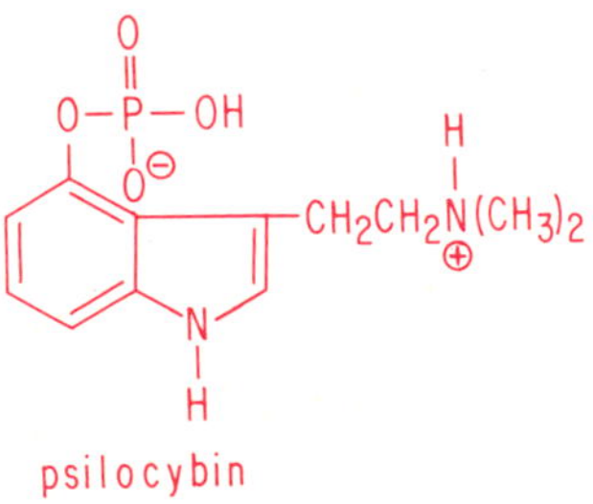

psilocybin

It is speculated that this phosphate linkage can be cleaved *in vivo* to release another active hallucinogen, psilocin (p. 173). Both psilocybin and psilocin are contained in the mushroom, *Psilocybe mexicana* Heim. The Aztecs referred to it as *teonanacatl*, "the food of the gods". Mushroom cults have been traced from 1500 B.C. to the Aztec and Mayan civilizations. As noted in an earlier section on ololiuqui and peyote, the Christian missionaries were enormously intolerant of the use of these plant substances. The Mexicans were forbidden the use of them even in their religious rites. Thus, the mushroom cult and the ololiuqui and peyote users were driven underground. Just over a decade ago, the Wassons, those outstanding husband-wife amateur mycologists, re-discovered the cult of the magic mushroom in Mexico. The Wassons observed the secret rites of the Aaxacan Indians and correctly placed the religious ceremony historically. They were able to obtain samples of the "sacred" mushrooms. R. Heim, a well-known authority on mushrooms, identified the specimens as coming from the genus *Psilocybe*. This species of mushroom occurs widely in Mexico, especially on wet hillsides in cow dung. Lastly, A. Hofmann of LSD fame, isolated the active principles, psilocybin and psilocin, proved their structures and accomplished a commercially feasible synthesis. Three stages were designed in the synthetic approach.

STAGE I   Synthesis of 2-methyl-3-nitrophenol

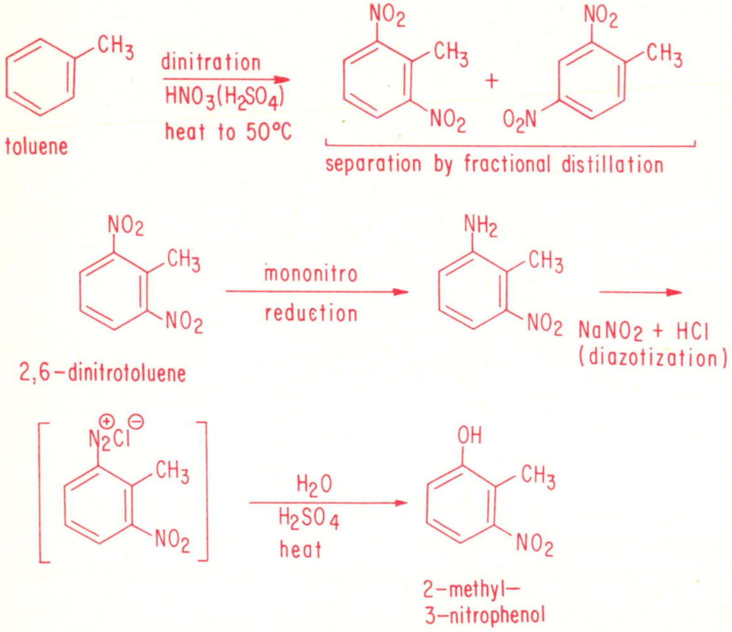

STAGE II   Synthesis of 4-benzyloxy-substituted indole

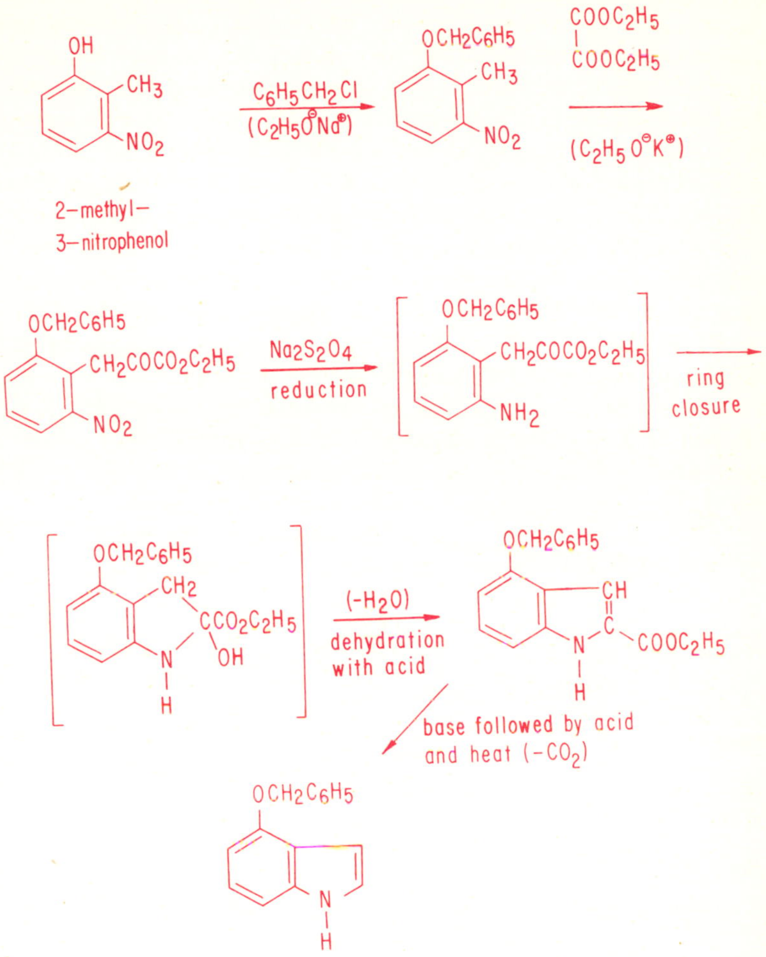

2-methyl-
3-nitrophenol

Initially, Hofmann and his associates prepared 2-methyl-3-nitrophenol. Next they designed the synthesis of the properly substituted indole. From this compound they were able to obtain psilocin and psilocybin, the desired end products.

STAGE III   Synthesis of psilocin and psilocybin

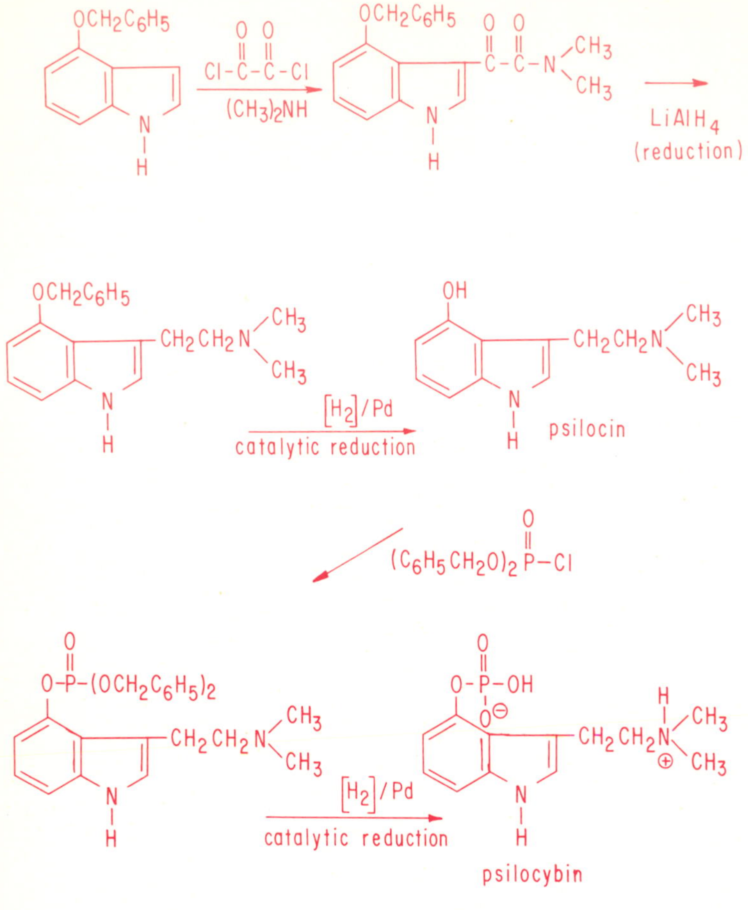

The synthetic scheme began with toluene which is a multimillion pound-a-year petrochemical. Treatment of this liquid with nitric acid and with a small amount of sulfuric acid led to two dinitration products, 2,4-dinitro-toluene and the desired 2,6-dinitrotoluene which could be separated by careful fractionation. If nitration conditions were much more drastic (heat and high pressure) the explosive trinitrotoluene (TNT) would result. After separation, the 2,6-dinitrotoluene is reduced carefully with zinc and ammonium chloride so that only one nitro group is converted to the amine function. Following this, the 2-methyl, 3-nitroaromatic amine is treated with nitrous acid to convert the amino group to a diazonium salt. Such reactions have long been known in chemistry. When the diazonium salt is heated in water with a trace of sulfuric acid, the first stage is completed with the preparation of 2-methyl-3-nitrophenol.

Stage II begins with the protection of the phenolic hydroxyl group of 2-methyl-3-nitrophenol. This is accomplished by an S_N2 displacement reaction using the phenolate anion to attack the easily displaced chlorine of benzyl-chloride. The phenolate anion is generated by allowing the strong base sodium ethanolate to react with the phenol. (This reaction is known as the Williamson ether synthesis.) Following the protection, a base-catalyzed addition to a carbonyl group is accomplished. The hydrogens of the methyl group of the protected 2-methyl-3-nitro phenol are slightly acidic. Very strong base and heat can generate the carbanion.

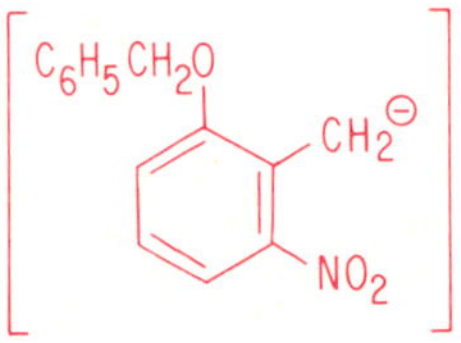

carbanion generated by potassium ethylate

This compound can react with diethyl oxalate by a nucleophilic addition-elimination reaction yielding a keto ester.

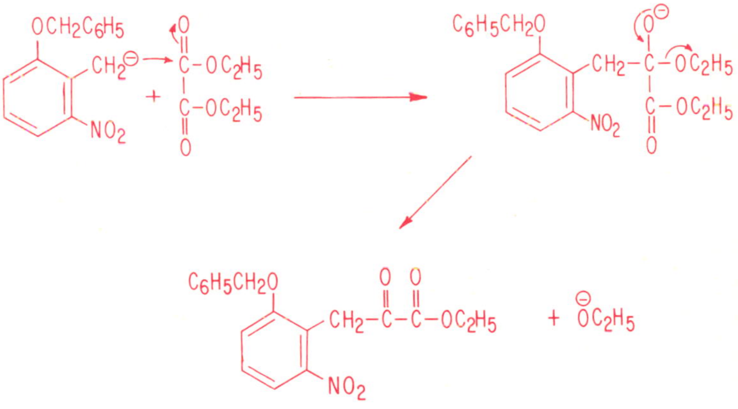

keto ester

The nitro group of the keto ester is reduced by sodium dithionate which immediately cyclizes by virtue of the proximity of the amino group to the ketonic group. The cyclized product is not isolated but is heated with a trace of strong acid to dehydrate the molecule.

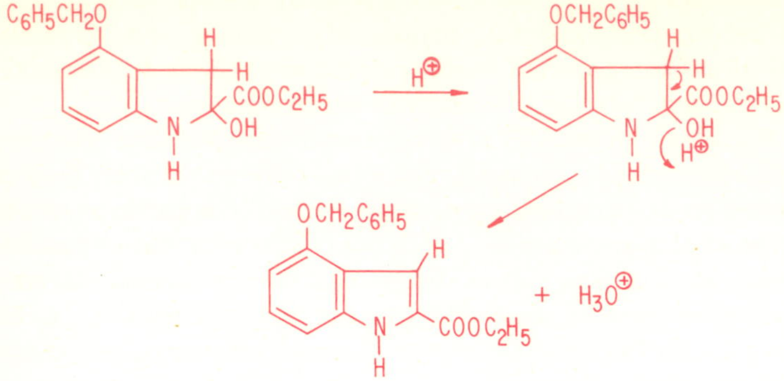

$C_6H_5CH_2O$ ... H, COOC$_2$H$_5$, OH, N, H $H^{\oplus}$ ⟶ $OCH_2C_6H_5$ ... H, COOC$_2$H$_5$, OH, $H^{\oplus}$, N, H

$OCH_2C_6H_5$ ... H, N, COOC$_2$H$_5$, H $+$ $H_3O^{\oplus}$

The second stage is completed by the decarboxylation of the ester. The ester is hydrolyzed by base and acidified. Heating the acid drives off carbon dioxide (see p. 175).

Stage III commences with an aromatic acylation of the 4-benzyloxy-substituted indole by oxalylchloride followed by amide formation with dimethylamine.

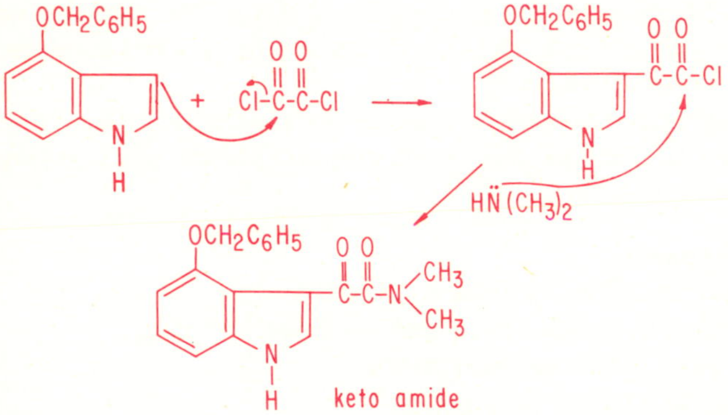

$OCH_2C_6H_5$ ... N, H $+$ $\overset{O\ O}{Cl-C-C-Cl}$ ⟶ $OCH_2C_6H_5$ ... $\overset{O\ O}{C-C-Cl}$, N, H

$H\ddot{N}(CH_3)_2$

$OCH_2C_6H_5$ ... $\overset{O\ O}{C-C-N}\overset{CH_3}{\underset{CH_3}{}}$, N, H   keto amide

The ketoamide is reduced by the strong reducing agent, lithium aluminum hydride. The protecting group for the phenol function is removed by noble metal (palladium or platinum) catalytic hydrogenation (reduction) to yield one of the desired products, psilocin. From psilocin, the psilocybin is prepared by phosphorylation of the phenolic function via dibenzylphosphorylchloride. The benzyl groups on the phosphate group are removed by catalytic reduction to give the final product psilocybin (see p. 176).

The initial reports on mushroom rites of Mexico were written by Bernadino de Sahagun, the Franciscan missionary of peyote fame. He provided us with a description of the coronation of Montezuma in 1502. The entire population celebrated with songs and dances. Special priests presided over the consumption of *teonanacatl*, while chanting periodically, dancing and clapping hands rhythmically. Even captured enemy princes were allowed to partake of the "food of the gods" during Montezuma's coronation feast.

The Wassons (see pp. 153; 174) not only investigated the mushroom cult from ethnobotanical and religiohistorical standpoints but they actually participated in a mushroom agape or love feast. The participants ate up to thirteen pairs of mushrooms over a few hours. The mushrooms tasted terrible, like rancid fat. Visions came in successive stages, geometric patterns, landscapes, intense abstract color designs not unlike fireworks. A euphoria developed for the Wassons. Clearly they were experiencing a psychotomimetic reaction from the mushrooms.

Although the psilocybin and psilocin structures have only recently been unraveled, the Mexican mushroom has received extensive attention by the medical profession. Typical doses of 150 µg/kg of body weight were administered under controlled conditions. Few toxic reactions developed even at higher doses. On one hand it appears that both psilocybin and psilocin are relatively nontoxic for man. On the other hand, cases of *Psilocybe* poisoning have been reported. In one such incident, two adults and two children consumed a *Psilocybe* species of mushroom that was later shown to contain both psilocybin and psilocin. All hallucinated but the children developed very high fevers and one actually died.

Hollister in his book on chemical psychoses lists the pattern of behavior and description of the mental state of normal subjects given average doses of psilocybin.

First 30 minutes:
    Dizziness, light-headedness or giddiness;
    Weakness, muscle aching and twitching, shivering;
    Nausea, abdominal discomfort;
    Anxiety, tension, restlessness;
    Numbness of the tongue, lips or mouth;
    Heaviness or lightness of the extremities;
30 to 60 minutes:
    Blurred vision, brighter colors, longer afterimages, sharp definition of objects, visual patterns (eyes closed);
    Increased acuity of hearing;
    Yawning, tearing, facial flushing, sweating;
    Dreamy state, loss of attention and concentration, slow thinking, feelings of unreality, depersonalization;
    Incoordination, difficult and tremulous speech;

60 *to* 90 *minutes:*
> Increased visual effects (colored patterns and shapes, generally pleasing, sometimes frightening, most often with eyes closed, occasionally superimposed upon objects in field of vision);
> Undulation or wavelike motion of viewed surfaces;
> Distance perception impaired;
> Euphoria, general stimulation, ruminative state;
> Slowed passage of time;

90 *to* 120 *minutes:*
> Continuation of many of the above effects in varying degrees, especially the introspective state;
> Increased bodily sensations and mental perceptions;

120 *to* 180 *minutes:*
> Waning of previously described effects;

180 *to* 300 *minutes:*
> Nearly complete resolution of drug-induced effects.

Later, subjects felt fatigued and many experienced headaches. Less common effects included uncontrollable laughter, breathing difficulties and strongly depressed appetites.

The metabolism of radioactive psilocin has been studied. Radioactive labels serve as tags for the molecule or its metabolic products. Rats given oral doses showed that the radioactivity is transported to the gastrointestinal tract within a half hour after ingestion. The concentration then decreased and appeared in most organ tissues, primarily the kidneys and the brain. Except for the first hour, the adrenal glands showed the greatest concentration of radioactivity. Greater than 90% of the radioactivity was eliminated in 24 hours, 65% in the urine and the remaining fraction through the feces and bile. Analysis of the metabolic products indicated that only 15% was demethylated, or oxidized, and about 25% was eliminated unaltered. After 7 days, radioactive metabolic products remained easily detected throughout the tissues of the test animals.

Hollister demonstrated that many alterations of metabolic patterns occur following ingestion of psilocin. It has been suggested that these materials inhibit adrenalin or serotonin or may block acetylcholinesterase in a manner not unlike that described for LSD and mescaline noted above (see p. 164).

As shown on p. 173, bufotenine is isomeric with psilocin. The former contains a 5-hydroxyl function while the latter has a 4-hydroxyl group. It has been demonstrated that bufotenine is one of the main constituents of the plant seeds of *Piptadena peregrina* from which the pre-Columbian Caribbean and South American Indians developed narcotic snuff (Cohoba or yopo).

Ramon Pane, who sailed with Columbus on his second voyage, wrote about the ways of the New World Indians. He noted that natives of Haiti used snuff and according to him: "This powder they draw up through the nose and it intoxicates them to such an extent that when they are under its influence, they know not what they do." Others wrote in detail about the manner of inhalation, the nature of the tubes inserted in the nostrils, and the preparation of the

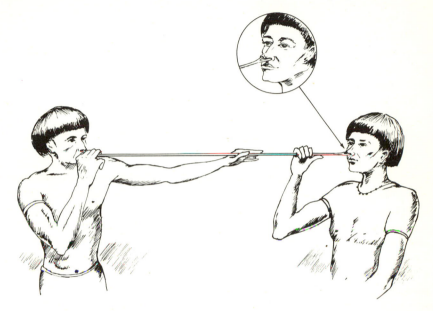

FIGURE 8.4 *A Potentially Lethal Blow*. Men of northwestern Brazil and Colombia partake of hallucinogenic snuff. One man inserts the bifurcated end of a bamboo tube into his nostrils; another man blows at the other end. Severe intoxication from the several tryptamines in the snuff results almost immediately. Often the blower used too much force and drove the snuff deep into the cranial tissue of the snuff taker; hemorrhaging and death frequently ensued.

ground seed which included mixing the seed with quick lime. For some tribes in South America a technique arose of using a tube with a bifurcated end and two men. The tube was loaded with the intoxicant powder. One man inserted both sides of the Y end of the tube in his nostrils. The second man blew as hard as he could to drive the snuff as far up the nasal passages of the first man as possible (see Figure 8.4). The ensuing severe intoxication led to uncontrollable convulsions followed by violent rages and/or deep sleep. The habit of taking these hallucinogens as snuff is truly fraught with great danger. Numerous deaths have been reported among Cohoba and yopo users.

Bufotenine was initially identified as occurring in the skin glands of toads (e.g., *Bufo vulgaris*); hence the name. It also occurs in mushrooms of the *Amanita* group; *A. muscaria mappa* and *A. m. pantherina*. The Chinese knew that dried secretions of a species of toad (Ch'an Su) were biologically active. These extracts have been shown to be rich in bufotenine.

It has been shown that bufotenine inhibits cholinesterase activity more than serotonin. In the 1950s Bumpus and Page collected evidence for the presence of bufotenine in human urine. Other researchers have found that schizophrenics do not have bufotenine or other methylated indoles in their urine. This has led to the hope that a diagnostic test could be developed to uncover potential schizophrenics long before overt symptoms appear. Through such findings, early treatment and control of the disease may be accomplished.

The synthesis of bufotenine was accomplished by Stoll and Hofmann in a manner identical to that shown earlier for the synthesis of psilocin and psilocybin (pp. 173–178). They, of course, began their synthesis by preparing 2-nitro-5-hydroxytoluene.

4.   Cannabinoids from *Cannabis sativa*

The oldest and most broadly occurring hallucinogen comes from the plant family, the Cannabinaceae. This plant group has served as the source of hemp fiber and oil in addition to its use as a psychoactive substance. The Chinese made hemp fiber more than 3500 years ago. Indians have long fashioned candles from the oil extracted from the seeds. Zoroastrians and other Asian peoples utilized the plant as an hallucinogen more than 2500 years ago. In the beginning of this chapter, we included other historical aspects of the use of cannabis. There are many more such examples.

At present, it is estimated that nearly 400 million people partake of some form of cannabis, making it second to opium as the most widely employed "mind altering" plant material used today. Numerous names have been applied to cannabis such as bhang, charas, dagga, garija, grass, guage, hashish (or hash), joint, marihuana, muta, pot, reefer, tea and many more. Although cannabis has such a long history and its occurrence is so ubiquitous, comparatively little is known about its biological effects on long term usage. Only recently has it been shown that the primary active principle in *Cannabis sativa* (the most common form of the Cannabinaceae) is probably Δ^1-tetrahydro-cannabinol.*

Δ^1-tetrahydrocannabinol

Major advances in the structural elucidations and syntheses of the cannabinoids have recently been reported. We owe much to Dr. R. Mechoulam of the Hebrew University for his definitive chemical investigations and to the Israeli police who continually supplied him with the hashish with which he worked.

More than twenty compounds related to the active Δ^1-tetrahydrocannabinol have been isolated and characterized from *Cannabis sativa*. Most of these materials are psychotomimetically inactive. By and large it is believed that most of the hallucinogenic activity resides in the Δ^1-tetrahydrocannabinol.

Chemically the cannabinoids are substituted terpinoids. The term "terpene" comes from the turpentine tree which exudes many oils. The chemical nature of plant flavors and fragrances are closely related to plant essential oils. The term terpene in its broadest sense now includes all of these compounds and many others involved with vitamin, steroid and pigment formation. It was noted many years ago that terpenes can be divided into groups of parent molecules constructed of multiples of isopentane units:

	Number of isopentane units	Number of carbons
Monoterpenes	2	10
Diterpenes	4	20
Triterpenes	6	30
Tetraterpenes	8	40

* The term tetrahydro refers to the fact that ring "a" in the formula contains only one double bond. It is possible to have two more double bonds in that ring. Formalistically, four hydrogens would have to be removed to achieve the maximum double bond content of ring "a", hence tetrahydro. The Δ^1 refers to the fact that the double bond in ring "a" exists between carbons 1 and 2. There are two asymmetric centers in Δ^1-tetrahydrocannabinol located at carbons 3 and 4. Of the four possible optical isomers, only three are geometrically attainable.

† The *n* stands for normal and indicates that what follows is a straight chain hydrocarbon (see also *n*-pentyl resorcinol, p. 185).

The monoterpenes as a class contain such compounds as β-ionone (a component of vitamin A₁), geranial or citrala (scent of geraniums), citronella (oil of citronella), limonene (oil from lemon, orange, etc.), camphor (camphor tree) and many others.

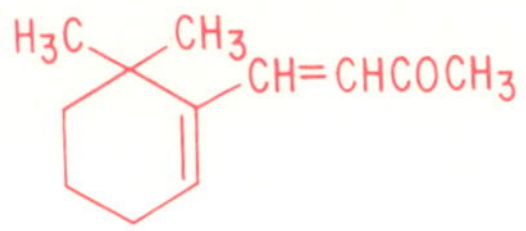

β-ionone

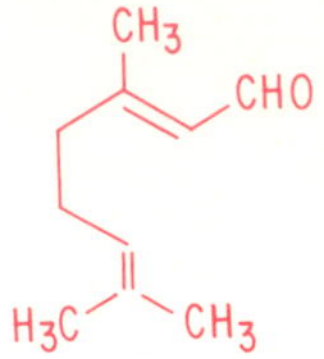

geranial

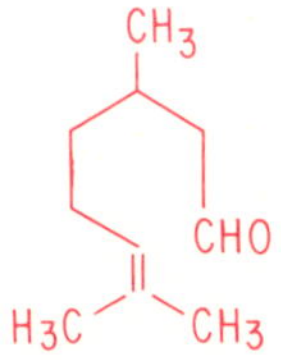

citronellal

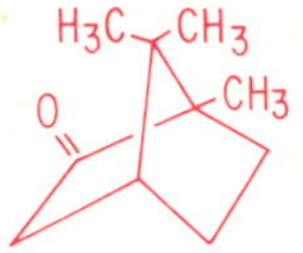

camphor

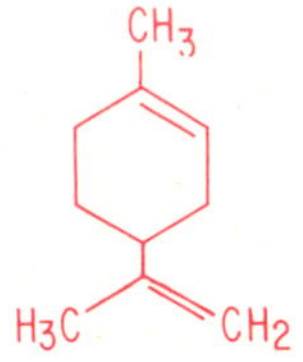

limonene

Among the large group of diterpenes are compounds such as vitamin A₁ (see Chapter IX) and abietic acid, the principle component of rosin which is an important commercial product (more than a billion pounds a year). Rosin is used in paints, varnishes, sizing of paper, on violin bows, and in synthetic rubber production.

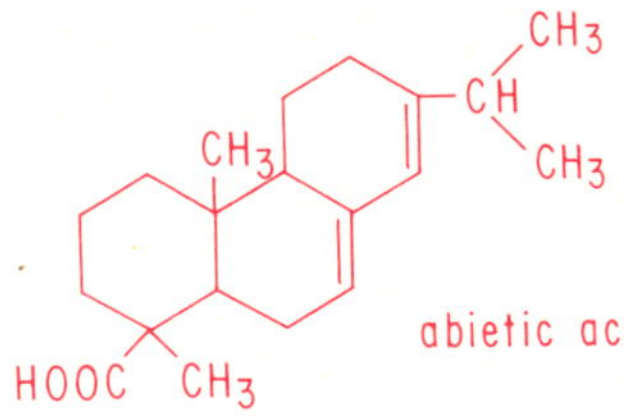

abietic acid

The triterpene squalene has been shown to be a precursor of steroids (Chapter VII). Lastly, the carotenoids are actually tetraterpenes.

This digression allows us now to consider the synthetic approaches to prepare the dl-Δ¹-tetrahydrocannabinol. Mechoulam allowed geranial (a monoterpene) to react with a properly substituted benzene derivative to achieve the synthesis of racemic desired product. This route shows clearly why the cannabinoids can be considered to be substituted monoterpenes.

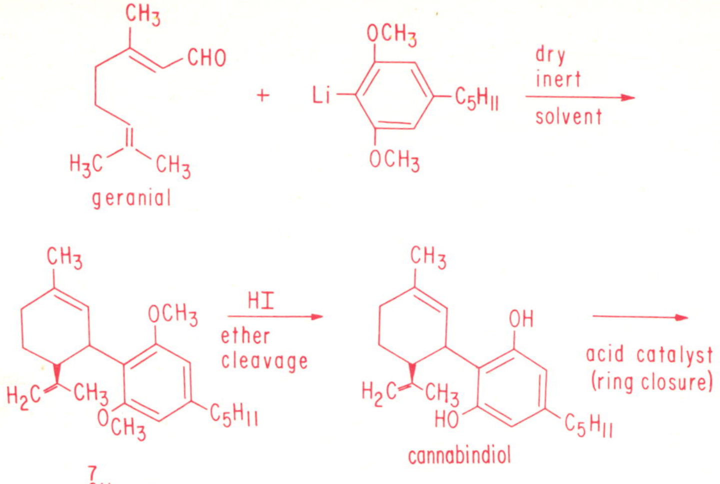

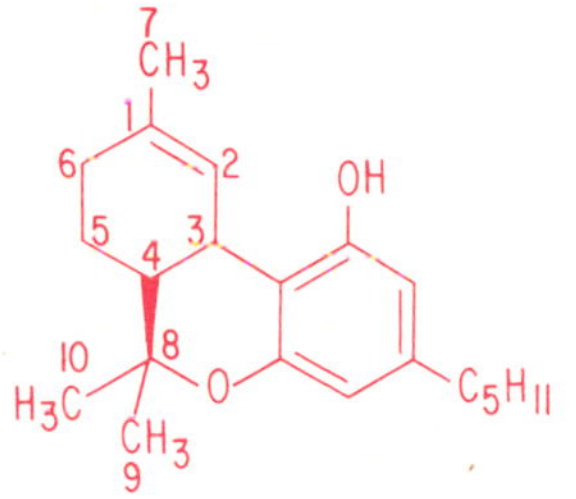

geranial

cannabindiol

The asymmetric centers
are located at carbons
3 and 4.

dl-△1–tetrahydrocannabinol (racemic)

Mechoulam was able to synthesize the naturally occurring, optically active, negatively rotating Δ^1-tetrahydrocannabinol by the following sequence. He began with the optically active monoterpene, verbenol, which was allowed to react with n-pentyl resorcinol in the presence of boron trifluoride as a catalyst.

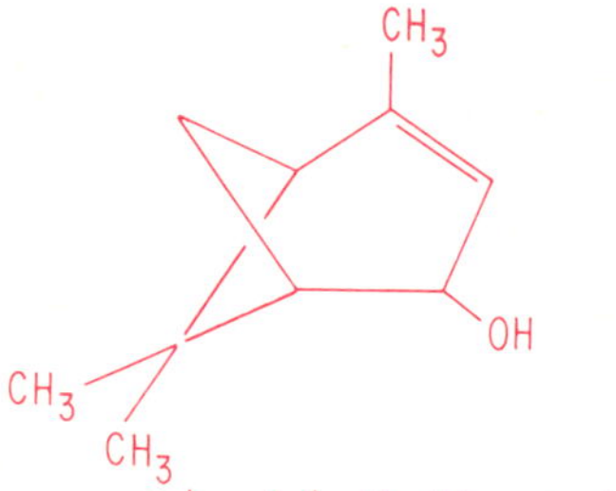

verbenol (optically active)

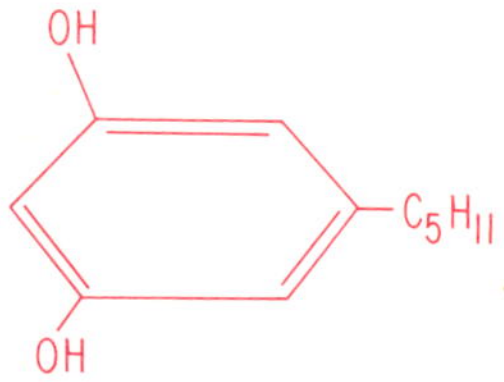

n-pentyl resorcinol

This gave an optically active isomer of the desired Δ^1-tetrahydrocannabinol, namely, the $\Delta^{1(6)}$-tetrahydrocannabinol* in 85% yield.

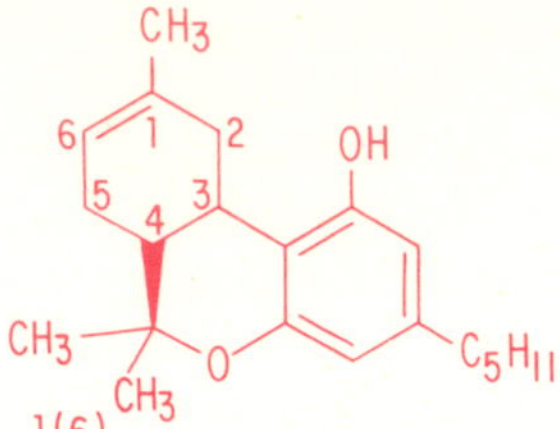

$\Delta^{1(6)}$-tetrahydrocannabinol

The $\Delta^{1(6)}$-isomer was treated with hydrogen chloride in toluene at $-15°C$ with zinc chloride as a catalyst which gave l-chlorohexahydrocannabinol.

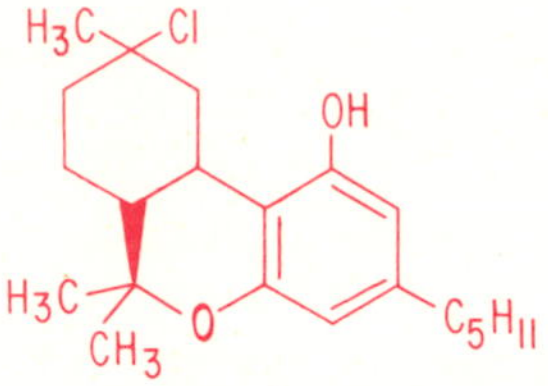

l-chlorohexahydrocannabinol

This product was dehydrochlorinated using sodium hydride in tetrahydrofuran as a solvent. Both optically active $\Delta^{1(6)}$ and the desired Δ^1-tetrahydrocannabinols result which can be separated by careful chromatography on a Florisil adsorbent column. Other routes to the synthesis have also been reported.

The availability of the synthetic material and the development of accurate techniques to establish Δ^1-tetrahydrocannabinol content in various plant specimens allow scientists today to undertake meaningful pharmacological, biochemical and toxicological studies. It should also be possible to place the psychotomimetic experiences on a more quantitative basis.

The problem of measuring response to psychotomimetic drugs quantitatively is not as simple as it sounds. The cannabinoids are fundamentally nonpolar molecules, insoluble in aqueous systems. When they are smoked, a variable fraction is volatilized and lost into the air either prior to inhalation or during the process of exhalation. Between 20 and 80% of the cannabinoids are

---

* The designation $\Delta^{1(6)}$ indicates that the double bond in ring a proceeds from carbon 1 to carbon 6.

utilized by smokers depending on their techniques of smoking. Oral doses are three times larger than the smoking cannabinoid contents for equivalent response. It appears that the gastrointestinal system and/or the liver quickly deactivates much of the dose. Although intravenous injections seem highly reproducible from the standpoint of drug delivery, it is unrealistic from the standpoint of mimicking the actual conditions of drug use.

Very little is known about the metabolism of the cannabinoids in man. Less than 1 % of unchanged Δ^1-tetrahydrocannabinol is found in the urine. Tracer studies have been carried out on rats using C^{14} labeled Δ^1-tetrahydrocannabinol. Within a week, a substantial fraction of the radioactivity was eliminated in the feces and urine. Nearly 50 % of the radioactivity was retained after 1 week in the form of metabolic products of Δ^1-tetrahydrocannabinol. Similar tracer studies have commenced in which doses of about 6 micrograms per kilogram were given to human subjects intravenously. Two different rates of metabolism of the Δ^1-tetrahydrocannabinol in plasma were observed: a fast rate where reactions occurred over 30 minutes and a slow rate where the transformations took as long as 60 hours. The nature of the metabolic products remains mostly obscure. Much more work must be carried out. Knowledge of the metabolic pathways is essential to an understanding of the effect of these drugs.

It is interesting to note that the fast rate correlates well with the rate of appearance and extent of the psychotomimetic experience. Even with this agreement, it is difficult to relate laboratory tests directly to effects described by marihuana users in social settings. Scientists try to use established methods that allow valid comparisons of both groups. Hollister in an article in *Science* describes a questionnaire that was employed with laboratory and social marihuana users. The results indicated that "The most common symptoms and signs reported were: paresthesia, floating sensations and depersonalization; weakness and relaxation; perceptual changes (visual, auditory, tactile); subjective slowing of time; flight of ideas, difficulty in thinking and loss of attention; loss of immediate memory; euphoria and silliness; sleepiness. Other common symptoms which are not verifiable in the laboratory were claims of increased insight and perception, as well as increased sexual desire, performance and enjoyment". Many of the psychological reactions noted here for marihuana are similar to those we've seen before for LSD, mescaline, amphetamines, psilocybin and psilocin. It does not appear as if typical use of cannabinoids leads to the violent outbursts or severe chemical psychoses observed with some of the other hallucinogens. People have even claimed substantial medical and therapeutic value for cannabis. In Western society, there has been great pressure for the legalized use of marihuana. Before commenting on this question, it is important to consider a brief comparison of marihuana and alcohol and also to indicate what is known about psychological dependency on marihuana.

Alcohol (ethanol) is a simple organic molecule, easily and quickly metabolized to normal products of respiration. Its dosage and concentration in the blood stream can be exactly determined. Excessive use of alcohol can lead to serious destructive consequences including cirrhosis and psychological addiction; but alcohol can be used in moderation without intoxication. It is extremely difficult to establish dosage for marihuana (as noted above). The object of marihuana use is clearly to achieve the hallucinatory effects. Frequent use of marihuana can certainly lead to psychological addiction in a manner not unlike the dependencies some people feel for alcohol or cigarettes.

As more research results are accumulated, the similarity of psychological response between Δ^1-tetrahydrocannabinol and LSD becomes obvious. The effects of long term use of both drugs are not known. Marihuana or its active principles represent a unique drug in that controlled experimentation with human subjects is proceeding at an ever accelerating pace. From these studies, we should learn enough about marihuana or the other forms of the cannabinoid drugs to make valid judgments about legalizing its use. It is likely that without such knowledge legal penalties for possession are bound to be counterproductive. The consequence of these severe penalties will be to drive the social users underground in a manner not unlike the "mushroom cults" of Mexico. Illicit traffic in the cannabis drugs will increase and our ability to implement scientific findings will be seriously diminished.

SUMMARY AND SOME CAUTIONARY REMARKS

In this chapter, we attempted to place the origins and chemistry of certain "mind altering" molecules in perspective. We included a substantial analysis of the organic reactions and structural assignments for numerous hallucinogens. To this was added some information on biological reactions derived from these drugs. Much research remains to be carried out.

Unfortunately, the widespread use of psychotomimetics in today's Western society has created a whole new set of problems. Advocates of nonmedical use of these drugs claim new intense sensations and insights. Opponents of the "drug culture" describe the extremes of asocial behavior among users as if they were typical effects. Both positions are oversimplified. To us, it is clear that hallucinogens are ingested to achieve hallucinations. Associated with this effect are numerous additional sensations among which is euphoria. While under the influence, organized thought becomes severely impaired. People withdraw into themselves and seek emotional satisfactions. It is interesting that

Eastern religions such as Zen do not employ drugs. They teach that self-realization comes through enormous mental self-discipline.

Hollister in his book on *Chemical Psychoses* sums up his worries about the people who take these "molecules of mysticism". He states: "I can only regard the psychedelic drug movement as a new brand of antiintellectualism. Just as the ancients were fearful of the unknown mysteries of nature, many people today are fearful of the information explosion which is so difficult to understand and assimilate. Fear of reality, known or unknown, has always been the major impetus for seeking escape through drugs. The prophets of the new drug cults are convinced that modern man needs to think less and feel more. Drug-taking is a sensual, not an intellectual experience. Indeed, the title of a recent psychedelic "happening" starring the cult's leader was appropriately titled *The Death of the Mind*. Psychedelic drugs are solvents of the logical, formal mind, ordinarily used by intellectuals, and words and numbers are supplanted by random, unorganized and often meaningless images. As the prospective Christ of the cult has so aptly described it, LSD is a powerful nonverbal, metaintellectual agent".

We agree with Dr. Hollister that statements such as this are not only metaintellectual, they are clearly antiintellectual. This is a time when the problems of the world are overwhelmingly complex—overpopulation, the urban decay, imbalance in the use of natural resources, pollution of the air and water and many more. We need clear thinking, rational planning and sound implementation. The notion of following a person "strung out" on these hallucinogens or narcotics in order to find the answers to society's needs is ludicrous. It is always easy to abdicate responsibility and seek rapid personal gratification. It is far more satisfying for us to attempt the accumulation and control of knowledge to achieve a more livable world.

SOURCE MATERIALS AND SUGGESTED READING

Allegro, J. M., 1970. *The Sacred Mushroom and the Cross*, Doubleday and Co., Inc., Garden City, New York.

Barron, F., M. E. Jarvik, and S. Bunnell, Jr., 1969. "The Hallucinogenic Drugs", p. 219 in *Science, Conflict and Society; readings from Scientific American*, W. H. Freeman and Co., San Francisco.

Hoffer, A., and H. Osmond, 1967. *The Hallucinogens*, Academic Press, New York and London.

Hollister, L. E., 1968. *Chemical Psychoses: LSD and Related Drugs*, Thomas, Springfield, Illinois.

Hollister, L. E., 1971. "Marihuana in Man: Three Years Later," *Science*, **172**, pp. 21–28.

Mechoulam, R., 1970. "Marihuana Chemistry," *Science*, **168**, pp. 1159–1166.

Schultes, R. E., 1969. "Hallucinogens of Plant Origin," *Science*, **163**, pp. 245–254.

Wasson, R. G., 1968. *Soma: Divine Mushroom of Immortality*, Harcourt, Brace and World, New York.

CHAPTER IX

MOLECULES OF GROWTH AND HEALTH
Vitamins

INTEREST IN vitamins and in the so-called "health foods" has grown rapidly during the last several years. What was confined not long ago to a select group of health faddists has spread to include a substantial segment of our citizenry. The little neighborhood health food store is a new but not uncommon part of the local scene now. This trend can have major beneficial results, if the concept of proper nutrition is emphasized and the cultism minimized.

The revolutionary suggestion has occasionally been made that doctors should be paid for keeping people healthy, rather than for curing them after they get sick. The idea is perhaps idealistic and impractical, but the underlying thought could well be applied by an informed individual to his own advantage. The easiest, least expensive, and probably most effective way to begin such a program would be to know the components of a balanced diet and to consume them. Specific vitamin shortages could be made up by the common low-dosage multiple-vitamin pills now seen in most supermarkets. The motive should match the act in modesty—merely to ensure a minimal daily intake of the few organic molecules which the body cannot synthesize, and without which the body cannot live for long. With vitamins probably more than anything else, the difference between none and little can be a lot. This chapter tries to show this, along with the ways some of these small molecules of health do their work.

Vitamin B$_1$ (Thiamine)

Toward the end of the last century, sailors on Japanese ships were suffering by the thousands, and many were dying hideous deaths, because of a mysterious disease called "beriberi". One of the mysteries of beriberi was that marine personnel of other nations were apparently immune to the disease.

Then in 1882, a marvelous piece of medical detective work by a Japanese navy doctor named Kanehiro Takaki produced a dietary solution to the problem. He discovered that replacement of *polished* rice—the Japanese seamen's staple—by *unpolished* rice not only cured the disease but also prevented it. Dr. Takaki postulated that some essential protein in the rice husk had been lacking in the sailors' former diet.

Subsequent work has shown that this is partially correct and partially incorrect. An essential dietary factor is indeed present in the rice husk, but this factor is not a protein. It is a small organic molecule called thiamine, which acts as a coenzyme—that is, it acts in conjunction with an enzyme (a protein) to catalyze specific metabolic chemical reactions.

One of the essential metabolic reactions which requires thiamine is the decarboxylation of pyruvic acid. We have seen (pp. 35, 37) that this is the reaction

$$CH_3-\underset{\underset{\text{pyruvic acid}}{}}{\overset{\overset{O}{\parallel}}{C}}-COOH \quad \xrightarrow[\substack{\text{thiamine as} \\ \text{coenzyme}}]{\text{enzyme} +} \quad [CH_3CHO] \quad + \quad CO_2$$

acetaldehyde

$$\downarrow \substack{\text{biooxidative} \\ \text{pathway} + \\ \text{coenzyme A}}$$

$$\text{citric acid cycle} \quad \longleftarrow \quad CoA-\overset{\overset{O}{\parallel}}{C}-CH_3$$

which "feeds" the citric acid cycle—that important sequence of reactions which serves as a fount of energy for aerobic life systems. Little wonder that thiamine deficiency causes extreme cellular distress.*

Plants have the ability to synthesize thiamine, but animals including man do not, and neither do some bacteria and fungi. For animals then, thiamine is a dietary necessity, and as it is neither protein, nor carbohydrate, nor fat, it is called a vitamin and has been arbitrarily designated vitamin B_1. Good sources of this vitamin include eggs, meat, peas, and beans. Man's daily requirement is one to three milligrams.

Vitamin B_1 was first isolated in 1926 from rice husk. Ten years later, the structure was determined and the molecule was synthesized. The molecule consists of two cyclic components joined by a single carbon-atom bridge. In this sense, it is reminiscent of the quinine molecule. The cyclic components of thiamine, however, are different from those of quinine. Thiamine contains a pyrimidine ring (seen in nucleic acids, p. 42) and a thiazole ring. A thiazole ring resembles the five-membered ring in penicillin in that it contains a sulfur atom (thio) and a nitrogen atom (azo) (see p. 116).

---

* Thiamine is the only coenzyme known from which diagnosis of impending vitamin deficiency can be diagnosed chemically prior to the development of the full-blown disease— beriberi (or polyneuritis).

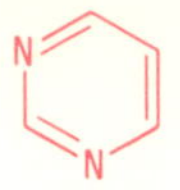

pyrimidine 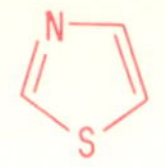   thiazole

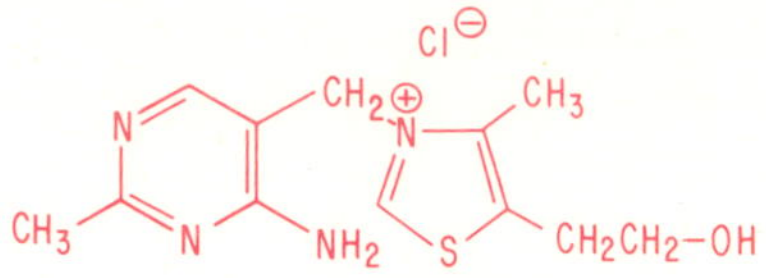

vitamin B₁ (thiamine chloride)

Determination of the structure of vitamin B$_1$ was facilitated by the fact that the molecule can be split into two parts under very mild conditions. A sulfite ion in nearly neutral aqueous solution attacks the bridging carbon atom freeing the thiazole ring from it.

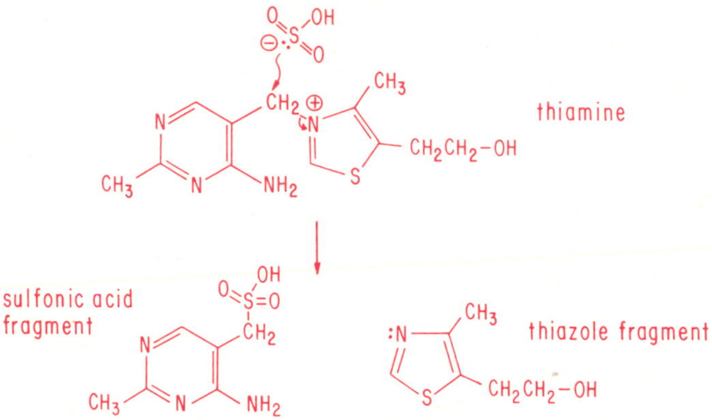

thiamine

sulfonic acid fragment   thiazole fragment

The bridging carbon atom in thiamine is vulnerable to nucleophilic attack because of the powerful electron-withdrawing effect of the positively charged nitrogen atom. This sulfonic acid group can be removed by treatment with sodium metal dissolved in anhydrous liquid ammonia.

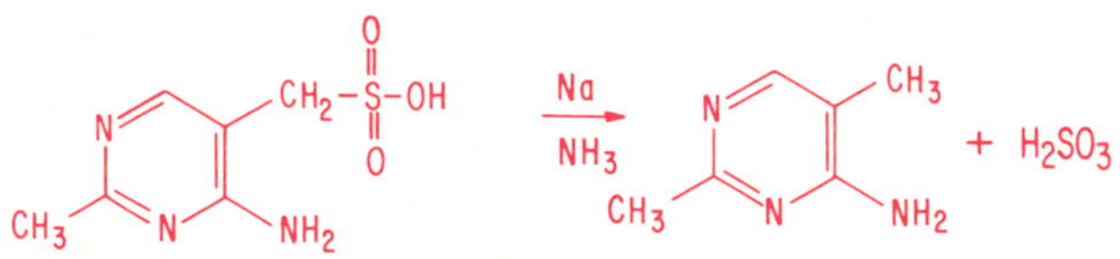

sulfonic acid fragment   4-amino-2,5-dimethylpyrimidine

The ultraviolet spectra of both the sulfonic acid derivative and its reduction product indicated that they might be pyrimidines, and their structures were subsequently proven by synthesis.

Ultraviolet spectra also suggested that the other portion of the molecule obtained from the sulfite degradation might contain a thiazole ring. When this other fragment was oxidized with nitric acid, it did actually produce a known thiazole derivative, 4-methylthiazole-5-carboxylic acid.

thiazole fragment   4-methylthiazole-5-carboxylic acid

The thiazole fragment before nitric acid oxidation did not contain a carboxylic acid group and therefore that carbon atom must have been changed during the reaction. It was also known from elemental analysis that, before oxidation, this fragment contained six carbon atoms and only one oxygen atom. This oxygen atom was known to be hydroxylic since it could be esterified. All this information pointed to two possible structures for the thiazole fragment—one with a primary alcohol group and the other with a secondary alcohol group.

primary alcohol   secondary alcohol

Primary and secondary alcohols can be distinguished by simple chemical tests (see p. 104). These indicated that the thiazole fragment contained a primary alcohol group. Also, the secondary alcohol structure contains an asymmetric atom and therefore should be resolvable into optically active fractions. The thiazole fragment could not be resolved. Synthesis of the primary alcohol structure confirmed the identity of the thiazole fragment.

This completes the structure determinations of both fragments of the thiamine molecules, but leaves unanswered the question of how the fragments are joined to each other. Two pieces of information indicate the correct mode of joining. First, the intact thiamine molecule does not contain a sulfonic acid group, so the carbon atom which becomes sulfonated by a sulfite ion must be involved in the joining. Second, thiamine was long known to be a quaternary ammonium compound, i.e., it exists as a salt. The only nitrogen atom which can fulfill this requirement and join the two fragments is the thiazole nitrogen atom. Hence the carbon-atom bridge must be attached to the thiazole nitrogen atom, as shown.

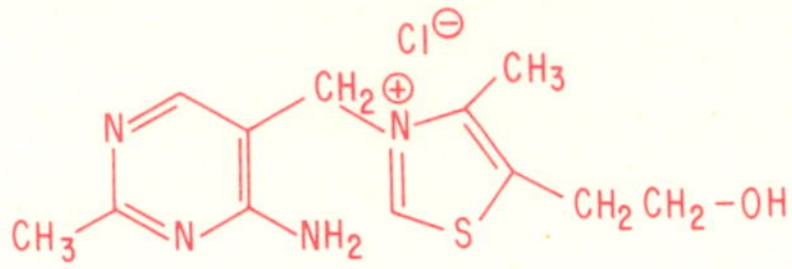

thiamine chloride

Knowing the structure of thiamine whets the desire to know how it works as a vitamin in the body. We have said that it is a coenzyme for the decarboxylation of pyruvic acid which feeds the all-important citric acid cycle. As a coenzyme, thiamine is present in all living cells as the pyrophosphate ester.

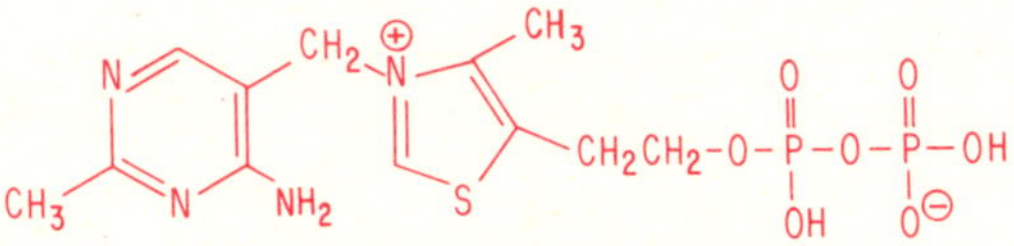

thiamine pyrophosphate
(cocarboxylase)

The coenzyme is called cocarboxylase because it acts in conjunction with the enzyme carboxylase, which catalyzes the decarboxylation of pyruvate. The entire enzymatic system consists of a protein with a molecular weight of about 150,000, one molecule of the coenzyme, and five atoms of magnesium.

The active site of the coenzyme is believed to be the carbon atom between the sulfur atom and the quaternary nitrogen atom.

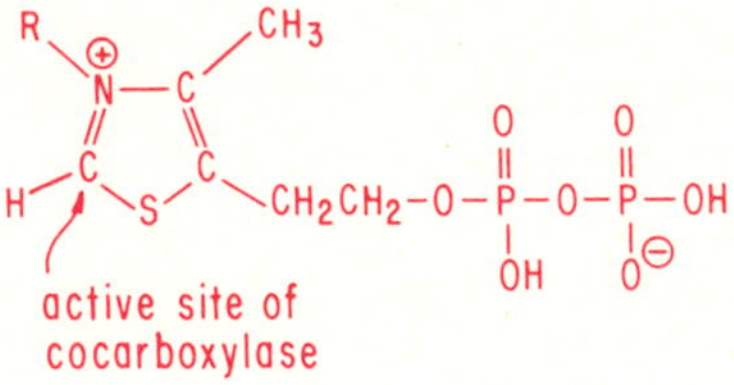

active site of
cocarboxylase

This carbon atom presumably loses a proton and becomes a negatively charged nucleophile.

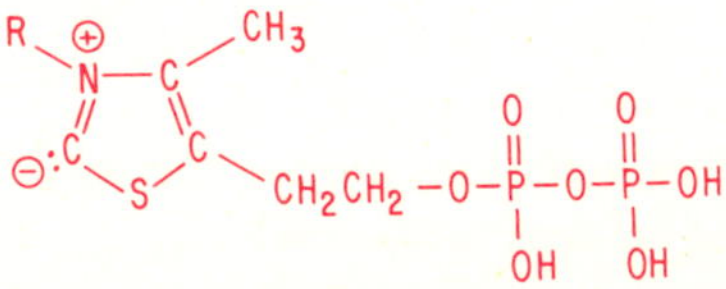

where R represents:

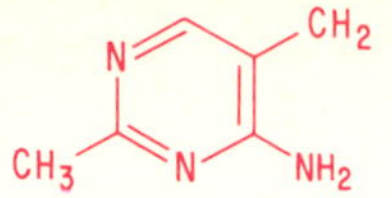

As a nucleophile, it can attack the ketonic carbon atom of pyruvate.

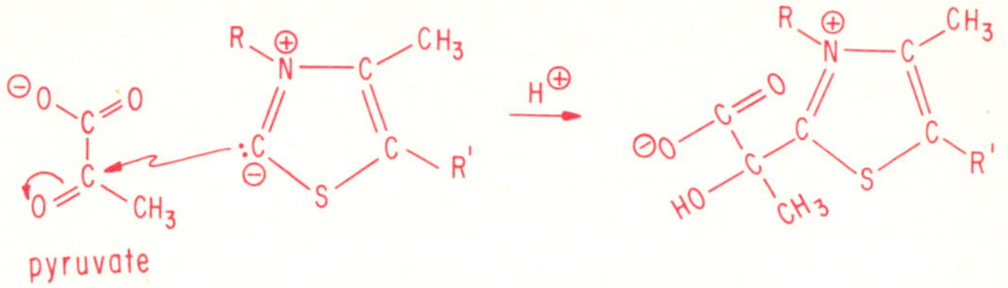

where R' represents

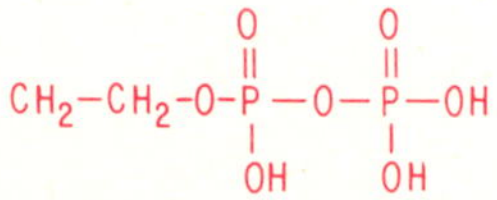

and R is as noted above.

The powerful electron-withdrawing effect of the positively charged quaternary nitrogen atom now facilitates the decarboxylation of the pyruvate anion.

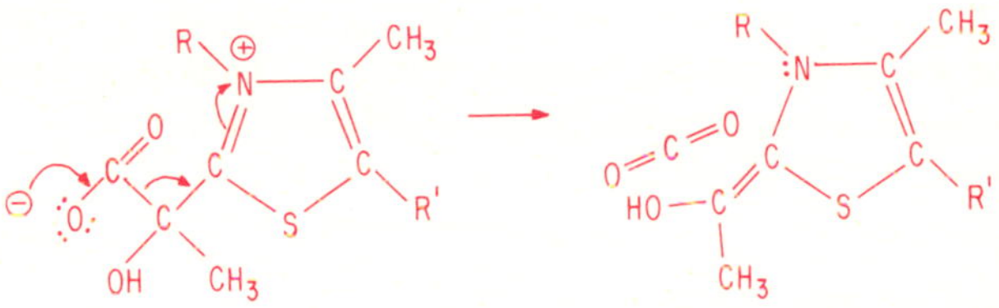

Re-establishment of the conjugated-double-bond system of the thiazole ring produces a stable carbanion.

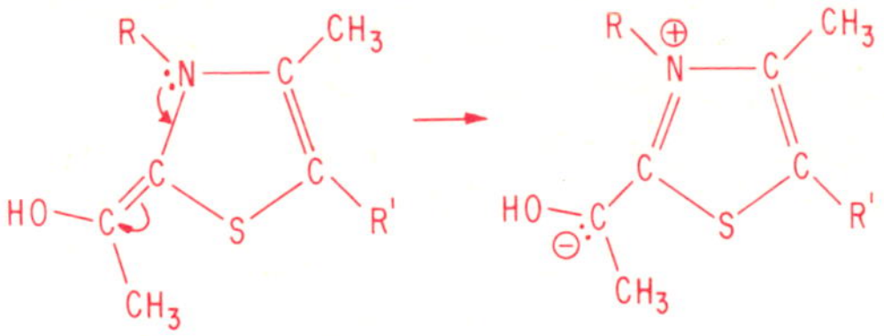

This molecule can shift a proton, release the decarboxylated product, and regenerate the original thiazole nucleophile.

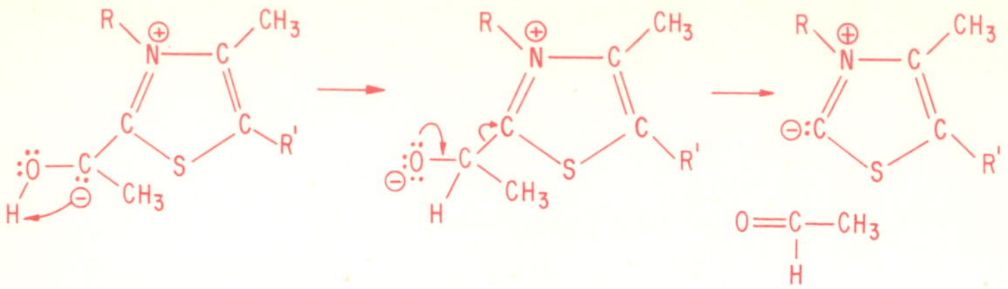

Two pieces of evidence indicate that this is really the way vitamin B_1 works. First, one of the above intermediates, hydroxyethylthiamine, was synthesized artificially and was found to be almost as active biologically as vitamin B_1 itself.

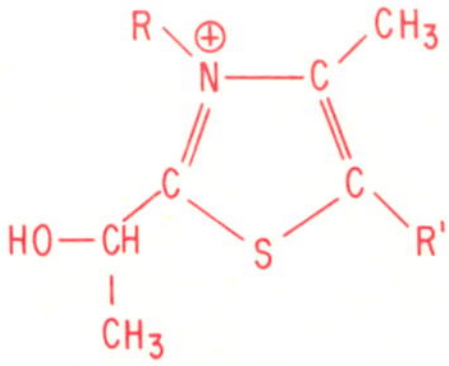

hydroxyethylthiamine (protonated form
of the carbanion described above)

Second, this same intermediate has been isolated from microorganisms.

As shown in the initial overall reaction (p. 37), the acetaldehyde is then oxidized and taken into the citric acid cycle.

Vitamin B_2 (Riboflavin)

The first vitamin preparations were obtained as concentrates of extracts from various natural sources. The intention at the time was to name these preparations systematically in alphabetical order according to the chronological sequence of discovery. Unfortunately, these early materials were later found to contain more than one substance. This in itself would not have caused any difficulty other than purification, but unfortunately some of the "impurities" turned out to be vitamins in their own right.

To clarify matters, a subscript system was begun, giving rise to vitamin B_1, vitamin B_2, and so on. Then even some of these were found to contain more than one vitamin. This led to some individual vitamins becoming known by name and not by letter, for example niacin (or nicotinic acid) found in early vitamin B_2 preparations. However, the designation vitamin B_2 now refers exclusively to the molecule riboflavin, which has already been seen as a component of flavin adenine dinucleotide (FAD) (see pp. 37, 38).

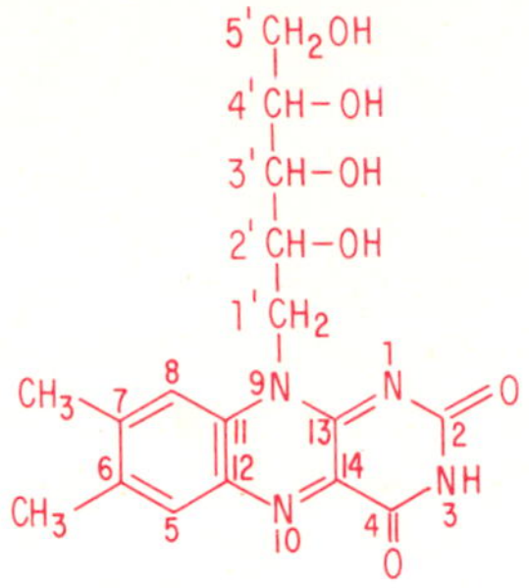

riboflavin

The parent ring system for this vitamin and related structures is called iso-alloxazine. Thus, riboflavin can be called 6,7–dimethyl-9-(1′D-ribityl) isoalloxazine.

FAD was seen earlier as one of the hydrogen-transfer molecules in the respiratory chain. It was pointed out at the time that the flavin portion of the molecule is actually the site of the hydrogen transfer function. Riboflavin itself can function in an oxidation-reduction system at a metallic electrode with an electropotential $E_0 = -0.185$ volt (pH 7, 20°C).

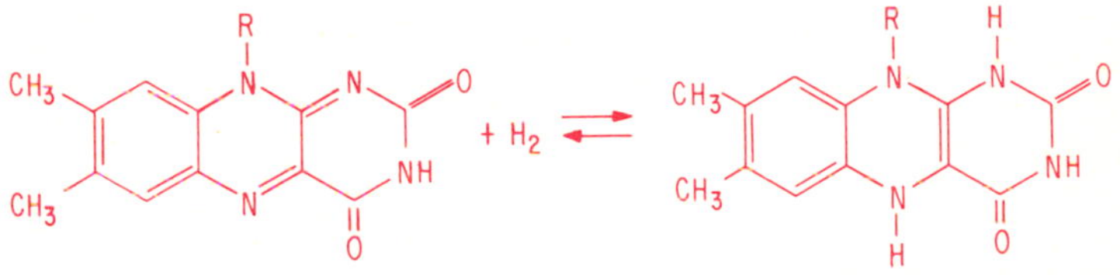

FAD*   FADH2
                            (reduced flavin)

This is one of the reasons the riboflavin molecule is so important to normal human metabolism and, since man cannot synthesize it, he must obtain

---

* The complete structure of the flavinadenine dinucleotide is:

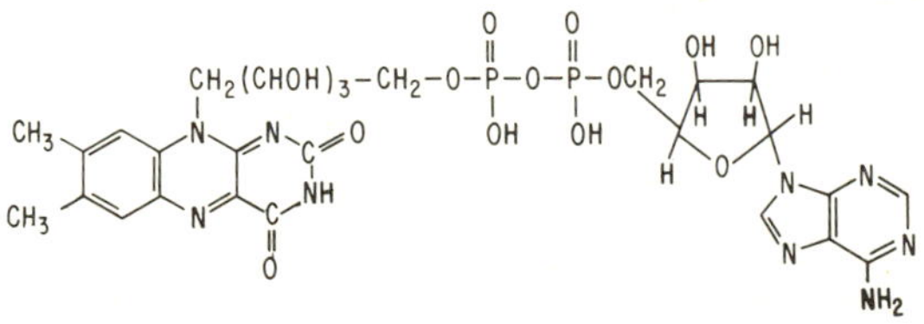

it from external sources. Some of the best sources are liver, wheat germ,
yeast, egg yolk, milk, fish and green vegetables. Man requires an average
daily intake of two to three milligrams of vitamin B_2. Deficiency causes symptoms such as cracking of the lips and corners of the mouth and, in extreme
cases, corneal opacity.

Interestingly, some animals play host to intestinal bacteria capable of
synthesizing the vitamin. These animals absorb enough riboflavin from this
internal "external source" to get along on a B_2-free diet.

Riboflavin was first isolated in 1933 from whey, the curd-free watery
portion of sour milk. The structure of the vitamin was determined independently by two groups of workers in 1935.

The molecule is unstable to light which complicated the structural determination somewhat. Finally, however, examination of the photolysis products
provided the key to the structure, which was then confirmed by synthesis.

In alkaline solution, photoirradiation of riboflavin produces lumiflavin.

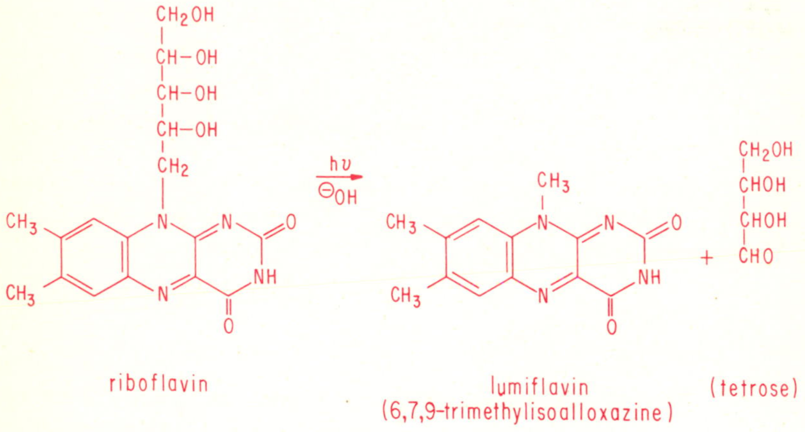

riboflavin   lumiflavin (6,7,9-trimethylisoalloxazine)   (tetrose)

Hydrolysis of lumiflavin produces urea and a carboxylic acid.

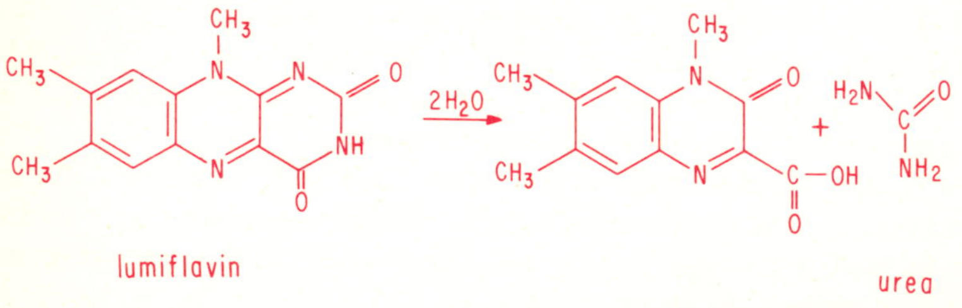

lumiflavin   urea

This carboxylic acid decarboxylates easily, which suggests that it is a β-keto acid.

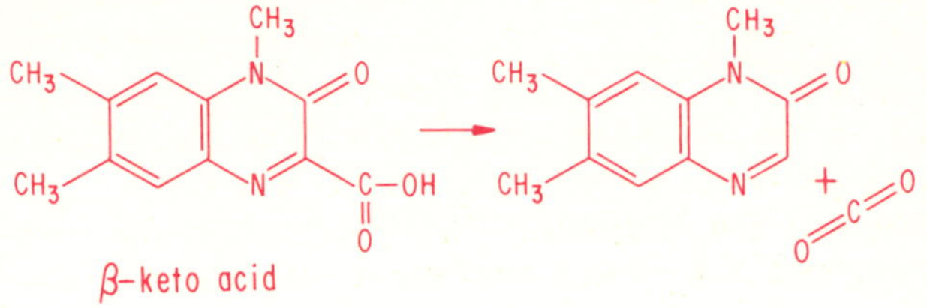

β-keto acid

Finally, the decarboxylation product, when heated in strong base, hydrolyzes and splits out glyoxylic acid to become identifiable as a phenylene diamine.

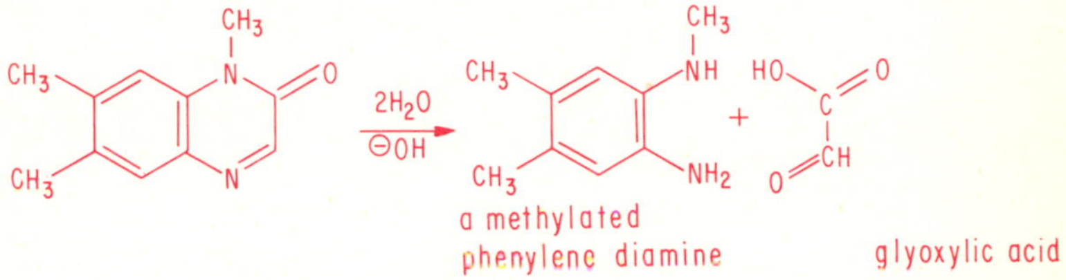

a methylated phenylene diamine   glyoxylic acid

Working backwards through this series of degradation fragments led to a good guess at the structure of lumiflavin and from there to riboflavin, which was subsequently synthesized.

Vitamin B₆ (Pyridoxine)

About 1 year after the isolation of riboflavin, another substance was found in the B complex of vitamins and was designated vitamin B_6. This new vitamin was shown to be essential for normal growth in rats. A deficiency of it in the rat diet caused a kind of dermatitis. It was long presumed to by physiologically important in man as well, but not until 1962 was it shown to be essential to human life by the discovery of a B_6-dependent anemia.

Vitamin B_6 occurs widely in nature in both plants and animals. Man's daily dietary requirement has not been accurately established but dosage recommendations are from one to two milligrams.

The vitamin B_6 molecule is simpler than those of thiamine and riboflavin. It consists of a single pyridine ring substituted with a methyl group, a hydroxyl group, and two hydroxymethyl groups. The structure of B_6 suggested its more descriptive name—pyridoxine.

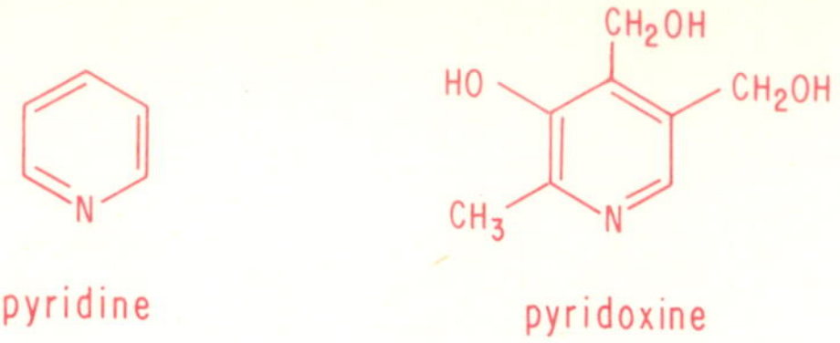

pyridine   pyridoxine

The term vitamin B_6 has come to include not only pyridoxine, but also pyridoxal and pyridoxamine, since all three of these are interconvertible in living cells.

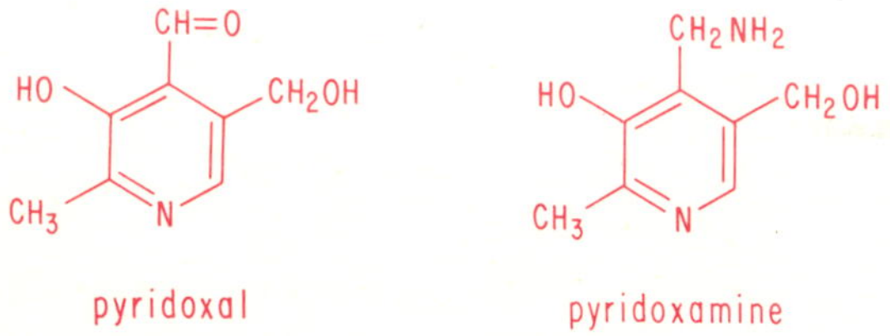

pyridoxal   pyridoxamine

Even more important, pyridoxal and pyridoxamine (as phosphates) are the "working forms" of the vitamin. These molecules perform a variety of essential metabolic functions, many of which deal directly with transformations of amino acids. One of these functions is called transamination. This is an equilibrium reaction between an α-amino acid and an α-keto acid. An example is the interconversion of alanine and pyruvic acid.

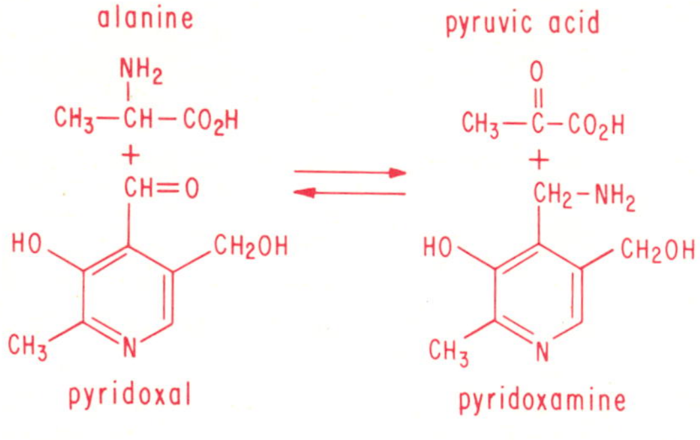

alanine   pyruvic acid

pyridoxal   pyridoxamine

This rather neat molecular sleight-of-hand occurs easily through the intermediacy of a pair of tautomeric Schiff bases (see appendix).

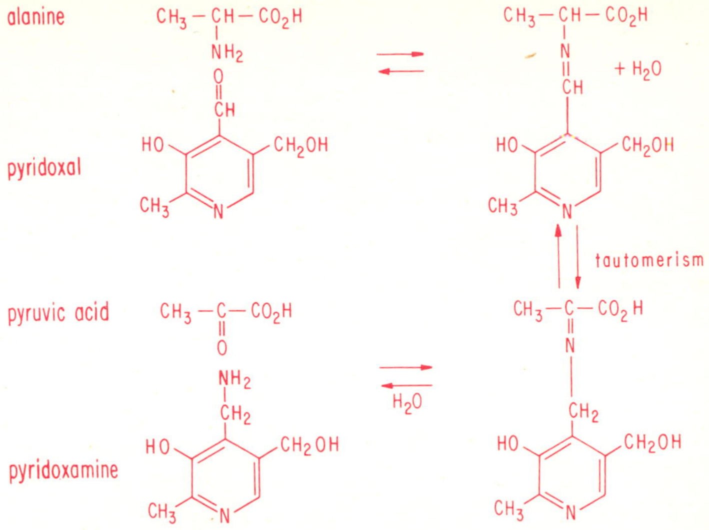

The transamination reaction provides cells with the ability to obtain amino acids for protein synthesis from intermediates in the citric acid cycle, and vice versa. For example, oxaloacetic acid (see pp. 35, 37) and aspartic acid are directly interconvertible by this means.

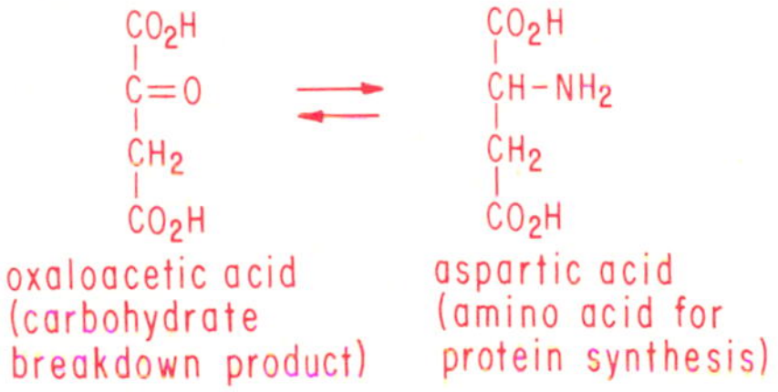

oxaloacetic acid (carbohydrate breakdown product)   aspartic acid (amino acid for protein synthesis)

This is only one of many examples of crossover reactions between "separate" metabolic pathways.

Nicotinic Acid (Niacin) and Nicotinamide

Among the early recognized vitamin-deficiency diseases, pellagra was a more serious problem in the United States than was scurvy or beriberi. As late as 1927, more than 120,000 cases were reported here, and these mainly in the south among poor people whose unvaried diets depended heavily on maize as the staple. Pellagra begins with dermatitis and abnormal skin pigmentation, and proceeds through nervous and digestive disorders, loss of memory, and ultimately causes insanity and death.

It was known to be a deficiency disease since 1914 but, like beriberi, it was thought at first to be due to inadequate protein. Then, during the 1920s, preparations of vitamin B were found effective in preventing pellagra. This unknown component of the complex was called the PP (pellagra-preventing) factor. When the PP factor was finally isolated in 1937, it turned out to be nicotinic acid, a simple pyridine carboxylic acid known since 1870. Its name arose from its synthetic origin—the oxidation of nicotine.

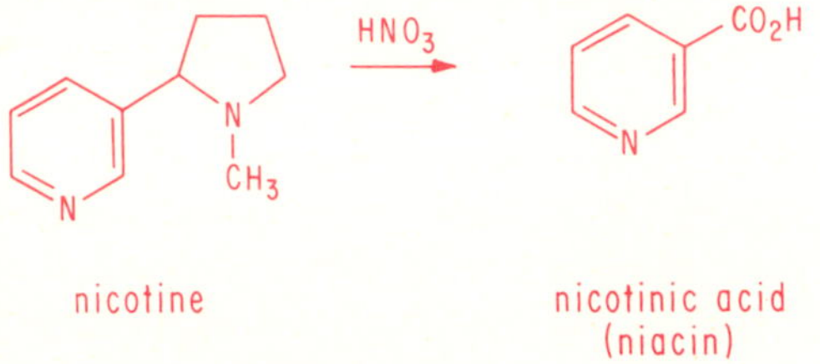

nicotine   nicotinic acid
                                        (niacin)

In view of its newly-discovered role in preventive medicine, however, the more respectable name niacin was immediately applied to the molecule. The best natural dietary sources of niacin are liver, wheat germ, and yeast. As much as ten to twenty milligrams is the average adult daily requirement.

Dietary nicotinic acid (niacin) is absorbed by the body from the gastro-intestinal tract. It then becomes converted to the amide, nicotinamide.

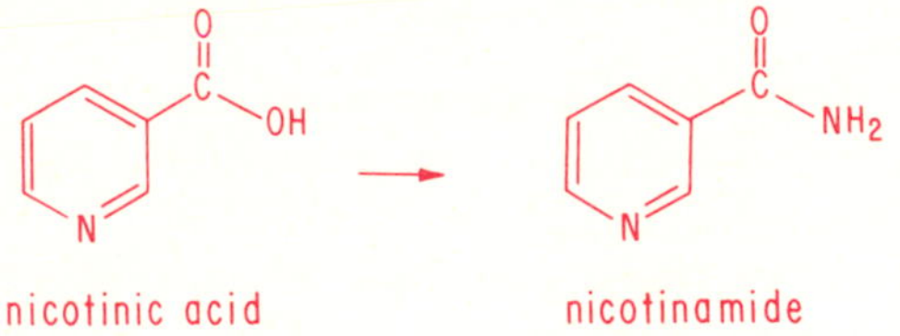

nicotinic acid   nicotinamide

The nicotinamide molecule then takes on its essential role as the active component of the hydrogen-transfer molecule NAD^+ (nicotinamide adenine dinucleotide) which we discussed earlier (see p. 24).

Vitamin C (Ascorbic acid)

The dramatic story of vitamin C harks back to the early 1600s when merchant seamen of the British East India Company sucked citrus fruit to prevent scurvy. This regimen provoked the nickname "limey", which has dogged the English to this day. The nickname, however, is far easier to bear than the horrors of the deficiency disease.

At the outset, scurvy causes bleeding, in the gums, under the skin, and in the joints. Skin softens and joints swell. Even touching a victim can cause agony. Without vitamin C, death comes none to soon.

Today, scurvy is not a serious problem, but even mild vitamin C deficiency can be dangerous. It can interfere with wound healing and with the body's ability to fight infection. The vitamin has recently reentered the "limelight" as a possible protection and even as a cure for the common cold.

Ascorbic acid is different from other vitamins in that only a few animals (such as man) cannot synthesize it and require external sources for the vitamin. Several groups of scientists including Linus Pauling and his associates claim that regularly administered doses of between 0.2 and 5 grams a day for typical adults are effective in preventing the common cold and also protect against other diseases including pneumonia, tonsilitis and acute rheumatism. They recommend oral doses of more than 10 grams a day at first appearance of cold symptoms, this level to be reduced when the cold has disappeared. Not only are the effects of the cold diminished according to the advocates but recovery rates are substantially enhanced.

Dr. Pauling and his colleagues carried out experiments in which normal and schizophrenic subjects received oral doses of ascorbic acid, niacinamide and pyridoxine. Schizophrenic patients were found to be low excretors for one or more of these vitamins. Regular ingestion of large amounts of these vitamins converted the schizophrenics into more normal excretors of the vitamins. The researchers deduced that vitamin therapy can help prevent and/or treat schizophrenia.

These claims and results have led to much scientific controversy. Many scientists doubt Pauling's statistics and conclusions. Recently, twenty-one prison volunteers were subjected to a double blind study.* Eleven men received three grams of vitamin C daily while the remaining ten were given an equal dose of a placebo. After 14 days, all subjects were exposed to rhinovirus 44, a common cold causing virus. For the next week both groups continued their regimen, i.e., vitamin C or the placebo. Within one or 2 days, after inoculation of the virus, all the men came down with colds. The symptoms did not differ between the group taking the vitamin C and that taking the placebo.

The prisoners were not the best group of subjects for scientists to have chosen in their attempt to prove the efficacy of vitamin C in fighting the common cold. These subjects were under the emotional strain of prison life and probably

---

* In a double blind study, subjects are given equal amounts of the test material or a placebo. None of the subjects knows which one he is being given.

came from backgrounds replete with malnutrition and related problems. Perhaps those factors made all of the subjects more susceptible to colds than the general populace. Obviously more research is necessary on programmed intake of vitamin C.

As in the case of nicotinic acid, the vitamin C molecule was known before its vitamin role was ascribed to it. In 1928, a crystalline acidic substance containing six carbon atoms, six oxygen atoms, and eight hydrogen atoms was isolated from orange juice. This molecule was named hexuronic acid. Four years later, the scurvy-preventing factor was isolated from lemon juice, and discovered to be hexuronic acid. Once again, in view of its newly discovered identity as the antiscorbutic factor, hexuronic acid was renamed ascorbic acid. The designation vitamin C had already become popularized before the actual isolation.

As an organic acid, ascorbic acid is somewhat unusual in that when two hydrogen atoms are removed from the molecule, the product (dehydroascorbic acid) is not an acid. Therefore, ascorbic acid cannot be a carboxylic acid. The disappearance of acidity on dehydrogenation indicates the presence of an ene-diol function (which is acidic) being converted to a diketone (which is neutral).

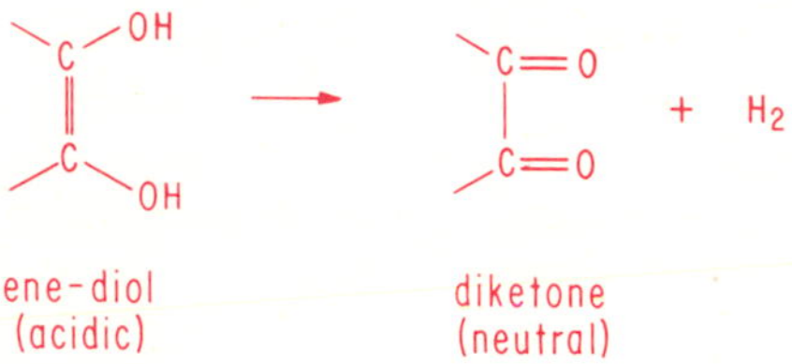

ene-diol
(acidic)

diketone
(neutral)

This accounts for two of the six oxygen atoms in the molecule. Two more of the six must also be in hydroxyl groups because ascorbic acid forms a tetra-acetate. This second pair is not enolic, however, since dehydroascorbic acid is not acidic.

Methylation of the ene-diol group with diazomethane produces a dimethyl ether which is not acidic.

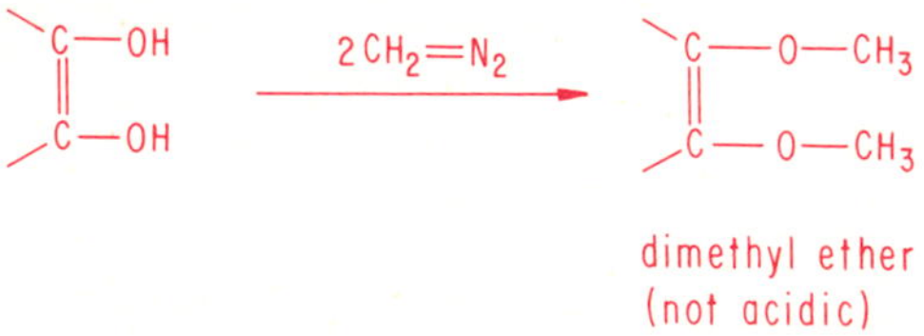

dimethyl ether
(not acidic)

Now, when this dimethyl ether is treated with base, and then neutralized, acidity returns to the molecule but *without* the loss of either methyl group.

This indicates that the hydroxide ion has opened a lactone ring (a cyclic ester) and freed a carboxylic acid.

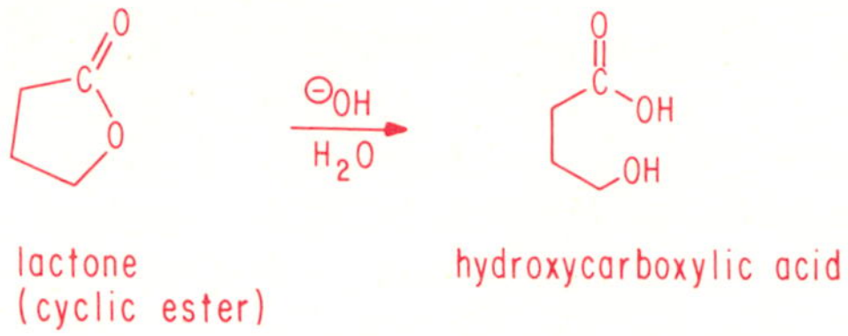

lactone
(cyclic ester)

hydroxycarboxylic acid

This would account for the last pair of oxygen atoms in the ascorbic acid molecule. We can now propose some possible structures for the molecule, using the most probable five or six-membered ring sizes for the lactone ring.

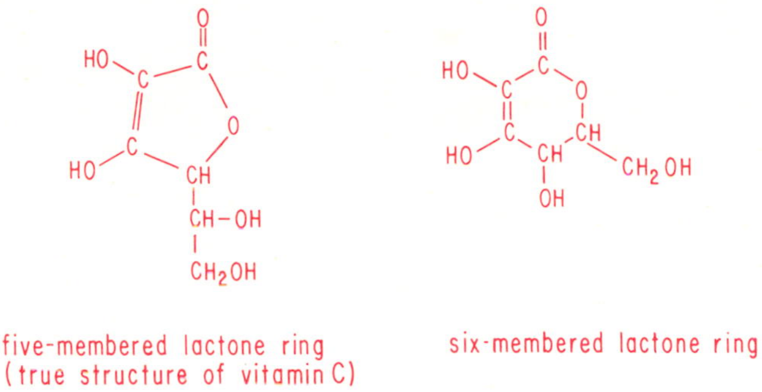

five-membered lactone ring
(true structure of vitamin C)

six-membered lactone ring

Additional chemical evidence, x-ray studies, and finally synthesis have shown that the five-membered ring structure is the true structure of the vitamin.

Vitamin A (Retinol)

The B complex of vitamins and vitamin C are all water-soluble. Vitamins A, D, and E are not. They are fat-soluble vitamins, which means that their structures must be predominantly hydrocarbon in nature.

We have seen (p. 140) that this is so in the case of vitamin D, which can be considered a derivative of cholesterol. Actually, the similarity among these three fat-soluble vitamins goes even deeper for, as vitamin D can be traced back through cholesterol to the fundamental hydrocarbon building-block isoprene, so can vitamin A and most of the vitamin E structure. All three can be built from isoprene (except for a benzene ring in vitamin E).

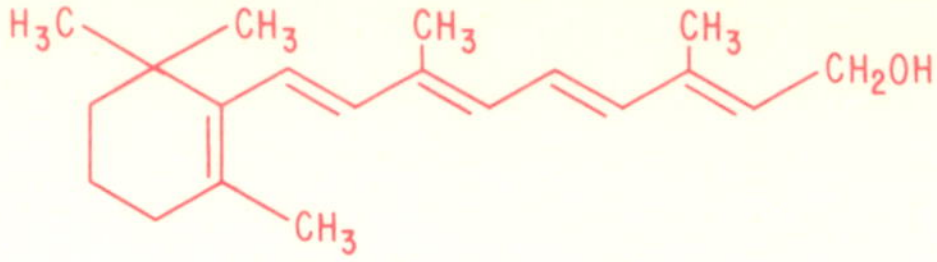

vitamin A (retinol)

The vitamin A skeleton can be imagined as being formed by head-to-tail bonding between four isoprene units, followed by a cyclization.

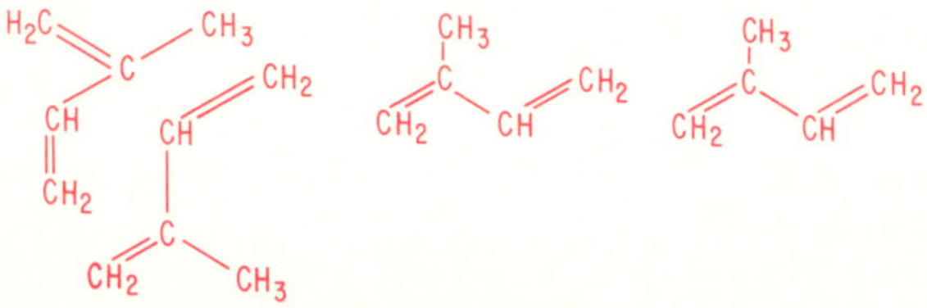

Subsequent hydroxylation leads to retinol.

As in the case of vitamin D, vitamin A was first isolated from fish liver oil. It was later named retinol because of its role in the sight-producing chemical reactions within the photo-receptors of the retina of the eye. An early indication of vitamin A deficiency is impaired vision in dim light, known as night blindness. A healthy adult requires a dietary daily average of about two-thirds of one milligram of the vitamin. One milligram is equivalent to 4500 international units of vitamin A.

An interesting difference has been observed between salt-water fish and fresh-water fish. Liver oil from fresh-water fish contains retinol and dehydroretinol (vitamin A_2), whereas that from salt-water fish contains only retinol (vitamin A_1). The extra double bond in vitamin A_2 is in the ring.

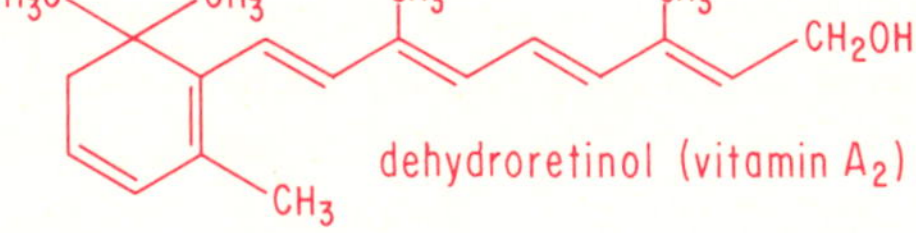

dehydroretinol (vitamin A_2)

Syntheses of structural variants of vitamin A show that all eleven carbon atoms and all four double bonds in the side chain are necessary for vitamin activity. Furthermore, each one of these conjugated double bonds must be *trans*. Also, the conjugated system must be in conjugation with the double bond in the ring. Finally, the carbon skeleton of the ring must be intact, and the side chain must have methyl groups at the third and seventh carbon atoms. With the exception of the hydroxyl group then, the entire molecule is required for vitamin activity.

Although vitamin A occurs only in animal tissue, man can obtain vitamin A from dietary precursors known as carotenoids, which are widespread in both plants and animals. Several carotenoids can be used by man as a source of the vitamin, but the one most closely related structurally to vitamin A is called β-carotene.

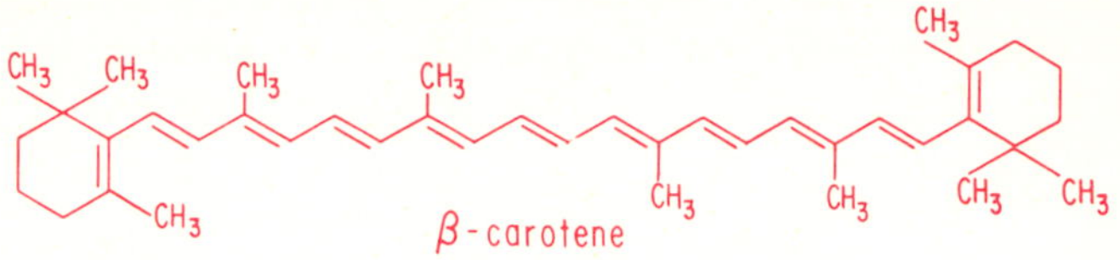

β-carotene

If β-carotene were snapped in half, each half would possess the carbon skeleton of retinol.

The E Vitamins (Tocopherols)

The vitamin E family is mysterious in that little is known about what these vitamins do in the human body or how they do it. At the same time, the E vitamins are fascinating in that many wonderful works have been attributed to them. These include the ability to dispel the alcoholic's craving for demon ethanol, the ability to produce fecundity in the supposedly barren, and ability to slow the degenerative process known as aging.

Actually, in the human, there is no specific disease which has been linked causatively to vitamin E deficiency. In other animals, deficiency symptoms vary with species, although tissue degeneration is a common feature. In dogs, skeletal muscles degenerate. In cattle and sheep, heart muscle depreciates and, in rats, the usual symptom is sterility.

The E vitamins are called tocopherols because they all contain the fundamental carbon skeleton of the molecule known as tocol.

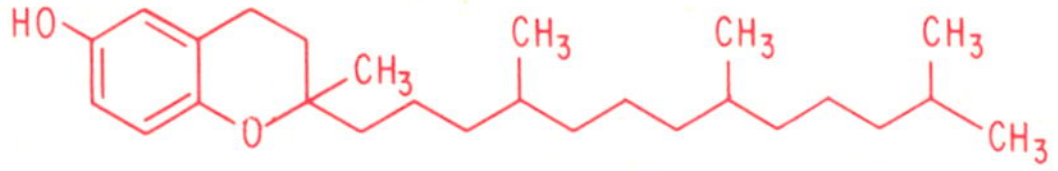

tocol

Tocol is formally a derivative of isoprene and hydroquinone (a difunctional phenol).

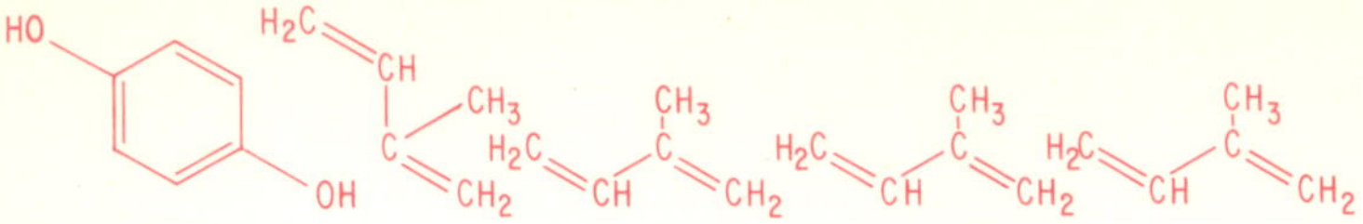

hydroquinone   isoprene units

The tocol side-chain is similar to the vitamin A side-chain in that both are derived from head-to-tail isoprene units. They differ, however, in two important ways. First, one is tail-terminated and the other is head-terminated. Second, the vitamin A side-chain contains conjugated double bonds, and the tocol side-chain is saturated.

Most of the eight known tocopherols contain the saturated tocol side-chain. An example is α-tocopherol, found in wheat germ.

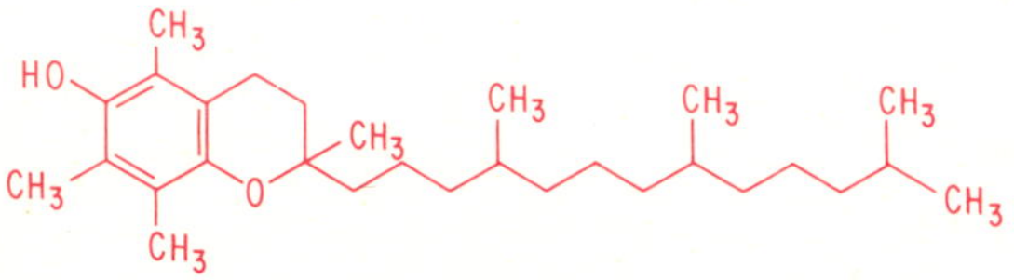

α – tocopherol

Tocopherols of this type differ from each other in the number and positions of methyl groups on the aromatic ring.

Another type of tocopherol contains three *non-conjugated* double bonds in the side chain. An example is ε-tocopherol, found in wheat bran.

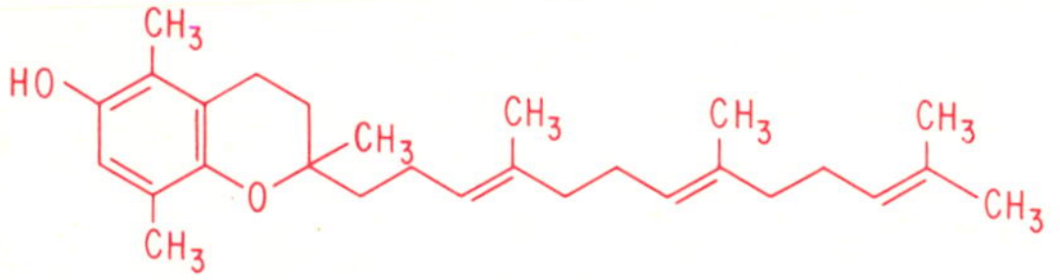

ε – tocopherol

Obviously, these double bonds are not required for vitamin activity. This contrasts sharply with the situation in vitamin A. Also, the stereochemistry of the two asymmetric centers in the tocol side-chain is not important to vitamin activity. The presence of the side-chain is required, however, since analogs with shorter side-chains are inactive.

One of the alleged roles of vitamin E which is receiving a good deal of attention at present is that of an elixir for protracted youthfulness. Current theories on the cause of the gradual degeneration of tissue and tissue function seen in the aging process lay heavy blame on free radicals. Free radicals form naturally within the body and are very reactive. One of the things they do readily, if their concentration is high enough, is couple with each other, as we have seen earlier.

$$R\cdot \ + \ R\cdot \ \longrightarrow \ R—R$$

This kind of thing going on between protein molecules or between nucleic acid molecules would of course lead to malfunction of these molecules, which are essential to cellular well-being. The gradual and eventual result would be degeneration and death of the cell, and then of the organ, and finally of the entire organism. Hydroquinone, and presumably vitamin E, act as radical scavengers. If vitamin E does indeed slow the aging process, it may very well do so by keeping the radical concentration low enough to discourage inter-radical coupling.

SUMMARY

The molecules discussed in this chapter were discovered primarily by scientists seeking to cure specific diseases arising from diet. Beriberi, facial tissue deterioration and related forms of dermatitis, pellagra, scurvy and night blindness all arise from vitamin deficiencies. The structure and chemistry of these molecules of growth and health differ substantially from each other, yet they appear to play an important role as cofactors in regulating proper enzyme action.

We discussed some controversial aspects of vitamin C usage and even vitamin E, the mysterious molecule whose functions are not clear at all. In both cases, more research is necessary to understand the mode of action and extent of effect for these vitamins.

SOURCE MATERIALS AND SUGGESTED READING

Dyke, S. F., 1965. *The Chemistry of the Vitamins*, Interscience Publishers, New York.

Pauling, L., 1970. "Evolution and the Need for Ascorbic Acid", *Proc. Natl. Acad. Sci.*, **67**, pp. 1643–1648.

Pauling, L., 1970. *Vitamin C and the Common Cold*, W. H. Freeman and Co., San Francisco.

Sebrell, W. H., Jr., and R. S. Harris, eds., 1967. *The Vitamins; Chemistry, Physiology, Pathology, Methods*, Volumes, I–V, second edition, Academic Press, New York and London.

CHAPTER X

MOLECULES OF THE SENSES
Taste, Odor and Attraction

TASTE AND SMELL represent closely interrelated responses to molecular stimuli. For higher animals, especially man, it is possible to separate the two on the basis of the activation of appropriate receptors. However, the phenomena remain closely allied. Who among us cannot recall the effect of head colds or simply blocked nasal passages on our sense of taste? All higher animals exhibit responses identifiable as separate taste and smell responses. When lower animals such as invertebrates are considered, taste and smell distinctions become less clear. We must rather describe the effects more as behavioral responses such as attraction or repulsion.

Taste

We tend to classify taste responses into four basic reactions: salty, sour, bitter and sweet, even when discussing the complex taste patterns exhibited by human beings. Clearly these represent a great oversimplification. The tongue (Figure 10.1) contains taste receptors which are organized into regions of specific sensitivities. When a wine connoisseur partakes of a great wine, the bouquet is initially savored by a smell sensation. The wine is then carefully brought in contact with the front of the tongue to produce a rapid taste response. The true flavor of the wine is appreciated by moving the wine to the back and underside of the tongue while breathing in through the mouth over the wine. This creates a maximum of effect and response, the basis of which is a combination of taste and smell.

Taste stimulation occurs through receptors which are located in structural entities called taste buds shown in Figure 10.2. For all taste effects, electrical neural responses are involved. Take the case where various salts are applied to the tongue. Easily measured electrical effects can be recorded as indicated in Figure 10.3.

Taste represents a complex phenomenon based upon molecular interactions. We know a great deal about taste molecules. Yet complete understanding of

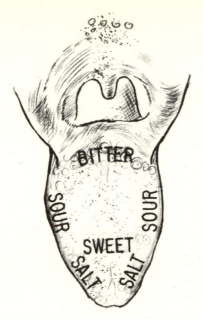

FIGURE 10.1   "Taste areas" on the tongue based on physiological data. [Courtesy of Dr. Robert Henkin]

the mechanism of taste remains unattained. Sweetness is probably the most studied and best understood taste effect from a molecular standpoint. We know that all sugars do not exhibit the same taste properties. The taste of D-glucose, sucrose and fructose can be easily differentiated from each other and from other sugars. Their relative level of sweetness varies widely (see Table 10.1). Even such minor structural differences as the configuration about the anomeric carbon atom of the cyclic form of a hexose affects its taste. For example, α-D-mannose is sweet while β-D-mannose is bitter.

mutarotation

β–D–mannose
Bitter

α–D–mannose
Sweet

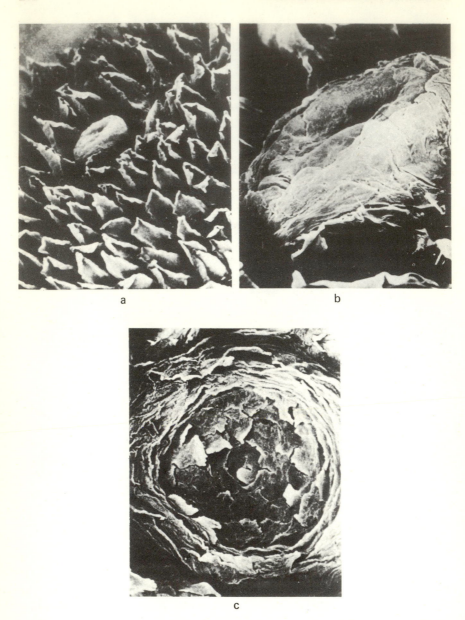

FIGURE 10.2   (a) Taste bud, or papilla, of a rabbit (×79); (b) electron micrograph of taste bud of a rabbit (×4,444); (c) view into the pore of a taste bud of a rat (×578). [Courtesy of Dr. L. Beidler]

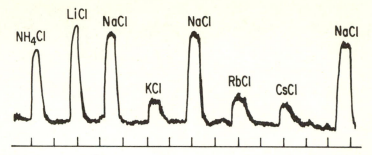

FIGURE 10.3   Electrical response of the taste nerve of a rat to 0.1 M concentration of salts supplied to the surface of the rat's tongue. One division on the horizontal scale equals 20 seconds. [From L. M. Beidler; see Source Material and Suggested Reading at end of chapter]

TABLE 10.1

Sugar	Sweetness in solution	Sweetness as "crystalline" material
β-D-Fructose	100–175	180
Sucrose	100	100
α-D-Glucose	40–79	74
β-D-Glucose	30–40	82
α-D-Galactose	27–67	32
β-D-Galactose	—	21
α-D-Mannose	59	32
β-D-Mannose	Bitter	Bitter
α-D-Lactose	16–38	16
β-D-Lactose	48	32
β-D-Maltose	32–46	—
Raffinose	23	1
Stachyose	—	10

ᵃ From R. S. Shallenberger (see "Source Material and Suggested Reading" at end of chapter).

These structures are related to each other by a process called mutarotation. If a cyclic form such as the β-D-mannose ring opens, it is possible for it to ring close from the aldehyde form to the α-D-mannose form. The only difference between the two structures is the configuration about the anomeric (or number 1) carbon atom. Fructose in the β-D-pyranose form is probably the sweetest of all the common sugars. Its relative sweetness decreases markedly with increasing temperature of the test solution and time of storage before tasting.

Such effects can be explained by the complex mutarotation which fructose can undergo. Thus, β-D-fructopyranose can be converted to the less sweet β-D-fructofuranose form and even to some small concentration of the less stable α-D-furanose isomer.

β-D-fructopyranose
(chair form)

β-D-fructofuranose

It is clear from Table 10.1 that the relative sweetness of all sugars differs. Much depends on how the taste sensation is measured. Near the threshold level for detection, sucrose is said to be bitter. As the concentration of sucrose is increased, the sweet taste becomes dominant. Chemical complications for other sugars such as mutarotation have been discussed for fructose and mannose. Such reactions are possible for numerous other sugars, some of which are listed in Table 10.1, including the monosaccharides, glucose and galactose, and the disaccharides, maltose and lactose.

Certain amino acids and peptides exhibit sweet tastes. Glycine and certain D isomers of normal amino acids tend to be sweet. Table 10.2 shows that general rules are difficult to apply. The L-isomers of alanine, serine, proline, threonine and even valine can be considered to taste sweet; almost all of the D-isomers are sweet-tasting. Clearly there must be a specific interaction possible between these unnatural isomers and the receptor sites on the taste bud. It is interesting to speculate that induced conformational effects on the proteins of the receptor sites caused by D-amino acids may be the basis of their sweet taste. Certainly the receptor site is asymmetric and may prefer to interact to form a favored D-amino acid-L-receptor site enantiomeric transitory complex.

A fascinating, novel finding has recently appeared. Dipeptides such as L-aspartyl-L-phenylalanine methyl ester and certain related compounds are nearly 200 times sweeter than sucrose. This synthetic material is most attractive

L-aspartyl-L-phenylalanine methyl ester

TABLE 10.2
Taste of Amino Acids[a]

Amino Acid	Taste of L-isomer	Taste of D-isomer
Alanine	Sweet	Sweet
Asparagine	Tasteless	Prolonged sweet taste
Glutamic acid	Special complex quality	Tasteless
Glycine[b]	Sweet	
Histidine	Tasteless	Sweet
Leucine	Slightly bitter	Very sweet
Methionine	Tasteless	Sweet
Phenylalanine	Faintly bitter	Sweet with bitter aftertaste
Proline	Sweet	—
Serine	Sweet	Very sweet
Threonine	Sweet to some, bitter to others	Sweet
Tryptophan	Tasteless	Sweet
Tyrosine	Bitter	Sweet
Valine	Slightly sweet, but with a bitter cotaste	Very sweet

[a] Adapted from R. S. Shallenberger (see "Source Material and Suggested Reading" at end of chapter).

[b] Recall that glycine does not have an asymmetric carbon and therefore does not have optical isomers. See appendix.

since it is composed of natural amino acids and is quite low in calories. Hydrolysis or enzymatic cleavage reactions lead to the easily metabolized natural amino acids, L-aspartic acid and L-phenylalanine, and the coproduct methanol. Mazur and his associates at G. D. Searles Laboratories found that substitution of any other amino acid for the L-aspartyl residue, even including such closely related structures as L-glutamyl or D-aspartyl residues, leads to tasteless products. They did find that the phenylalanine can be replaced by methionine or tyrosine in the dipeptide methyl ester and still maintain sweetness. The dipeptide free acids (such as L-aspartyl-L-phenylalanine) of these methyl esters are not sweet. The sweetness of the three dipeptide esters is lessened when an ethyl replaces the methyl group in the ester.

The synthetic organic chemist has long known that a molecule with a sweet taste cannot be easily modified and still maintain its sweetness. Saccharin is sweet while N-alkylated derivatives are tasteless. The alkali metal salts of cyclohexyl amine sulfate (cyclamates) are sweet while the same salt of aniline sulfate is nearly tasteless. Alkyl nitroanilines vary enormously in taste, depending on their substitution patterns.

Three Representative Synthetic Sweet-tasting Molecules
and Structurally Related Tasteless Molecules

1

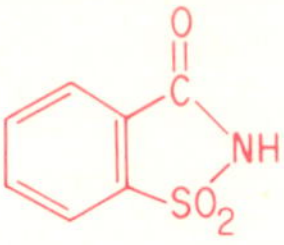

saccharin
sweet

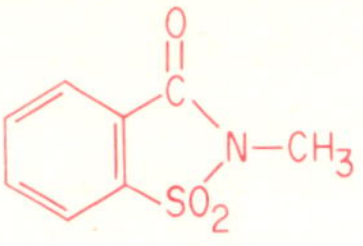

N-methyl saccharin
tasteless

2

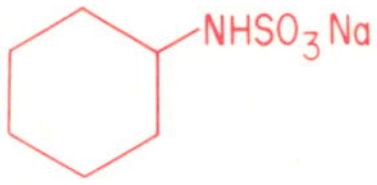

sodium cyclamate
sweet

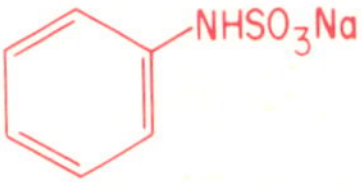

sodium aniline sulfate
tasteless

3

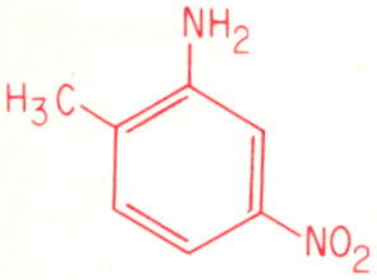

3-nitro-6-methylaniline
sweet

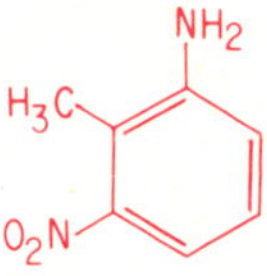

5-nitro-6-methylaniline
tasteless

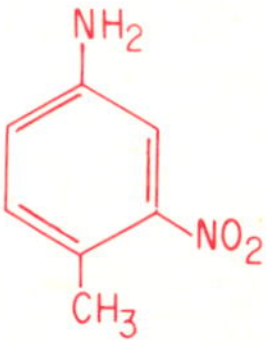

3-nitro-4-methylaniline
tasteless

When we add these to the sugars and amino acids noted above, it is difficult to draw structural generalizations concerning the mechanisms of taste. The receptor cells are able to respond to these molecular stimuli in a definite and reproducible manner. Molecular recognition must play a part. Certain taste responses are genetically controlled. Phenylthiourea is bitter-tasting to about 75% of the population while the remaining 25% taste nothing with this compound even at high concentration.

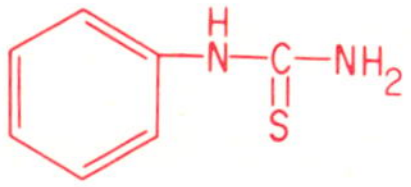

phenylthiourea

It is indeed baffling that such widely differing structures as those we have considered can elicit sweet-tasting responses. Several theories have been offered to explain the molecular basis of taste. These involve such considerations as the nature of the functional groups in the molecules being tasted; the chemical and physical properties of said sweet molecules; specific intra-molecular hydrogen bond formations; and/or the presence of a gate-keeper protein which governs access to the receptor sites in the taste bud.

As long ago as 1914, specific sweet-causing molecular groups were defined as sapophoric units. The structures often were comprised of polyhydroxy or amino acid residues. By analogy with dye chemistry, glucophores were described as groups able to form sweet-tasting molecules when combined with an auxogluc, an otherwise tasteless group. Some workers indicated that factors such as the following are important in producing the sweet taste:

a) Modification of sweet molecules alters the taste.

b) Active hydrogen-containing compounds have discernible taste.

c) Optical antipodes usually possess different tastes.

It was postulated that three-atom enolization is involved in the sweet taste of saccharin.

This tautomerism mechanism would explain why N-alkylated saccharins are not sweet. They cannot undergo enolization. The greater sweetness of α-anomeric hexoses and disaccharides has been explained on the basis of the fact that α-anomeric forms contain *cis* hydroxyl groups on the anomeric and adjacent carbon atoms. This generalization does not always hold since β-D-lactose is sweeter than the α-form, and β-D-mannose, which possesses the desired *cis* structure, is bitter.

Hansch and his associates related the sweetness of a series of nitroaniline derivatives to their Hammett σ-values.

$$\log RS = k\pi + k'\sigma + k''$$

where RS is the relative sweetness, σ is the Hammett constant which describes the electron withdrawing power of a group, and π, another linear free energy substituent constant related to σ but more descriptive of the hydrophobic character of a group. Hydrophilicity or hydrogen bonding ability is necessary for sweetness. Along with this, a hydrophobic region of the structure is essential in order to gain access to the receptor site of a taste bud. Using this approach, Hansch explained why 3-nitro-6-methylaniline is sweet while the other isomers are not.

Certain researchers found that the magnitude of electrophysiological response is related to the concentration of the sweet-tasting molecule. These hypotheses are based on the assumption that the hydrogen bonding of a sweet molecule to the receptor site causes a conformational change in a protein on the surface of the receptor site. Indeed some proteins have been isolated from taste buds that exhibit specific, weak interactions of the hydrogen bonding type with sweet molecules. Bradley and Henkin discovered that administration of thiol drugs lowers taste acuity, while metals such as copper(II) and zinc(II) return the taste sensitivity to normal levels. These findings suggest that scission of protein disulfide bonds by the sulfhydryl-disulfide interchange can occur by the following mechanism:

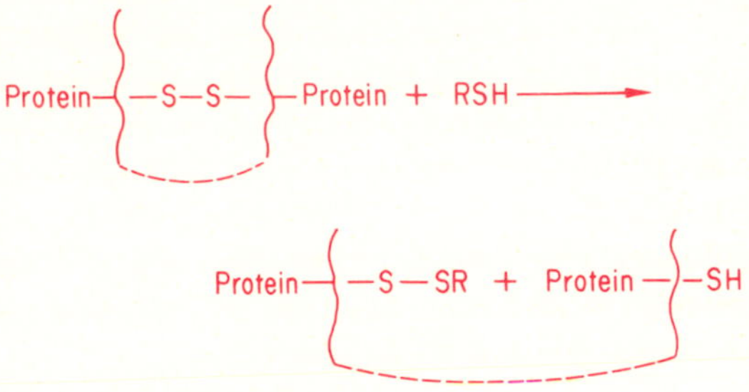

$$\text{Protein}\!-\!\{-S\!-\!S\!-\}\!-\text{Protein} + \text{RSH} \longrightarrow$$

$$\text{Protein}\!-\!\{-S\!-\!SR + \text{Protein}\!-\!\}\!-SH$$

The reversal of the effect by copper and zinc can be explained by chelation of the thiol derivative with the metal ion:

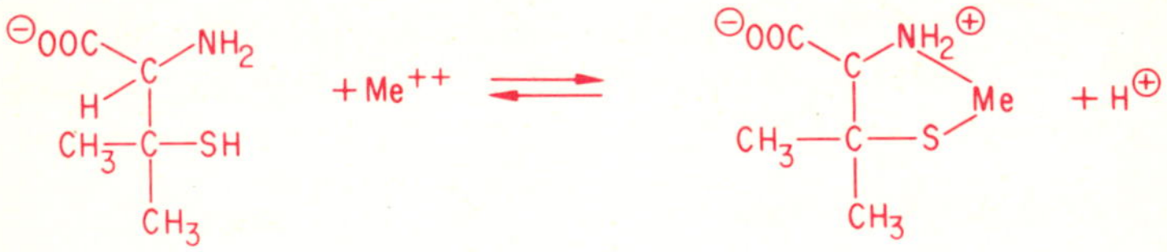

D–penicillamine

Continuation of this process can lead to the following:

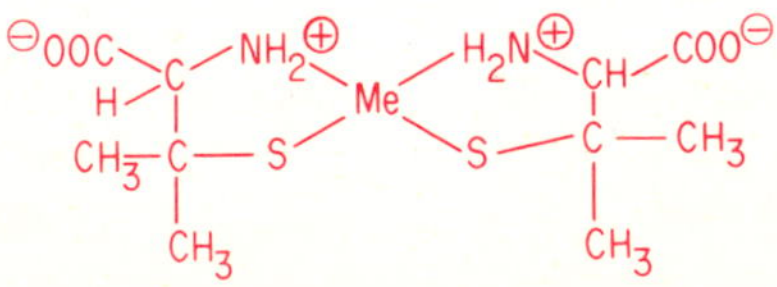

Reduction–oxidation reactions may also be involved:

$$2RSH + Cu^{++} \rightleftharpoons 1/2\ RSSR + RSCu^{\oplus} + 2H^{\oplus}$$

The binding of a sweet-tasting molecule to a metal and sulfhydryl-containing protein can produce a conformational change about the receptor site. Bradley and Henkin designated the protein which controls and regulates the passage of the sweet tasting molecules to the receptor site as the gate-keeper protein. When the gate-keeper protein reduces the effective size of the pore following administration of thiol drugs, the taste thresholds should of course rise substantially. The fact that they do can be taken as evidence in favor of this explanation.

Shallenberger developed a different theory which incorporates many of the known facts about molecular sweetness. He defined the basic nature of a sweet compound's saporous unit. This unit, in turn, must meet specific configurational and conformational requirements. The initial phase of the sweet taste reaction is thought to involve intermolecular hydrogen bonding between the saporous unit of a sweet compound and the taste bud receptor site. These properties are in turn related to a compound's physical and geometric properties. He found that many sweet compounds possess an AH,B system where A and B represent electronegative atoms or groups. The distance between the AH proton and the B orbital was calculated to be an average value of about 3 Å. When compounds meet these requirements, they possess a sweet taste. Some examples of molecules whose sweet taste can be explained using the Shallenberger approach are shown in the formulae below:

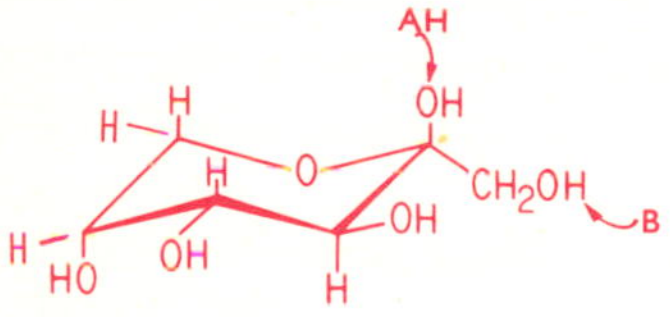

β -D-fructose (pyranose form)

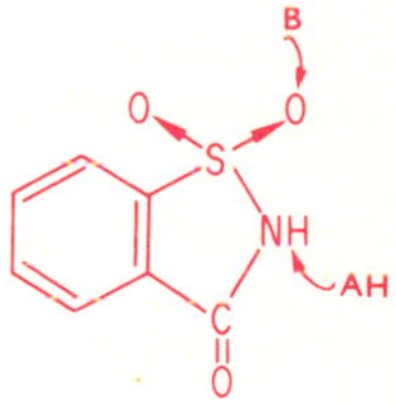

saccharin

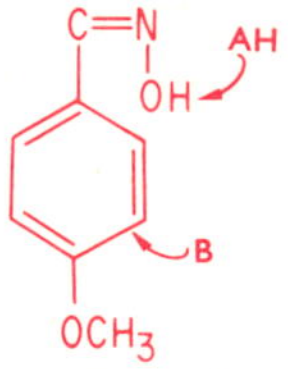

α–anisaldehyde oxime

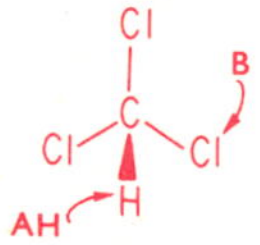

chloroform

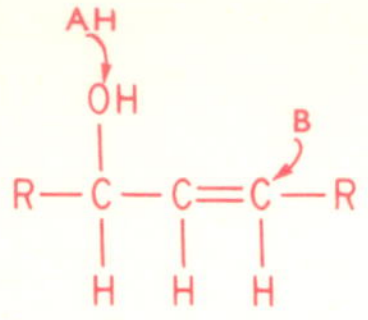

unsaturated alcohol

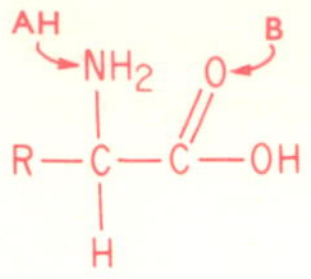

α − amino acid

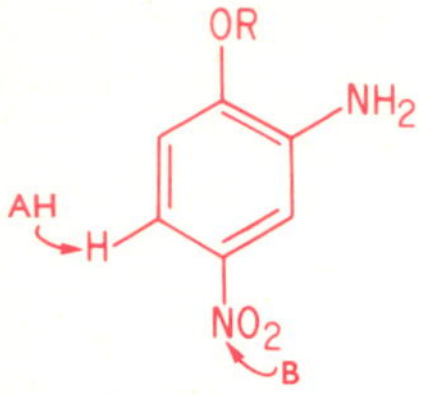

1−alkoxyl−2−amino−4−nitrobenzene

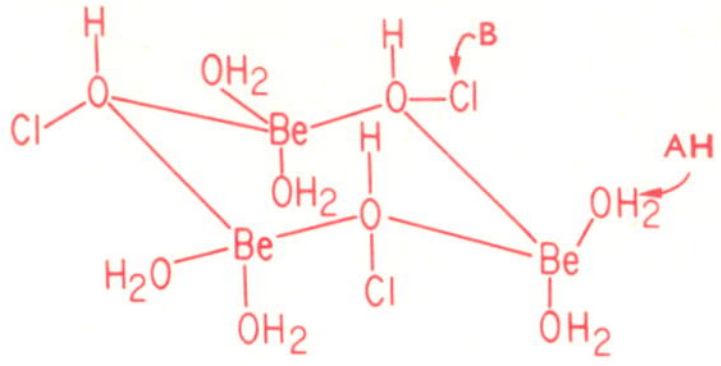

beryllium hydroxo−chloride

It is difficult to understand configurational effects such as between α and β anomers and D-L amino acids by the Shallenberger theory. Other molecular aspects of sweet taste are interpretable by the AH,B interaction system. For example, the β form of anisaldehyde oxime is not sweet because the AH group is far from the B region.

The above mechanisms and explanations for the sweet taste are not applicable for all cases. Further research must be carried out in order to expand our knowledge on the mechanism of taste. Complex chemical reactions appear to be involved in the preneural and perineural stages of the taste effect. We do not yet know the broad implications of structure, topology and desorption aspects of molecular interactions during the taste sensation. The proposals discussed above each contain special insights into the problem. A unified and comprehensive theory cannot be proposed at this time.

Odor

Before proceeding to consider the fascinating chemistry of attractants and stimulants (with specific reference to insects), we will examine olfaction as a related response to taste. Human beings can describe smells in most significant though complex terms. General classifications are easily recognized. Certain molecules produce odors (i.e., react with the olfaction receptor sites) in an easily describable manner. The molecules listed in Table 10.3 fall into general categories; however, the table cannot convey the complexity of value judgment used by people to characterize odor. These odors do not comprise all of the scents known to man. The categories are merely devised to cover cleanly reproducible responses from test panels.

TABLE 10.3
Primary Odors for Human Beings[a]

Camphoraceous	Pungent	Ethereal
Borneol	Acetic acid	Acetylene
tert-Butyl alcohol	Allyl alcohol	Carbon tetrachloride
D-Camphor	Cyanogen	Chloroform
Cineol	Formaldehyde	Ethylene dichloride
Pentamethyl ethyl alcohol	Formic acid	Propyl alcohol
	Methyl isothiocyanate	

Floral	Peppermint	Musky
Benzyl acetate	tert-Butylcarbinol	Androstan-3α-ol (strong)
Geraniol	Cyclohexanone	Cyclohexadecanone
α and β ionones	Menthone	17-Methylandrostan-3α-ol
Phenylethyl alcohol	Piperitol	Pentadecanolactone
Terpineol	1,1,3-Trimethylcyclo-5-hexanone	Muscone (3-methyl cyclo-pentadecanone)

Putrid
Amylmercaptan
Cadaverine
Hydrogen sulfide
Indole (when concentrated; floral when dilute)
Skatole

[a] According to J. E. Amoore, *Proc. Sci. Sec., Toilet Goods Assoc. Suppl.*, **37**, 1 (1962).

The structure of odorous molecules can be readily established. It is far more difficult to determine how these molecules interact with olfactory receptor sites (Figure 10.4). Volatility obviously plays an important role in that the odorous molecule must be delivered to the olfactory sites in an airborne state. What is generally not appreciated is how incredibly sensitive the smell sense can be. For a particularly high potency odorant, the average person can detect one part in a trillion parts of air. After a small number of odorous molecules reach the olfactory receptor site (Figure 10.4), they are adsorbed on the receptor surface.

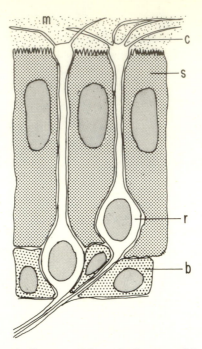

FIGURE 10.4 Schematic diagram of the main features of the olfactory epithelium. m is the mucus sheet; c, the cilia; s, the supporting cells; v, the receptor cells; and b, the basal cells. [Adapted from D. G. Moulton, 1969, "Detection and recognition of odor molecules," in *Gustation and Olfaction*, G. Ohloff and A. F. Thomas, eds., Academic Press, London and New York]

Functional groups alone do not establish molecular odor characteristics. Acetic acid, for example, has a pungent odor, while butyric acid possesses a rancid aroma. Valeric and caproic acids smell like sweaty locker rooms while higher fatty acids have little if any odor. That geometry plays an important role can be seen from the xylenols listed below:

faint odor   musty odor   oil of wintergreen odor

It is also well established that hydrocarbons possess odors strongly dependent on chain length. Thus methane is odorless; n-pentane gives off a characteristic lighter-fluid aroma and alkanes in the octane and nonane range exhibit gasoline-like aromas. Esters are known to be fruity but their specific odor qualities depend on the nature of their acid and alcohol components. The γ-lactones provide a particularly interesting variation of odor depending on substitution. The 5-n-pentyl derivative possesses a coconut odor, while the 5-n-hexyl derivative smells like peaches.

coconut odor   peach odor

A closely related γ-lactone emits an extremely strong aroma of beef bouillion.

beef bouillion odor

Ethyl butyrate smells like pineapple while n-octyl acetate gives off an orange scent.

How are odors discriminated? What are the steps involved in the olfactory process? Certainly part of the effect resides in the nature of the odorous molecule itself, while the remaining contribution must come from the stimulus to the nervous system of the organism.

As with taste, electrical effects can easily be measured. The interaction between odor-producing molecules and a receptor site must be weak in order to account for flow rate effects and the desorption rates. It has been suggested that a weak complex occurs between the odorant and β-carotenoids of the olfactory sensory cilia.

Carotenoids are semiconductors. They may interact or complex with odorants; this complex may cause the conductivity to increase. The increase in conductivity would then result in the depolarization of the membrane about the receptor site. In this manner, a signal would be generated simultaneously with the desorption of the odorant. Another more traditional explanation is that weak complexes are formed between the odorant and proteins of the olfactory sensing cells. This is followed by an increase of electrical conductance of the membranes about the receptor site and an impulse is generated in an analogous fashion to that described for carotenoids.

Evidence that the complex is probably between odorant and proteins rather than carotenoids comes from work with a group-specific protein reagent, N-ethylmaleimide.

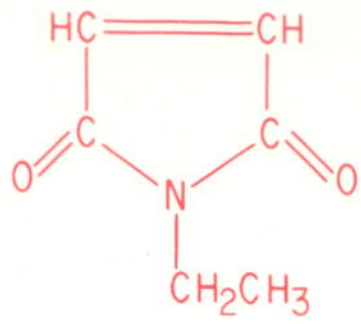

$$HC=CH$$
$$O=C \quad\quad C=O$$
$$N$$
$$CH_2CH_3$$

This reagent reacts primarily with sulfhydryl groups on proteins.

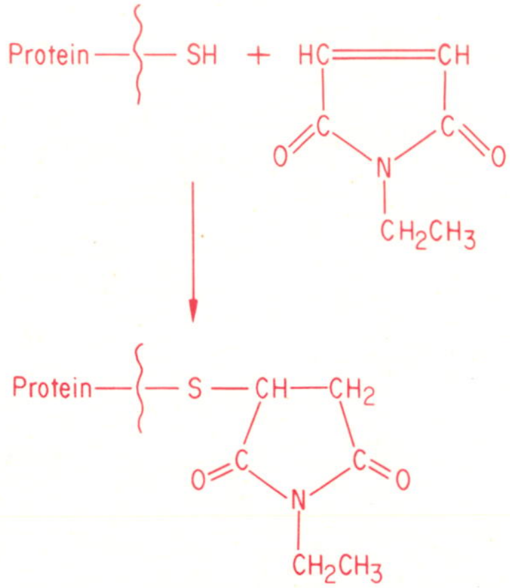

$$Protein-SH \;+\; HC=CH$$
$$O=C \quad\quad C=O$$
$$N$$
$$CH_2CH_3$$

$$\downarrow$$

$$Protein-S-CH-CH_2$$
$$O=C \quad\quad C=O$$
$$N$$
$$CH_2CH_3$$

This reaction irreversibly blocks the sulfhydryl group. Frog olfactory epithelia completely lose their ability to respond to odorants after treatment with N-ethylmaleimide. It was found, however, that if the odorous molecule, ethyl butyrate, is administered in sufficient concentration to saturate the receptor sites prior to and during treatment with N-ethylmaleimide, normal response can be reestablished for ethyl butyrate and closely related compounds after washing and waiting some time for tissue recovery to occur. Responses to all other types of odorants were completely blocked. These results indicate not only that the complex occurs between proteins and odorants but that different receptor sites exist for different odor sensations within olfactory cilia.

Suggestions have been made that the quality of odors of molecules depend on their characteristic low frequency infrared vibrational bands. Perfume chemicals have been studied from the standpoint of empirical vibrational correlations. All possess far infrared absorption bands between 100 cm^{-1} and 500 cm^{-1}. It appears that the correlation involves not only the presence of certain bands but the absence of others. A relationship seems to exist between the lowest wavelength infrared band for a given molecule in the vapor state and its olfactory threshold. The lower the wavelength for this band, the lower is the threshold for detection. Butyl mercaptan (skunk odor) possesses an infrared band below 200 cm^{-1} and can be detected by humans at below 10^{-12} molar. Methanol shows no infrared band below 1000 cm^{-1} and can only be detected at concentrations greater than 10^{-3} molar.

All of these results have provided a basis for the highly speculative vibratory theory which holds that the molecular vibrations of the odorant set up specific resonances in the receptor sites which lead to electrical responses noted above. Supporters of this hypothesis have shown that four of six test subjects were able to distinguish the difference in the odor of ordinary naphthalene (IR frequencies 363 and 183 cm^{-1}) and fully deuterated napthalene (IR frequencies 331 and 169 cm^{-1}). In addition, for most cases, no difference in olfactory response exists between optical antipodes. Such results can be considered to be consistent with the vibratory theory, since mirror image forms have the same dipole moments and vibrational frequencies.

An alternative explanation is based on a stereochemical approach. Molecular silhouettes are produced and orthogonal views examined. To do this, a conformation that is most likely to interact with the receptor site is chosen for each compound of a series. A standard is selected to which all related compounds are compared. For fatty acids, using isovaleric acid as the standard, correlations were obtained between molecular shape and degree of specific inhibitions of smell (anosmia). For benzaldehyde (the standard) and substituted benzaldehydes, correlations were obtained between almond odor and molecular shapes. Thus o-methyl benzaldehyde possesses the same almond odor as benzaldehyde, while the o-ethyl, o-isopropyl and o-t-butyl benzaldehydes exhibit much lower levels of almond scents.

The theories proposed above are most likely great oversimplifications. In the cases of the androsterols, small changes in structure alter odor substantially without materially affecting the far infrared vibration spectrum or the molecular silhouette.

Δ^{16} – androsten–3α–ol
strong musk–like odor

Δ^{16} – androsten–3β–ol
weak musk–like odor

Even more perplexing are the results from the so-called *ortho* musks. A simple methyl group substitution alters the odor enormously.

weak musk odor   strong musk odor

Pyrazines have been isolated from roasted peanuts, cocoa, and coffee. The derivative 2-methoxy-3-methylpyrazine smells like roasted peanuts. The closely related compound 2-methoxy-3-isopropylpyrazine has a potato-like aroma while the 2-methoxy-3-*n*-hexylpyrazine possesses a pepper-like odor. These three compounds vary substantially with respect to their specific odor thresholds.

Thus we have seen cases where very small changes of structure can alter the quality and/or potency of odor-producing molecules. The theories to date are not adequate to explain such subtle factors in olfaction. When we add the complexities resulting from psychological factors in smell, it is no wonder that many scientists working in this field feel like organoleptic idiots. Much more research must be undertaken before meaningful generalizations on olfaction can be made.

Attraction

Karlson and Butenandt initially devised the word "pheromone" in 1959 to describe substances secreted by animals to affect the behavior of other animals. The term comes from the Greek "pherein", meaning to carry, and "horman", meaning to stimulate. We will use pheromone to denote those chemical substances which allow for intra-species communication. In this section, our main emphasis will be on insect and other lower animal attractants.

It has been demonstrated that certain insect species react to pheromone molecules in such incredibly low concentrations as 10^{-17} molar. Effects such as these are among the most highly sensitive known to science today. These chemical attractants govern many of the activities of insects, such as the seeking of food, mating behavior and copulation, and location of egg-laying domains. Other responses may include alarm, trail marking, intra-group recognition and swarming. The workings of these pheromone molecules may not involve a simple stimulus and a unique response. Rather, more complicated patterns may be involved where pheromones induce modification of the physiology of an organism. In this way, they prepare the organism for a secondary stimulus, which may or may not involve pheromone molecules.

The first complete characterization of an insect pheromone was carried out by Butenandt and his associates, who worked on the female silkworm moth, *Bombyx mori*, which exudes a sex attractant. In order to accomplish their work, the Butenandt group extracted the pheromone from the terminal abdominal section of more than a half million virgin female silkworm moths. Careful fractionation and monitoring of the activity ultimately produced an active material which was identified as *trans*-10-*cis*-12-hexadien-1-ol. This material is known as "bombykol".

In order to prove this structure, spectroscopic studies and chemical degradative reactions were carried out. Ultimately, of course, the complete and total synthesis of the proposed structure was undertaken and careful comparison between synthetic, and natural products carried out. In this way the structure of the *Bombyx mori* sex pheromone was established.

The gypsy moth, *Porthetria dispar*, represents another fascinating tale of the isolation and characterization of an insect pheromone. This insect was accidentally brought to Massachusetts in the middle of the last century and has proved to be an enormous pest, destroying many forest lands in the northeastern part of the United States. It is now spreading to other areas with dire consequences as an ultimate possibility. Research as long as fifty years ago established that isolated virgin gypsy moths could somehow attract the male gypsy moth. It was deduced that the female exuded some chemical which could be recognized by the male. After many years of research, it was thought that the molecule is 10-acetoxy-*cis*-7-hexadecen-1-ol. About ten years later the same research group proved that the actual pheromone has the structure of an epoxide of a nineteen carbon alkane, 2-methyl-7-epoxyoctadecane. This compound has been given the name "disparlure".

10-acetoxy-*cis*-7-hexadecen-1-ol   2-methyl-7-epoxyoctadecane (disparlure)

In research areas such as those involving insect attractants, erroneous structures for active principles are not uncommon. Many hundreds of thousands of insects must be collected, specific anatomical sections separated, and chemicals extracted. Chemical fractionation on micro scale follows with constant monitoring for activity. Chromatographic techniques are employed to continue the process of fractionation and purification which usually yields only a few milligrams of the pheromone. Once comparatively pure fractions are obtained, the material is subjected to careful spectroscopic examination and chemical degradation reactions in order to make tentative structural assignments. Total synthesis and comparisons between the natural and synthetic compounds provide the ultimate proof of the structure.

As an example of this technique, consider the case of the male beetle, *Trogodernia inclusium*. More than 100,000 virgin female beetles were ground and extracted with benzene in a Waring blender. The solvent was removed by lyophilization (freeze drying) in order not to destroy or damage any of the insect pheromones. The residue was distilled under high vacuum. Acidic compounds were extracted with dilute base. The remaining neutral substances were subjected to chromatographic separation on a silica gel column followed by analytical and preparative gas chromatography. Two compounds were isolated, compound A (3 mg) and compound B (1 mg).

Spectroscopic analysis and specific chemical reactions showed that compound A was a primary alcohol containing one *cis* double bond. Ozonolysis produced an aldehyde which eluted from a standardized gas chromatographic column between *n*-octanal and *n*-nonanal. From these results, the researchers postulated that the aldehyde was a methyl substituted octanal. Hydrogenolysis also proved very informative. It should be recalled that hydrogenolysis involves not only addition of hydrogen to multiple bonds but also cleavage of carbon to oxygen bonds. From this reaction, 3-methylhexadecane was obtained.

$$CH_3CH_2CH(CH_2)_{12}CH_3$$
with a CH_3 branch on the third carbon

Compound B was studied by the same techniques. Ozonolysis gave the same aldehyde as that isolated from compound A which established the location of the double bond and the methyl branch. Total synthesis proved the structure of each compound as follows: diborane was allowed to react with 2-butene to produce a reactive organoboron intermediate:

which was added to acrolein in a Michael type 1,4 addition.

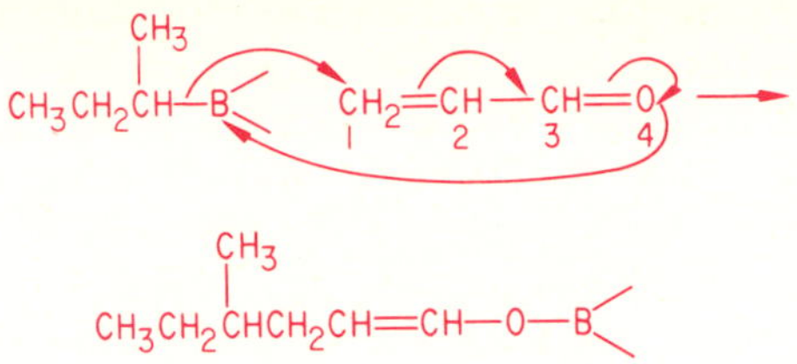

$$CH_3CH_2\overset{\overset{\displaystyle CH_3}{|}}{CH}-B< \qquad CH_2\!\!\overset{1}{=}\!\!CH\overset{2}{-}CH\overset{3}{=}\overset{4}{O} \longrightarrow$$

$$CH_3CH_2\overset{\overset{\displaystyle CH_3}{|}}{CH}CH_2CH\!=\!CH\!-\!O\!-\!B<$$

Treatment of this intermediate with water produced the desired initial product:

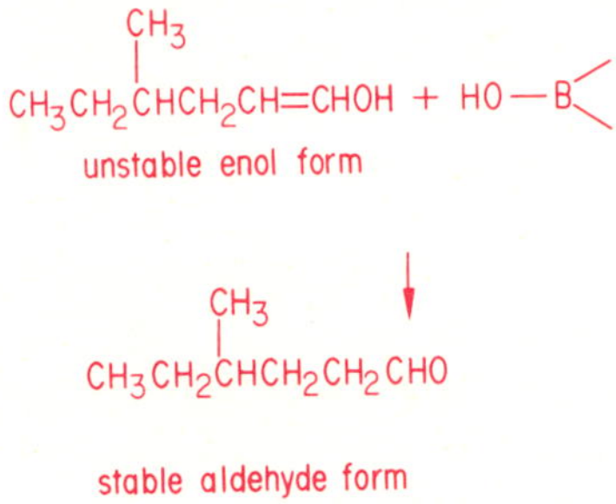

$$CH_3CH_2\overset{\overset{\displaystyle CH_3}{|}}{CH}CH_2CH\!=\!CHOH \; + \; HO\!-\!B<$$

unstable enol form

$$CH_3CH_2\overset{\overset{\displaystyle CH_3}{|}}{CH}CH_2CH_2CHO$$

stable aldehyde form

A Wittig reaction was then carried out on this aldehyde:

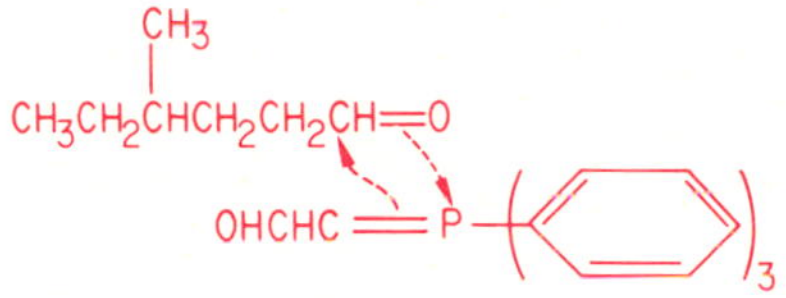

$$CH_3CH_2\overset{\overset{\displaystyle CH_3}{|}}{CH}CH_2CH_2CH\!=\!O$$

$$OHCHC =\!\!= P\!-\!\!\left(\!\!\left\langle\!\!\bigcirc\!\!\right\rangle\!\!\right)_3$$

through the following unstable intermediate:

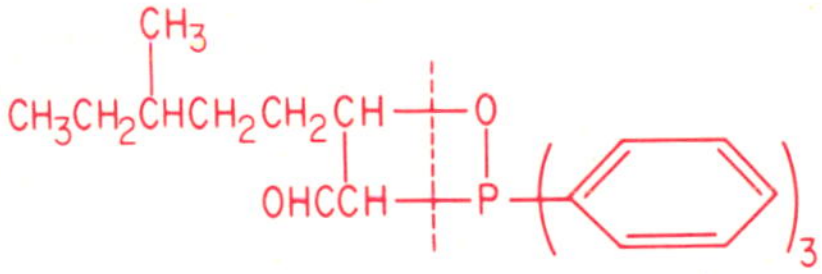

$$CH_3CH_2\overset{\overset{\displaystyle CH_3}{|}}{CH}CH_2CH_2CH\!-\!\!-\!O$$

$$OHCCH\!-\!\!-\!P\!-\!\!\left(\!\!\left\langle\!\!\bigcirc\!\!\right\rangle\!\!\right)_3$$

which cleaved as shown to give an unsaturated aldehyde and triphenyl phosphine oxide.

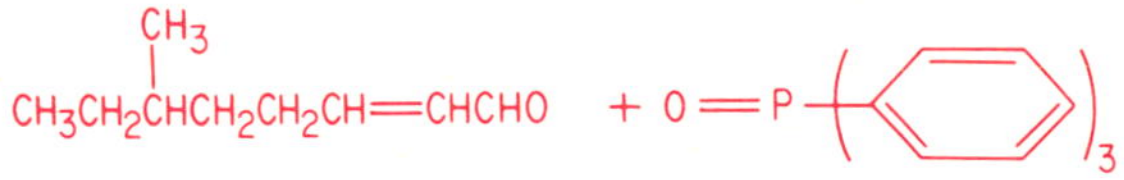

$$CH_3CH_2\overset{\overset{\displaystyle CH_3}{|}}{CH}CH_2CH_2CH\!=\!CHCHO \; + \; O\!=\!\!=P\!-\!\!\left(\!\!\left\langle\!\!\bigcirc\!\!\right\rangle\!\!\right)_3$$

This compound was reduced to the saturated alcohol by hydrogenation and then converted to the primary bromide by heating with hydrogen bromide.

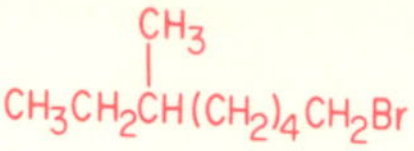

$$CH_3CH_2CH(CH_3)(CH_2)_4CH_2Br$$

Treatment of this halogen compound with triphenylphosphine

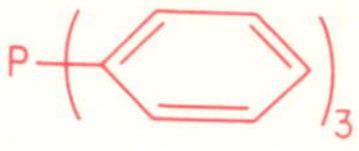

$$P{-}(\!\langle\;=\;\rangle\!)_3$$

led to a phosphonium salt

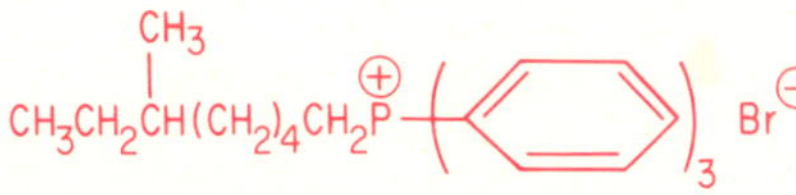

$$CH_3CH_2CH(CH_3)(CH_2)_4CH_2\overset{\oplus}{P}{-}(\!\langle\;=\;\rangle\!)_3\;\;Br^{\ominus}$$

which was allowed to react with the mono-aldehydosuberate methyl ester. [Suberic acid is the linear eight carbon dicarboxylic acid.]

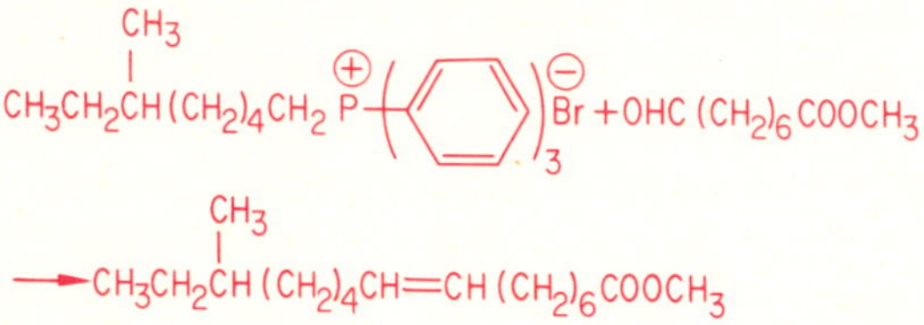

$$CH_3CH_2CH(CH_3)(CH_2)_4CH_2\;\overset{\oplus}{P}{-}(\!\langle\;=\;\rangle\!)_3\;\overset{\ominus}{Br} + OHC(CH_2)_6COOCH_3$$

$$\longrightarrow CH_3CH_2CH(CH_3)(CH_2)_4CH{=}CH(CH_2)_6COOCH_3$$

This *cis* isomer of this compound proved to be identical to compound B. Hydride reduction of this *cis* compound gave a *cis* unsaturated primary alcohol which was identical to compound A.

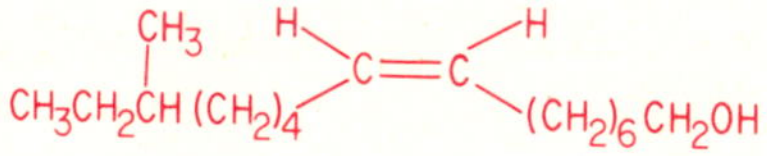

$$CH_3CH_2CH(CH_3)(CH_2)_4\underset{}{\overset{H}{C}}{=}\overset{H}{C}(CH_2)_6CH_2OH$$

The pink bollworm moth, *Pectinophora gossypiella*, exudes a sex pheromone of the following structure, 10-propyl-*trans*-5,9-tridecadien-1-ol acetate which has been designated "propylure".

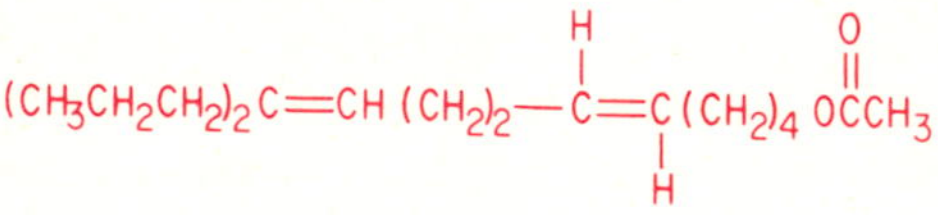

$$(CH_3CH_2CH_2)_2C{=}CH(CH_2)_2{-}\overset{H}{\underset{H}{C}}{=}C(CH_2)_4\,O\overset{O}{\overset{\|}{C}}CH_3$$

propylure

Of the many isomers possible for this structure, only one elicited the sexual excitement from the male bollworm moth. It was subsequently shown that propylure alone does not attract males in field traps although it proved highly exciting to the males in laboratory experiments. Scientists working on the problem discovered that N,N-diethyl-m-toluamide, which is present in the crude natural extract activates propylure so that males will be attracted in the field.

N,N-diethyl-m-toluamide

It must be stated, however, that this finding has recently been seriously questioned. It is interesting to note that N,N-diethyl-m-toluamide has been used as an extremely effective mosquito repellent, but this represents the first time this compound has been isolated from a natural source. It appears to be stored in relatively large amounts by the female pink bollworm moth.

Among insects, social organization or collectivization reaches its highest level in bees, wasps, ants, and termites. Clearly, the social organization among these insects is complex. It would be premature to assign chemicals as the prime causative agents which regulate the level and character of participation for individuals within the insect social structures. However, it is certainly possible at this stage to show that varied and specific pheromones are exuded by these insects and that highly characteristic responses ensue as a consequence. Arthropods, of which these insects are a part, exhibit the most diverse and complex pheromonal emissions known among the lower animals. The chemicals are usually produced in exocrine glands and stored in sacs. They may be discharged to accomplish any number of purposes. Primary among them are trail-marking in order for individuals within the group to find their way; alarm, to alert individuals to danger or to protect them once a predator has appeared; swarming, and other aspects of gathering together, to collect food, reproduce and/or to protect each other.

Trail-marking pheromones must be made up of short-lived molecules exuded by the insects. It has been shown that extracts from the fire ant can be stored in hexane at 4°C for at least three years without losing the trail-marking ability. At 25°C, however, the active principles become rapidly deactivated. Compounds such as n[cis-3,cis-6,trans-9]-dodecatrien-1-ol have been isolated as trail-marking pheromones for termites.

$$CH_3CH_2-C=C \overset{H}{\underset{CH_2}{}} \overset{H}{} C=C \overset{H}{\underset{CH_2}{}} \overset{H}{} C=C \overset{H}{} CH_2CH_2OH$$

Bees deposit droplets containing trail-marking pheromones to locate food sources and the return route to their nests. The trail-marking pheromones are excreted from the mandibular gland. When the scouts leave the nest, they deposit these mandibular gland secretions at specific distances from each other along their line of travel. After a food source is located, worker bees proceed directly to the food source following the aerial trial deposited by the scout bees by means of volatile components from the droplets. On their return to the nest, these worker bees reinforce the concentration of the trail-marking pheromones. Extracts from the mandibular gland of two species of bees have been analyzed and shown to contain methylketones and benzaldehyde. It is fascinating to speculate how bees recognize proper trail markings. Certainly the mixture of ketones specific to a given species will differ from those associated with other species, and even other populations of the same species. It is clear that disaster would occur within any given colony were the worker bees unable to follow the pheromone trail-marking language unique to that colony. Thus it is likely that the composition of pheromone molecules varies from colony to colony and from species to species.

Another means of communication between bees is the amazing "waggle dance", which is used to describe both the direction and distance from the hive where a new food source can be located. On returning from a scouting expedition, the bees perform a circular ritual bisected by a tail-wagging pattern. The angle of the bisect with respect to the sun indicates the direction of the food source; the number of waggles/second tells the other bees how far to fly.

The particular part played by benzaldehyde in pheromone activity is quite interesting. This compound which appears to enhance the short-term trail-marking character of the mandibular deposits turns out to be an important defensive agent in the secretions eliminated from ants. Thus it turns out that insects with a rather limited number of simple and relatively volatile organic molecules are able to construct various pheromone-like activities by using the same or similar combinations of molecules over and over for different purposes. Among the main alarm pheromones are compounds such as terpenes, esters, ketones and alkenes. The common denominator among these molecules is that they are quite volatile and long-lasting and have a potential irritating quality. In contrast to the trail-marking pheromones, these molecules are produced in relatively large amounts because they are involved in generating rapid signals without being enormously long-lasting.

Arthropods maintain a constant predator-prey relationship with other organisms. It is no surprise therefore that the land-dwelling representatives possess a most diverse chemical defense system. Terpenoids and related molecules are common types of molecules isolated from ants which most probably serve in their defense mechanisms. Some of the typical structures are listed below.

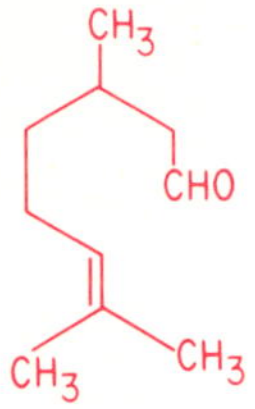

citronellal 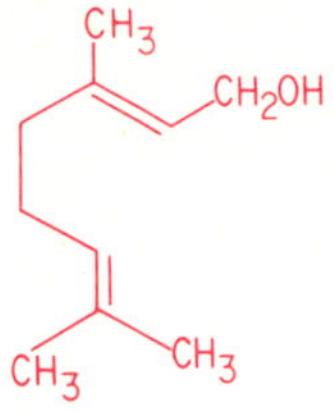   geraniol 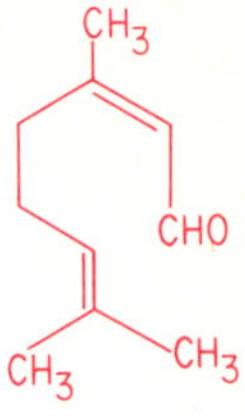   citral

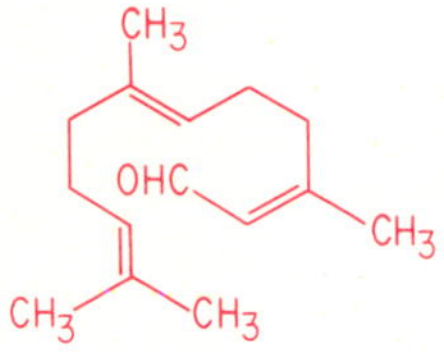

farnesal 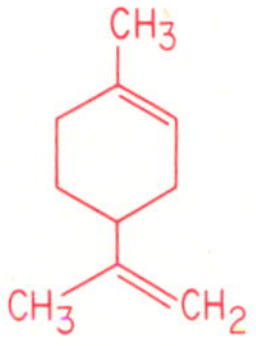   limonene 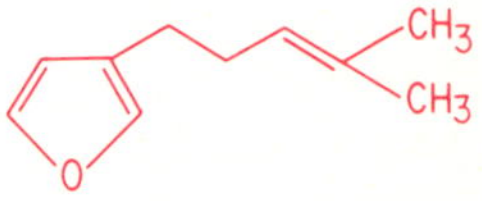   perillen

Benzoquinones are among the most widely distributed molecular types of arthropod defensive substances. They often are found with other compounds such as long chain hydrocarbons, aliphatic esters or aldehydes.

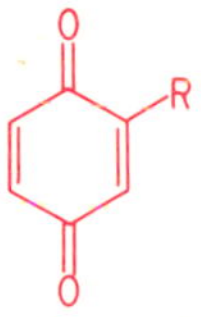

The quinones are thought to be stabilized by being dissolved in these nonpolar molecules. As a result, they increase the effectiveness of the secretion. Insects often secrete a combination of benzoquinone and toluquinone, or toluquinone and 2-ethyl-benzoquinone.

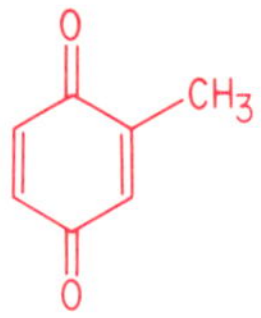

toluquinone 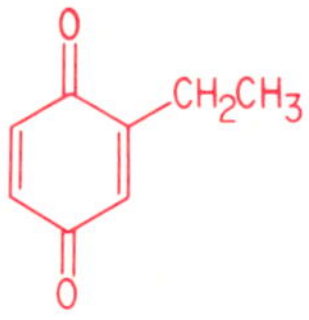   2-ethyl-benzoquinone

Certain millipedes, for example, secrete the benzoquinone derivatives
toluquinone and 2-methoxy-3-methyl-benzoquinone.

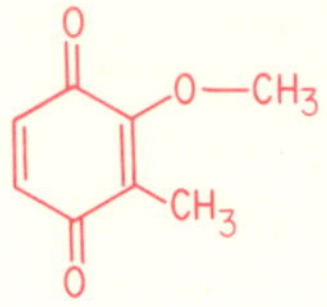

2-methoxy-3-methyl-benzoquinone

A route for the biosynthesis of these benzoquinones may stem from
preformed aromatic ring compounds such as the amino acids phenylalanine
and tyrosine.

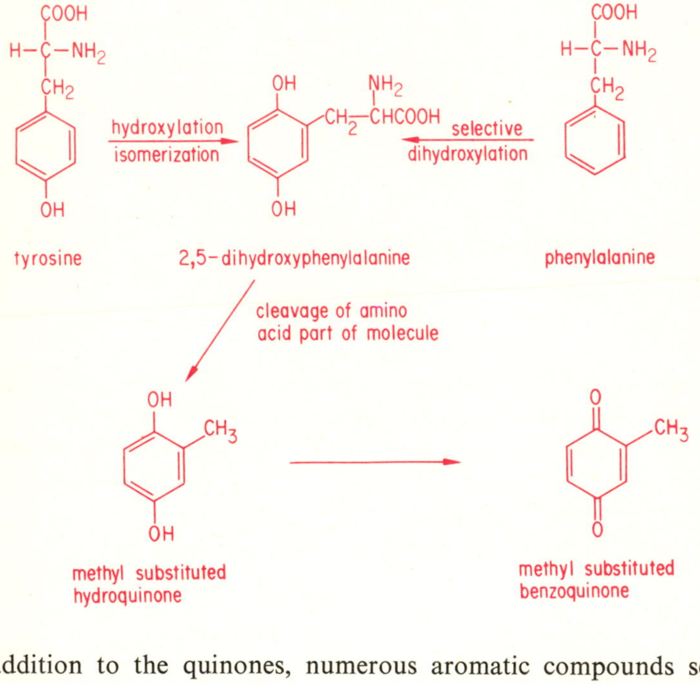

In addition to the quinones, numerous aromatic compounds serve as
arthropod defensive substances. Cresols, salicylaldehyde, p-hydroxybenzal-
dehyde, benzoic acids, benzaldehyde and others have been isolated. The diving
beetles from the family Dytiscidae possess two kinds of secretion glands which

produce defensive compounds. One is situated in the prothorax and produces steroidal compounds, while the other is located in the abdomen and secretes aromatic compounds such as benzoic acid, *p*-hydroxybenzoic acid and *p*-hydroxybenzaldehyde.

To this point we have presented the structure and some descriptions for typical insect pheromones. In several cases, mixtures of molecules are essential for activity. The male boll weevils, *Anthonomus grandis*, emit a complex mixture of terpenoid substances including the four listed below.

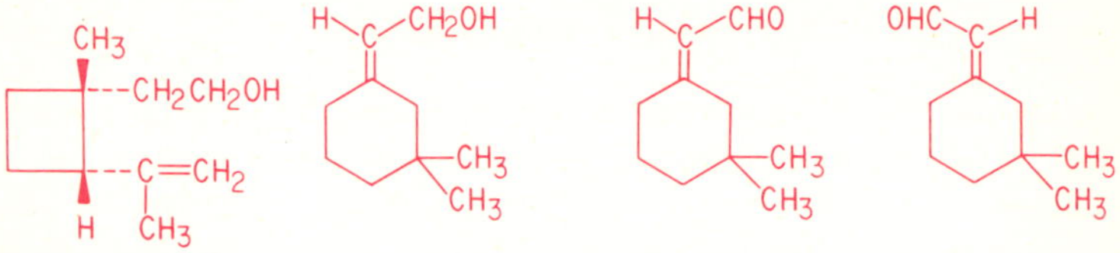

The combination of all four compounds is necessary to produce the attractive effect since fractionation of this mixture leads to inactive materials. Activity can only be restored when the four terpenes are recombined.

Pheromones containing nitrogen have also been reported. The husband and ated in the reaction. Also, the products of alkylation of benzene generally ketone from the male danaid butterfly, *Lycorea ceres ceres*, in addition to two long chain esters similar to the structures we have seen before.

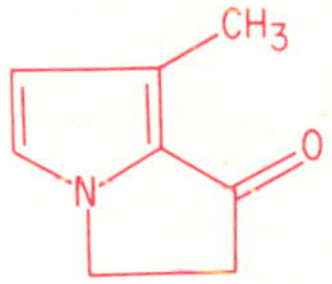

Synthetic organic chemists have entered the field of research on insect pheromones with the aim of controlling the destructive features of insect populations. Numerous attractants and repellents have been prepared. Trimedlure attracts the mediterranean fruit fly. Heptyl butyrate and 2,4-hexadienyl butyrate lure yellow jackets.

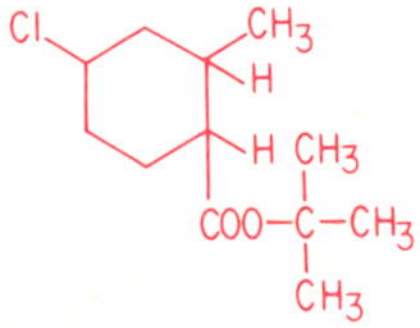

$$CH_3(CH_2)_5CH_2O-\overset{\overset{\displaystyle O}{\|}}{C}-C_3H_7$$

heptyl butyrate

trimedlure

$$CH_3CH=C-CH=CH-CH_2O-\overset{\overset{\displaystyle O}{\|}}{C}-C_3H_7$$

2,4-hexadienyl butyrate

Many compounds have been synthesized which act as repellents. Typically they contain benzyl benzoate which repels mites, *N*-butylacetanilide, which repels ticks, and 2-butyl-2-ethyl-1,3-propanediol which repels mosquitoes.

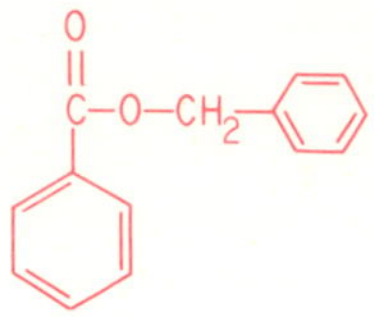

benzyl benzoate 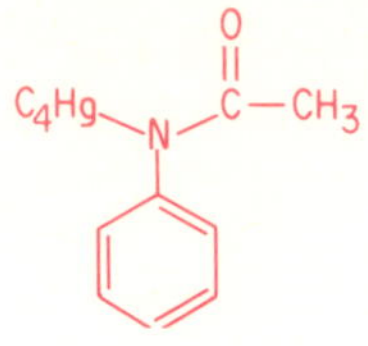   N-butylacetanilide

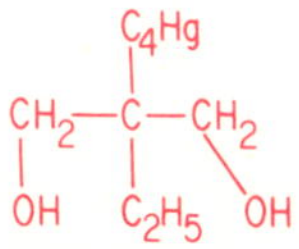

2-butyl-2-ethyl-1,3-propanediol

These compounds are effective, safe and highly specific for the assigned tasks. They must act for appropriate periods of time and not endanger other living systems. Attractants and repellents must be developed to protect human beings and other animals from dangerous insect-borne diseases (cf. Chapter VI).

The structures of some mammalian pheromones have also been elucidated. Since many primate species possess analogous behavior patterns to those discussed above for insects, it is no wonder that they secrete molecular attractants. Reproducible pheromonal activities for higher animals are much more difficult to establish because individuality within any group can mask interpretable responses. Human pheromones have been postulated. Even though none has yet been isolated and characterized, this area of research seems most exciting and potentially fruitful.

SUMMARY

In this chapter, we examined some aspects of the molecular basis for taste, smell and attraction. Our approach primarily concentrated on describing the broad range of molecular types which elicit such specific responses. We attempted to present current thinking concerning theories of taste and smell. Some examples of the elegant methodology used in pheromone research to identify active principles were presented. In addition we indicated directions research in these areas is taking. The subject matter of this chapter has broad social implications. Taste, smell and attraction are central phenomena in understanding chemical ecology. Utilization of knowledge about these molecules of the senses and synthesis of much needed new molecules can substantially raise the quality of human existence.

SOURCE MATERIAL AND SUGGESTED READING

Beidler, L. M., 1966, "Chemical excitation of taste and odor receptors" in *Flavor Chemistry*, Advances in Chemistry Series, R. E. Gould, ed., Vol. 56, Am. Chem. Soc. Publications, Washington.

Beroza, M., ed., 1970, *Chemicals Controlling Insect Behavior*, Academic Press, New York and London.

Ohloff, G., and A. F. Thomas, eds., 1971, *Gustation and Olfaction*, Academic Press, London and New York.

Pfaffmann, C., ed., 1969, *Olfaction and Taste*, The Rockefeller University Press, New York.

Shallenberger, R. S., and T. E. Acree, 1971, "Chemical structure of compounds and their sweet and bitter taste," in *Handbook of Sensory Physiology*, Vol. IV. Chemical Senses, Part 2, L. M. Beidler, ed., Springer-Verlag, Berlin, Heidelberg, and New York.

Sondheimer, E., and J. B. Simeone, eds., 1970, *Chemical Ecology*, Academic Press, New York and London.

CHAPTER XI

MOLECULAR STRUCTURE BY X-RAY VISION
Organic Molecules in 3-D

NOBEL PRIZES by themselves do not provide criteria by which to judge a field of scientific endeavor. It is unmistakable, however, that x-ray diffractionists have been singled out by the Swedish Academy far out of proportion to their numbers. In 1914 Max von Laue was awarded the Nobel prize for the discovery of x-ray diffraction. The following year the father-and-son team, William and Lawrence Bragg, were similarly recognized for their experiments elucidating the nature of x-ray beams diffracted from crystal surfaces. Three additional Nobel prizes were awarded over the next three decades to scientists who explained novel aspects of x-ray diffraction.

In 1954 Linus Pauling received the Nobel prize for his discovery of the α-helix, a configuration of the polypeptide chain found in certain proteins. This ushered in the modern era of x-ray crystallography in which the three-dimensional structures of many important and complex organic molecules have been unraveled. Three-dimensional analysis is important because biological function depends very specifically on the spatial arrangements of the atoms of a molecule. Since then, Dorothy Hodgkin presented complete structures for vitamin B_{12} and insulin. (For the first, she received the Nobel prize in 1964.) John Kendrew and Max Perutz received the Nobel prize in 1962 for their work on the structure of the globular proteins, myoglobin and hemoglobin. In the same year, Francis Crick, James Watson, and Maurice Wilkins were similarly honored for discovering the double helix of DNA.

Many other molecular structures were elucidated without being singled out by the Nobel committees. Though less dramatic, these results have helped develop the major approach in science today for obtaining the three-dimensional structure of molecules.

The famous organic chemist Emil Fischer devised a convention (see appendix) to describe two-dimensional representations of molecules containing asymmetric carbon atoms. It took over fifty years until Bijvoet established the correctness of the Fischer convention by x-ray diffraction analysis. Molecular biology thrives on the findings of x-ray crystallography. Through the diffraction technique, it has been possible to explain the logic of active sites of enzymes, the genetic code of DNA, and even patterns of molecular evolution.

An x-ray diffraction pattern of a molecule is not like a photograph taken with an optical camera; the atoms of a molecule are much too small to see with visible light. X-rays are used because their wavelengths are of the order of atomic dimensions. X-rays are scattered by the electrons of the atoms of a molecule and under certain favorable circumstances, the resulting diffraction patterns can be interpreted by complex mathematical relationships.

Actually, as James Watson in his book *The Double Helix* clearly states, Pauling's great discovery came as much from his brilliant intuitive mind as from a knowledge of the mathematics of x-ray crystallography. Watson writes, "I was soon taught that Pauling's accomplishment was a product of common sense, not the result of complicated mathematical reasoning. Equations occasionally crept into his argument, but in most cases words would have sufficed. The key to Linus' success was his reliance on the simple laws of structural chemistry. The α-helix had not been found by only staring at x-ray pictures: the essential trick, instead, was to ask which atoms like to sit next to each other. In place of pencil and paper, the main working tools were a set of molecular models superficially resembling the toys of preschool children."*

Watson and Crick relied heavily on model building to unravel the structure of DNA. Following construction of plausible structures, they were able to reason backward as Pauling did and show these structures to be fully consistent with x-ray diffraction patterns and mathematical relationships.

Single Crystal X-ray Studies

Before proceeding to describe some immensely important contributions of x-ray crystallography to chemistry and biology, it is necessary to present an introduction to the language and principles of x-ray crystallography. As noted above, x-rays are used because their wavelengths are sufficiently small (~1 Å) to resolve molecular structure.

Let us define a crystal as a three-dimensional array or lattice of points with a regular repeating space between points. Mauritz Escher devised a graphic representation of a lattice as shown in Figure 11.1. Generally the smallest element of space which can be repeated and generate the lattice is called the unit cell.

The diffraction pattern itself is a lattice in which the spacings are not accidental. They are related to the spacings of the points in the original

---

* J. D. Watson, 1968. *The Double Helix*, Atheneum Press, New York, p. 50.

FIGURE 11.1   Representation of a lattice, entitled "Cubic space division."
Drawing by M. C. Escher, courtesy of Collection Escher Foundation—Haags
Gemeentemuseum—The Hague

crystal lattice. Both the placements and intensities follow complex mathematical rules for scattering from regularly placed atoms. There is a general
mathematical principle which states that a Fourier series can be used to represent a regular repeating function. The Fourier series of such a function is
expressed in terms of (1) the number of times the sines and cosines repeat within
a specific distance and (2) the amplitudes of the sine and cosine waves. A Fourier
synthesis is accomplished if the original function can be regenerated from
summation of the amplitudes of the sines and cosines.

Since a crystal is regular in three dimensions, the electron density can be represented by a Fourier series in three dimensions

$$\rho_{(x,y,z)} = \frac{1}{v} \sum_{h=-\infty}^{+\infty} \sum_{k=-\infty}^{+\infty} \sum_{l=-\infty}^{+\infty} F_{(h,k,l)} \exp^{-2\pi i (hx+ky+lz)}$$

where

$\rho_{(x,y,z)}$ represents the electron density at each point in the unit cell;

v is the volume of the unit cell; and

h, k and l are the components of the wave number for a given sinusoidal wave for all the principal axes of the crystal.

$F_{(h,k,l)}$ is called the Fourier coefficient or structure factor. It is the coefficient of the Fourier series and therefore is a complex number made up of magnitude and a phase (Figure 11.2). Every spot on a diffraction pattern conceptually yields one term in the Fourier synthesis. Its intensity is proportional to the square of the wave amplitude.

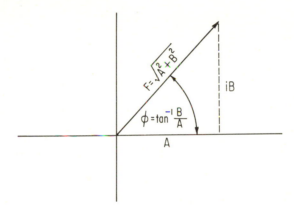

FIGURE 11.2 A graphic representation of the structure factor vector illustrating that it is composed of real (A) and imaginary (iB) components or equivalently a magnitude (F) and a phase angle (ϕ)

No information is available from the diffraction pattern which relates the phase of each wave to all the others. That this is important can be seen in Figure 11.3 (taken from a review by Dickerson) where the addition of two waves with and without phase shifts are shown.

The difficult task then for x-ray crystallographers is to solve the phase problem. An optical analogy may be useful to indicate why it is important to

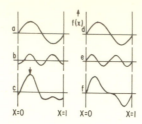

FIGURE 11.3   Fourier synthesis, addition of waves, and the effect of phase shifts.
(a)–(c). Build-up of a repeating function by addition of sinusoidal waves: (a) $f(x) = 10\sin 2\pi x$; (b) $f(x) = -5\cos 2\pi(2x)$; (c) $f(x) = 10\sin 2\pi x - 5\cos 2\pi(2x)$. Function c might conceivably represent a one-dimensional section through a structure with an atom at the peak marked by the arrow.

(d)–(f). Effect of a phase shift in one term upon the Fourier summation. Function b shifted by one-quarter cycle so that: (d) $f(x) = 10\sin 2\pi x$; (e) $f(x) = -5\cos 2\pi(2x + 1/4) = +5\sin 2\pi(2x)$; and (f) $f(x) = 10\sin 2\pi x - 5\cos 2\pi(2x + 1/4)$. Any resemblance to an atomic profile has now been largely removed. The entire character of a Fourier synthesis is drastically altered by a change of relative phases of the waves of which it is composed. (Taken with permission from *The Proteins*, Vol. 2, 2nd Edition, ed. by H. Neurath, R. E. Dickerson, Chapter 11, Academic Press, New York 1964)

know the phases of the diffracted waves. Figure 11.4 contains representations of objects (ducks) and their optical transforms. In part (a) we see that a single duck becomes a fuzzy series of concentric rings in which the form and features of the duck are not apparent. The intensity of the rings is dependent on the density of matter making up the duck. In part (b) we see two ducks at a distance x from each other. The optical transform now shows broken concentric rings with the distance $1/x$ between successive "smudges" on any given ring. Once again the features of the ducks are lost. Lastly, in part (c), we have four ducks at distances x and y from each other as shown. The optical transform shows smudges (not ducks) from which the distances $1/x$ and $1/y$ can be obtained. The distances in the transform are the reciprocals of the real distances in real space separating the objects. The periodicity in the transform is the reciprocal lattice. If the real distances are regenerated by Fourier synthesis without regard to phase, spots will be generated at the correct distances, but no ducks. The process of getting back the ducks is conceptually identical to generating a crystal structure from a diffraction pattern.

To begin with, the initial steps to solve a structure involve the determination of the size and shape of the unit cell and the calculation of the number of molecules in the unit cell. Following this, a search is made for "systematic absences" of diffraction spots (or reflections). These results allow one to place the crystal into a proper "space group".

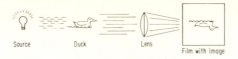

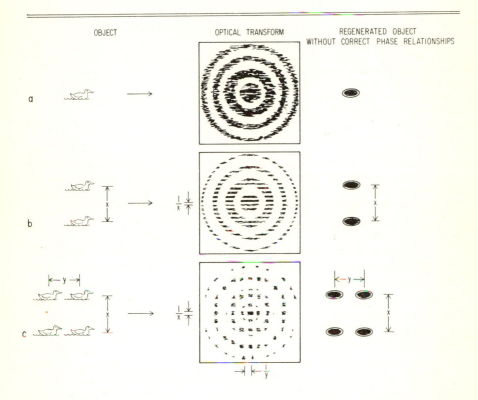

FIGURE 11.4 Optical analogy to x-ray diffraction. The optical transform represents the intensity distribution of the light scattered from the object as it appears in the lens. The relative phase relationships between the light waves scattered by the duck (top figure) are not altered by the lens. Since the long wavelength light waves can be collected together (focused), an image of the duck can be formed on the film. However, if the phases of the light waves were to be scrambled and an attempt to regenerate the ducks made, only simple blobs would result. Since the short wavelength x-rays cannot be focused (the refractive index for x-rays is unity for all materials), obtaining correct phases becomes the essential and definitive task in generating structures

The Bragg equation for x-ray diffraction is as follows:

$$d = \frac{n\lambda}{2 \sin \theta}$$

where d is the distance between parallel planes in a crystal; n is an integer; λ, wavelength of the x-rays; and θ is the angle of incidence of the x-ray beam with the crystal plane.

We initially assume that the scattering centers lay on parallel planes at a specific distance, d, apart and that each scattered wavelet will be in phase, cooperating with those scattered from successive parallel planes. In this case all diffracted beams in the diffraction pattern will result in spots of uniform maximum intensity. This is not valid for real structures, however. Inspection of a single x-ray photograph reveals that the intensities of the spots vary over a wide range. The variation of intensity arises because the scattering centers are not situated exactly on the crystal planes but are distributed throughout the unit cell. The problem is to determine the distribution of the scattering matter, i.e., the electrons of the atoms, from the observed intensities of x-ray reflections.

It must be remembered that $F(h,k,l)$ is a complex quantity; it represents not only the amplitude but also the phase of the diffracted beams. The observed intensities in an x-ray diffraction pattern enable us to calculate only the absolute value of intensities, but unfortunately provide no information concerning the relative values of the phase constants.

Thus, once again the application of the Fourier synthesis to the solution of crystal structure appears severely limited. However, the value of this powerful method in representing and refining the results of crystal analysis remains.*

A direct method approach proceeds in the following manner: an intuitive guess is made as to the position of the atoms. Values for $F(h,k,l)$ are then calculated. The observed intensities are collected and compared to the calculated intensities. If they agree reasonably well, then we can use them to calculate the electron density.

---

* If the crystal structure contains a center of symmetry, which is chosen as the origin for the coordinates, the possible phase angles are 0 or π. Therefore, the sign of the coefficients can be either positive or negative. For a centrosymmetric structure, the Fourier series may be written:

$$\rho(x,y,z) = \sum_{h} \sum_{k} \sum_{l}^{+\infty}{}_{-\infty} F(h,k,l) \cos \pi \left(\frac{hx}{a} + \frac{ky}{b} + \frac{lz}{c} \right).$$

In such simple cases, we must know the magnitude of $F(h,k,l)$. The structure problem can be solved by a trial-and-error procedure since both the intensities and phases can be obtained.

The electron density map will show maxima at positions somewhat different from the initial guess. Next an attempt is made to improve the atomic position and compute another set of $F(h, k, l)$, which is employed to calculate a new electron density map by a new Fourier synthesis. The Fourier series should provide a perfect representation of the crystal structure if all the coefficient and phase constants could be determined with great accuracy and if the measurement could be extended indefinitely. Practically, this is not the case, and the number of terms employed is limited by experimental conditions and the wavelength used. The prospect of success depends on whether the postulated positions are anywhere near the correct positions, giving some measure of agreement with observed intensities. If they are, the correct positions can be found by judicious small displacements of some or all the atoms from the original first positions. The utility of the direct approach falls off sharply with the size and complexity of the organic molecule. It has proven very valuable for low-molecular weight crystalline compounds. Structures obtained by direct methods for amino acid derivatives and dipeptides were used by Pauling to unravel fibrous protein structure.

Another way of getting around the phase problem involves the use of heavy atoms. Their scattering of x-rays dominates the diffraction pattern if the heavy atom contains an appreciable fraction of the electrons of the entire molecule.

$$I = |F|^2$$

where I represents the scattering intensity, and $|F|^2$ denotes the square of the amplitude of the waves, which is determined by the electron density. When the position of the heavy atom has been established, its phase can be calculated by a Fourier series. One then assumes that the entire structure is composed of this heavy atom. Of course, this is not true but is a close enough approximation to reveal some unambiguous aspects of structure. For most reflections the phase of the scattering from the entire molecule does not differ much from the scattering of the heavy atom alone. The calculated heavy atom phases are combined with observed intensities to achieve a Fourier synthesis. This stage yields an approximate structure. If positions of some light atoms are recognized, they are used to calculate better phase factors which are used once again in a Fourier synthesis. By successive improved phase relationships and Fourier refinements, the correct structure can be obtained (Figure 11.5).

Since proteins are so large, no heavy atom exists which can dominate the x-ray scattering. To solve protein structures, it is necessary to utilize the multiple isomorphous replacement method in which at least two heavy atom derivatives are prepared where the heavy atom does not alter the geometry of the protein crystal (isomorphous replacement).

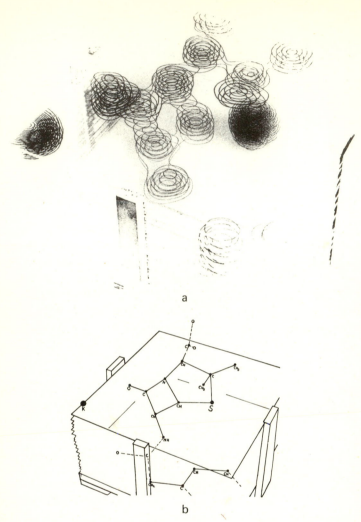

FIGURE 11.5 (a) Model of electron densities of potassium benzylpenicillin drawn on layers of plastic sheeting. (b) Chemical representation of the model deduced from part a. (Taken with permission from *The Chemistry of Penicillin*, Princeton University Press, Princeton, New Jersey, 1949)

There are extremely small differences between the protein diffraction patterns with or without the heavy atom. The total scattering (F_H) is the sum of the scatterings from the protein (F) and from the heavy atom (f).

$$F_H = F + f.$$

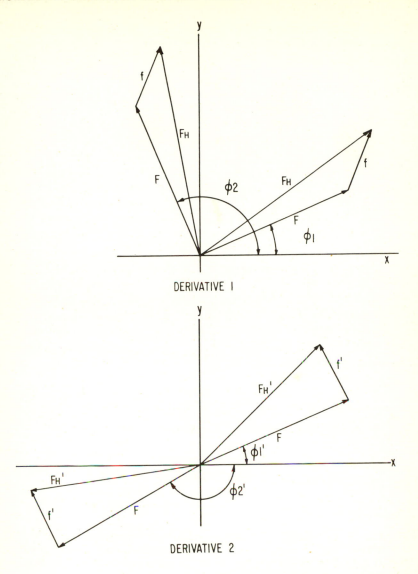

FIGURE 11.6 Representation of the vector diagrams illustrating the general need for multiple isomorphous replacements. The magnitudes of the structure factors for the parent protein (F) and the heavy atom derivatives, (F_H and $F_{H'}$) are observed vectors. The magnitude and orientation of the heavy atom contribution alone for the two derivatives can be calculated (f and f'). Two triangles can be closed for each derivative. Only one angle, the correct phase angle, is common to both derivatives (ϕ_1 and $\phi_{1'}$)

The intensities of the scattering are measured for the parent crystal and for the crystals containing the heavy atom. The phase problem reduces to a trigonometric question of finding the orientation of a triangle in Cartesian coordinates from a vector summation. For a single derivative, two phase angles exist which satisfy the vector equation. A second heavy atom derivative also provides two solutions, one of which is identical to the vector orientation of one of the solutions from the first isomorphous derivative (Figure 11.6). The phase angle that the two solutions have in common represents the correct solution to the vector equation.

Errors in heavy atom locations and intensity assignments for complex protein crystals dictate that more than two isomorphous heavy atom replacements are necessary. The accurate collection of reflections (diffraction intensities) is essential to structural analysis. Modern techniques involve automatic control of data collection by computer operation. The alignment and rotation of the crystal are automatic as is the storage of all intensities of reflection. The Fourier synthesis is also carried out by computer as are additional structural refinements.

The most serious problem in this method is that of finding good heavy atom derivatives in which one heavy atom occurs per molecule. Once such analogs have been produced, the solution of the three-dimensional structure is only a matter of hard work and time. Isomorphous crystals of a protein can be obtained by three methods:

1) Chemical modification of the protein followed by crystallization if the reagent is of a type which binds by a definite chemical reaction.

2) Crystallization in the presence of a reagent which does not form covalent bonds.

3) Diffusion of a reagent into a crystal.

All three approaches are based on trial and error. There are no general techniques or fundamental principles to follow in order to prepare successful isomorphous, heavy atom derivatives of complex organic molecules.

From the above discussion, we can outline the procedure to study the structure of globular proteins:

a) Obtain good crystals.
b) Prepare at least two isomorphous heavy atom derivatives.
c) Measure diffraction patterns for the parent protein and protein crystals containing heavy atoms.
d) Calculate the phase angles as shown in Figure 11.6.
e) Combine the phase and intensity values for a Fourier synthesis.

f) Construct an initial model.

g) Recalculate improved phases.

h) Repeat the Fourier synthesis.

i) Construct a more accurate model.

j) Repeat the Fourier synthesis and construct refined models.

The Fischer Projection and Vitamin B₁₂

Bijvoet was the first to establish the absolute enantiomorphy of a crystal structure (sodium rubidium tartrate). The differences between the magnitudes $F(h,k,l)$ and $F(-h,-k,-l)$ is known as the Bijvoet difference. These structural factors are geometrically related to each other by a center of symmetry. Once the phase angles of a structure are known, the Bijvoet difference can be calculated. If the sign of the calculated difference is opposite to the observed difference, the true structure is represented by a mirror image of the postulated structure. For sodium rubidium tartrate, the structure is consistent with the signs of the phase angles. This therefore confirms the correctness of the Fischer convention (see appendix).

From here, we proceed to a more complicated molecule whose structure has been solved by x-ray analysis. The structure of vitamin B₁₂ (chemical formula $C_{63}H_{88}O_{14}H_{14}PCo$) was elucidated in 1957 by Dorothy Hodgkin and her group at Oxford University. The first x-ray photographs of vitamin B₁₂ crystals taken in 1948 and 1950 showed that it was formally possible to solve the structure of these crystals by x-ray analysis. But the lack of complete information, at that time, about the chemical composition of the molecule made solution of the three-dimensional structure impossible. The actual solving of the structure required the combined skill of Dorothy Hodgkin's group of x-ray specialists and Lord Todd's organic chemists who worked on the chemical composition of vitamin B₁₂. Partial structures from chemical hydrolyses aided the x-ray researchers while pinpointing of specific groups by x-ray analysis erased ambiguities found by the organic chemists. The x-ray structure actually preceded the unraveling of the chemical composition of this important and elusive vitamin that plays such an important role in preventing pernicious anemia. Disturbances in the metabolism of proteins, carbohydrates and fats are observed in animals kept on vitamin B₁₂ deficient diets.

The presence of the relatively heavy atom of cobalt in the molecule, allowed the Hodgkin group to attempt to solve the structure by the heavy atom method (described above). Through the analyses of four different crystals of vitamin B₁₂ or related compounds (air-dried crystals of vitamin B₁₂, wet crystals kept in their mother liquor, crystals of a selenocyanate derivative of vitamin B₁₂, and crystals of a hexacarboxylic acid obtained by degradation of the vitamin)

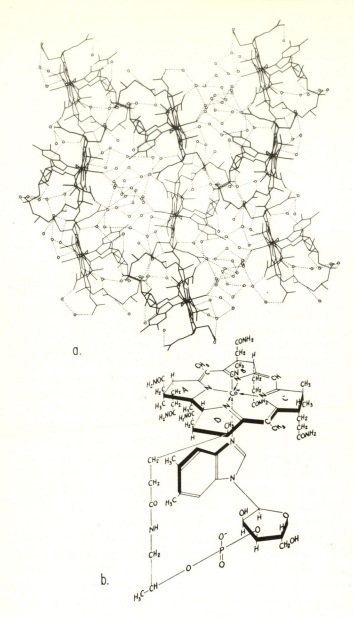

FIGURE 11.7   Vitamin B_{12}. (a) The crystal structure of wet vitamin B_{12}. Molecules of the vitamin are hydrogen bonded together by water molecules (shown by open circles). (b) Single molecule of the vitamin. [Photograph courtesy of Dorothy C. Hodgkin, Oxford University]

it was possible to gather partial information from each of these structures that led to the definition of the molecular and crystal structure of the molecule.

The cobalt atom lies in the center of the nucleus of four five-membered rings (Figure 11.7) and is linked to the benzimidazole group on one side of the plane and to the cyanide group on the other. The plane of the benzimidazole ring is nearly at right angles to that of the nucleus; the ribose group attached to it turns back toward a position nearly parallel with the nucleus.

Myoglobin and Hemoglobin

The first three-dimensional structural analysis of a protein using x-ray crystallographic methods was carried out by Kendrew and his associates on sperm whale myoglobin. The approach which has proven most successful in the study of proteins involves either multiple or single isomorphous replacements (cf. above section). Unfortunately, the search for such derivatives can not be systematic. As noted above, there is no way to predict whether a given derivative will be isomorphous with respect to the parent protein.

In 1954 Perutz found that horse oxyhemoglobin could be crystallized in the presence of p-chloromercury benzoate or silver ions. In this manner he was the first to obtain isomorphous replacements. Even with this result, it was not possible in the 1950s to interpret the x-ray diagrams by two-dimensional methods because of the overlapping of many atoms in any projection 60 Å thick. It was obvious that three-dimensional analysis was essential to the interpretation of protein x-ray diffraction diagrams. The solution of complex three-dimensional structures had to wait for the development of large computers. With the aid of these enormously fast instruments, the analysis of the hundreds and even thousands of reflections became possible.

Kendrew and his associates applied the technique of isomorphous replacement to myoglobin, a much simpler analog of the hemoglobin molecule. Myoglobin contains 153 amino acids and 1 heme group and has a molecular weight of approximately 17,000. Using trial and error techniques, Kendrew and his colleagues prepared several usable isomorphous derivatives of myoglobin. Some of these included a p-chloromercury benzene sulfonate, gold chloride and mercury diamine derivatives and a complex with mercuric iodide. Initially the three-dimensional analysis on myoglobin was carried out at 6 Å resolution.* This level of resolution gave 400 reflections. Using high-speed

---

* The level of resolution of an electron density map is defined by the number of terms used in the Fourier synthesis. The greater the number, the greater will be the details of molecular structure obtained. As one proceeds to finer and finer resolution, the number of terms and hence the number of reflections that must be measured rises enormously.

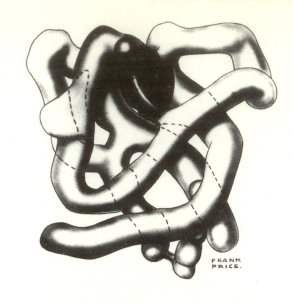

a.

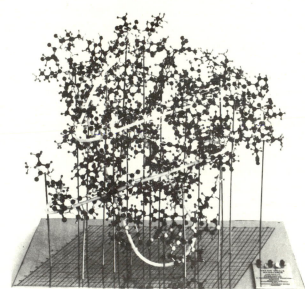

b.

FIGURE 11.8 Models of the three-dimensional structure of the myoglobin molecule. (a) At 6 Å resolution, only the outlines of the protein structure are seen. (b) At 2 Å resolution, it is possible to deduce amino acid sequences. The tubing shown defines the α-helical sections of the protein. [Photographs courtesy of J. C. Kendrew, Medical Research Council, Cambridge]

computers, Kendrew and his colleagues were able to interpret the electron density map. The analysis revealed the gross structure of the molecule and showed extensive α-helical content. A picture of the molecule deduced at this resolution can be seen from Figure 11.8a. The heme residues were analyzed by electron spin resonance which also provided the orientation of this group. Scouloudi worked on seal myoglobin. She prepared a different iso-morphous derivative and found that the three-dimensional structure of seal myoglobin did not vary substantially at the 6 Å level resolution from the sperm whale myoglobin which Kendrew and his associates studied.

The Cambridge group then proceeded to carry out a series of investigations at 2 Å resolution (Figure 11.8b). In this analysis, they encountered well over 9000 different reflections for the parent protein and for each of the four heavy atom derivatives. The most significant results of this high resolution 2 Å analy-sis was that the α-helical content deduced at 6 Å was clearly confirmed. They found that the segments of α-helix contain essentially the proper dimensions of pitch (5.4 Å) and residue translation (1.5 Å). (See section on fibrous pro-teins.) They concluded that the myoglobin molecule is quite compact; that the stabilizing forces for the globular structure probably come from proximity interactions of near-lying atoms rather than from polar forces; that the polar groups tend to be on the outside of the molecule while the non-polar groups tend toward the inside. The surface polar groups all appear to be highly solvated (see Figure 11.8).

It is no surprise that the work on hemoglobin proceeded more slowly, primarily because it is four times the size of myoglobin. It consists of two α and β chains. Each chain has one heme and at least one sulfhydryl group. Because of the sulfhydryl groups, hemoglobin possesses a definite site for heavy atom substitution. The beta chains were substituted with p-chloromercury-benzoate, dimercury acetate and a "beta mercurial" (1,4-diacetoxymercury-2, 3-dimethoxybutane). At a resolution of 5.5 Å Perutz and his associates obtained 1200 reflections for each of the heavy atom derivatives and for the parent compound. From these patterns, they were able to build up the gross features of the oxyhemoglobin molecule. In a manner similar to that discussed above for myoglobin, they then proceeded to obtain electron density data at 2.8 Å resolution. The molecule was seen to be roughly spheroidal with dimensions of $64 \times 55 \times 50$ Å. After study it could be seen that the molecule had lines of density which roughly corresponded to the chemically separable chains (i.e., two alpha and two beta chains). The general views can be seen in Figure 11.9. Remarkably, it is readily apparent that each subunit was essentially identical to the structure seen in myoglobin. (See p. 273 for the evolutionary significance of this).

Resolution at about 3 Å for human and horse deoxyhemoglobin has also

FIGURE 11.9   Model of the hemoglobin molecule. The bonding sites for oxygen
are indicated by O_2. [Photograph courtesy of M. Perutz, Medical Research Council,
Cambridge]

been obtained. The results indicate that the globular structures of the subunits
of human oxyhemoglobin closely resemble those of horse oxyhemoglobin.
Subunits have been shown to move as much as 7 Å with respect to each other
although the general shape of the subunits and the whole molecule remains
nearly unchanged.

Lysozyme

A complete structural determination of the protein lysozyme* at 2 Å resolution
was published in 1965 by D. C. Phillips and his associates at the Royal Institute,
London. The breakthrough in analysis of lysozyme occurred in 1960 when
Poljak was able to prepare heavy atom isomorphous derivatives. The pro-
visional structure at 6 Å resolution was confirmed in its general features.

* Biochemically lysozyme acts to break down mucopolysaccharides and to lyse some
bacteria such as *Micrococcus lysodeikticus*. It is found in egg white and in mucosal secretions
in man.

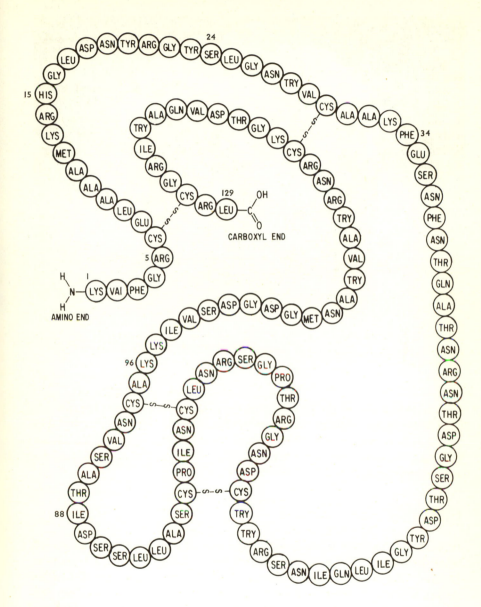

FIGURE 11.10 Sequence of amino acids in lysozyme with disulfide bridges shown. A three-dimensional model of this enzyme is shown in Figure 11.11. [Taken from an article by D. C. Phillips, *Scientific American*, November 1966, p. 78]

The molecule appears to be ellipsoidal with dimensions of $45 \times 30 \times 30$ Å and contains a marked cleft on its surface. The three isomorphous heavy atom derivatives which were used to produce the 6 Å electron density map were the complex ion involving $UO_2F_5^\equiv$, an ion derived from $UO_2(NO_3)_2$ and also an orthomercurihydroxytoluene-p-sulfonic acid derivative. With the exception of the last named, the derivatives used to obtain the data at 6 Å could be used for the 2 Å resolution analysis. A two-dimensional representation of the amino acid sequence and folding is shown in Figure 11.10. The research team from the Royal Institute employed the techniques essentially developed by Kendrew, Perutz and their associates in the analysis of myoglobin and hemoglobin. Their work, of course, involved a simultaneous study of the electron density maps and use of known polypeptide sequences to construct models.

Lysozyme consists of 129 amino acids. In the manner described above, the research workers were able to locate 114 of the amino acid residues clearly. The positions of the remaining residues were obtained by geometric deductions from electron densities which only showed a fraction of each of the amino acid side chains. Lysozyme contains three stretches of about ten residues each that are helically arranged; namely, residues 5–15, 24–34 and 88–96. These structural features are contained in the three-dimensional molecular array shown in Figure 11.11.

These short range helical arrays for lysozyme are actually somewhat distorted in that the hydrogen bonding arrangement is between the α and 10/3 helices.* An unusual feature of the lysozyme molecule occurs between residues 41 and 54 where the chain folds back on itself and forms an antiparallel pleated sheet. This type of arrangement was proposed much earlier for the structure of silk fibroin. (See section on fibrous proteins). Lysozyme is similar to myoglobin and the subunits of hemoglobin in that the polar side chains of the amino acid residues are distributed almost exclusively on the outside of the molecule. The non-polar side chains are primarily on the inside of the molecule.

Fibrous Macromolecules

To this point we have concentrated on some fundamental basic tenets of x-ray crystallography and on single crystal analysis. Fibers represent another potentially regular molecular domain. It is usual to transmit x-rays through a crystalline or partially crystalline fiber perpendicular to the fiber axis. The resulting diffraction pattern is used to calculate some of a fiber's molecular

---

* The α-helix contains about 18 residues in 5 turns while the 10/3 helix is a structure in which 10 residues repeat in 3 turns. We will consider the α-helix in the next section of this chapter.

dimensions such as repeat distance. The term crystallinity, when applied to a fiber, refers to the order within polymer chains and the regularity of chain packing. Conceptually, a fiber can have complete three-dimensional order if regular, infinitely long chains pack perfectly in the direction normal to the chain axes. Real polymers can only be partially crystalline because finite chain lengths lead to defects in order and packing at the chain ends. Fibers can be stretched to align many (but not all) chains, leading to substantial crystallinity parallel to polymer axes. Packing forces are less affected by mechanical orientation so that large regions of disorder normal to chain axes are common. Such systems made up of crystalline regions (crystallites) and disordered regions (amorphous zones) give diffraction patterns smeared out into areas about the center of the x-ray beam. As a result of the lower level of order, fibers produce x-ray diagrams with many fewer spots (reflections) than are typical of diffraction patterns from single crystals. Unfortunately, amorphous zones create background scattering that can interfere with the interpretation of fiber patterns.

The key to the understanding of fiber x-ray diffraction involves the realization that the location, distance and intensity of spots from the center of the x-ray picture bear direct relationships to structural features in the fiber. The vertical line through the center (origin) is called the meridian, while the horizontal line through the origin is called the equator (Figure 11.12).

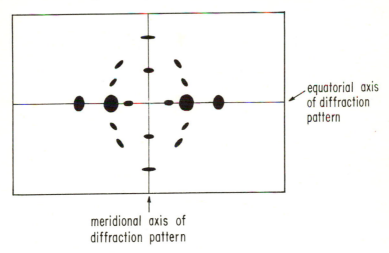

equatorial axis
of diffraction
pattern

meridional axis of
diffraction pattern

FIGURE 11.12   Schematic representation of part of a diffraction pattern of a fiber (held perpendicular to incident x-ray beam)

Reflections on the meridian provide structural information on order along the fiber axis. Equatorial reflections, on the other hand, show order perpendicular to the fiber axis.

A spot located at a distance d' from the origin along the meridian indicates a real repeating distance of d along the fiber axis:

$$d(\text{Å}) = \left(\frac{1}{d'}\right) \text{Å}^{-1}.$$

This was realized many years ago by K. Meyer and H. F. Mark who pioneered the study of fiber x-ray diagrams. They studied silk fibroin and found meridional reflections at $d' = 0.143$ Å^{-1}, which meant a repeating distance of 7 Å along the fiber axis (Figure 11.13).

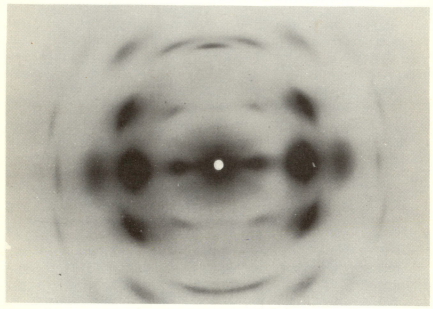

FIGURE 11.13   X-ray diffraction diagram of silk fibroin. Note the strong meridional reflections. [Photograph courtesy of R. E. Marsh, California Institute of Technology]

Mark proposed that this was just enough for a repeat of two fully extended amino acids (Figure 11.14). In addition, there is a very strong reflection at $d' = 0.285$ Å^{-1} on the meridian which corresponds to a repeat of 3.5 Å, enough for a single extended amino acid to repeat. Both are consistent with the same structural picture of fully extended polypeptide chains. In this way, the structure of the β-form of proteins was explained. Astbury interpreted the packing for β-keratin (a protein of skin, hair, etc.) on the basis of the equatorial reflections. He assigned the 4.65 Å spacing to the distance between hydrogen-bonded polypeptide chains within the same plane, and a 9.8 Å spacing to interplane stacking of the sheets of polypeptide chains (Figure 11.15).

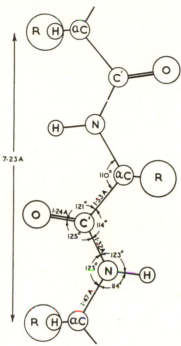

FIGURE 11.14   Diagram of a fully extended polypeptide chain, where the distance between amino acids is 7.23 Å. [Taken from a paper by L. Pauling and R. B. Corey, 1953, published in *Proc. Roy. Soc. B.*, **141**, p. 21]

As can be seen from the x-ray fiber diagram in Figure 11.13, spots occur only along a set of layers. A genuine crystallographic repeat spacing $d(\text{Å})$ will produce diffraction spots at multiples of the distance d' from the origin:

$$1d', 2d', 3d' \ldots \text{Å}^{-1}.$$

The coefficients are referred to as the orders of the spacing d in the real structure.

Most ordered fibrous structures are not composed of extended polymer chains. This is because of packing considerations or the nature of the side groups coming off the main chain. For example, although linear polyethylene can assume a zig-zag extended structure (Figure 11.16), isotactic polypropylene (see p. 75) cannot because in the fully extended chain the methyl groups

would actually interpenetrate each other. The only route open to such polymers in maintaining structural order is to form a helix. In fact, most ordered
vinyl polymers exist as helices in the solid state.

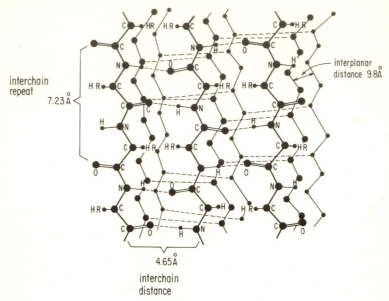

FIGURE 11.15   Sheets of fully extended polypeptide chains as in silk, hair, etc.
The reflections of x-ray diffraction diagrams (see Figure 11.12) lead to the indicated
interplanar and interchain distances and the interchain repeat

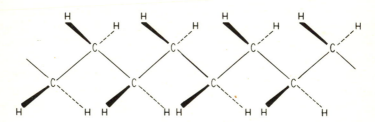

FIGURE 11.16   Linear polyethylene in a *trans*-staggered, zig-zag, extended
structure

Biopolymers also form helices. Most fibrous proteins and nucleic acids
fall into classes of polymers with helical symmetry. Repeat distances involve
rotations and translations about a helix axis.
Pauling developed the "equivalence postulate" where all monomeric units

occupy geometrically identical positions with respect to the polymer (crystallographic) axis. Since α-amino acid residues have no symmetry elements, the only route available to convert the coordinates of one residue to those of another involves screw translations along the chain axis. Such operations lead to helical structures. A helix with n repeating units per turn is said to have an n-fold screw axis.

In his protein studies, Pauling considered only those structures which leave the amide (peptide) bond planar with a *trans* configuration. This assumption is reasonable on resonance grounds and also from the x-ray results found for single crystals of such compounds as N-acetylglycine, glycylglycine, DL-alanine and L-threonine.

These findings were completely consistent with the generalized picture of the extended poly-α-amino acid chain discussed above. The hydrogen bonds from the amide-NH to the carbonyl are essentially linear, even though some deviations from these linear interactions are allowed. Out of this work came other discoveries which explained the structure of helical proteins.

Only a limited number of rotations per residue can be fitted to the restrictions of planar *trans* amide, standard bond lengths and angles, linearity and length of the hydrogen bond, and the equivalence postulate. In fact, a rotation of $97.2°$ falls out as an almost unique answer that embodies all these requirements. Pauling designated the structure resulting from this rotation, the α-helix (Figure 11.17).

Let us define some terms used to describe a helix. P is designated the helical pitch; h is the rise per residue (or axial translation) in the helical direction and n is the number of units per turn (P/h). Lastly, β is the angle of pitch (Figure 11.17). In the α-helix, the consecutive residues have an axial translation of approximately 1.5 Å; the pitch of the helix is about 5.4 Å; there are 3.67 residues per turn and the linear arrangement of carbonyls and NH groups forming the intrachain hydrogen bonds makes an angle of $12°$ with the helix axis (Figure 11.18).

Shortly after this prediction, Perutz obtained experimental evidence for the right-handed α-helix by finding a 1.5 Å spacing using a synthetic polypeptide, α-keratin and hemoglobin. The right-handed* α-helix has since been shown to be a common feature of many other proteins.

Following Pauling's proposal of the α-helix, several groups of scientists have worked out the details of x-ray diffraction by helical structures. Cylindrical coordinates for the real structure are used where r (the helical radius) and

---

* The handedness of a helix is determined by the direction of progression along the backbone of a chain. A helix is defined as right-handed when the direction of progression is clockwise.

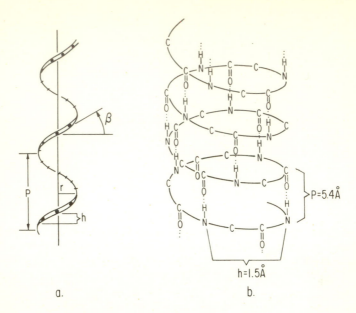

a.   b.

FIGURE 11.17   Schematic representations of a simple helix (a) and the α-helix (b). *P* refers to the helix pitch; *r*, the radius of the helix; *h*, the axial translation per residue; and *β*, the pitch angle

z represent coordinates in angstroms and ϕ denotes an angle between the plane containing the incident beam, the fiber axis and an atom situated on the radius (Figure 11.19a). The terms *R, Z* and ψ specify analogous coordinates in the diffraction pattern (Figure 11.19b). Diffraction spots are limited to values where $Z = l/P$ where *l* is the order of the layer which can have positive, negative or zero values and P is the pitch. (See Figure 11.17).

If a helix has an exact repeat after $Z = c$ over which distance there are an integral number of turns (*N*) and an integral number of monomeric units (**M**), then:

$$c = NP = Mh.$$

The diffraction pattern can be explained on the basis that $Z = l/c$.

Let us consider the α-helix in which 18 residues repeat in 5 turns (or 3.60 residues per turn) with a real repeat along the fiber axis of 27 Å:

$$Z = 1/27 \text{ Å}^{-1} = 0.0370 = l/c.$$

The indexing of layer lines would exhibit the first meridional reflection on the 18th layer line. Slight alteration of the twist leads to major changes that would

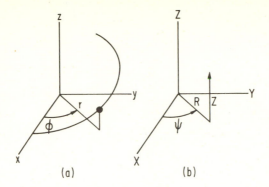

(a)   (b)

FIGURE 11.19   Cylindrical coordinates describing helical structures (a) and the x-ray diffraction pattern of the structures (b), where r (the helical radius) and Z are coordinates in angstroms and ϕ is the angle between the plane and incident beam. R, Z and ψ are analogous coordinates for the diffraction pattern

be expected for the Z values. A helix with 3.61 residues per turn would yield a pattern where 65 residues repeat in 18 turns over a distance of 97.5 Å.

Real helices actually allow such distortions of the pitch to accommodate packing factors and/or chain end defects.

How are molecular structures of fibers deduced from x-ray diffraction patterns? Each crystallographer has his own approach and each fiber presents its own unique problems. Let us consider isotactic polypropylene as a case in point. (See p. 75.) Natta and Corradini elucidated the structure only after extensive work on polymer synthesis; fiber preparation, measurements of degrees of crystallinity, geometric considerations and, finally, detailed examinations of x-ray diffraction patterns of the stretched (oriented) fiber.

As noted above, isotactic polypropylene must form a helix in its crystalline domain. The structural building unit of the polymer

$$\text{CH}_2\text{---CH}$$
$$|$$
$$\text{CH}_3$$

assumes specific geometries that allow a crystallographic repeat to occur further along the polymer chain. This can be accomplished by suitable rotations and translations of chemical bonds along the helix axis (Figure 11.20). Geometric considerations led the scientists to propose a model where the side chain methyl groups are arrayed as far from each other as possible. Such placements generate a helix in which the methyl groups assume positions 120°

from each other as one looks down the helix axis (Figure 11.20). After three monomeric groupings, the structure repeats.

FIGURE 11.20   Helical structure of isotactic polypropylene. A top-view of the helix axis shows that the methyl groups are spaced as far apart as possible—120°

The x-ray pattern of polypropylene fiber contains a prominent meridional reflection at 0.154 Å^{-1} which corresponds to a 6.5 Å repeat distance. This distance is completely consistent with an initial model where the helix contains three monomeric residues per turn.

The unit cell dimensions and density were then calculated. These values can be determined from the diffraction pattern. The meridional reflection at about 6.5 Å provided the unit-cell dimension along the fiber axis while reflections at or near the equator gave the other two dimensions. From the diffraction pattern, the values can be obtained for the angles made by the sides of the unit cell. Table 11.1 shows the results for polypropylene. Since two angles are 90°, the unit cell belongs to the monoclinic crystal class.*

---

* There are seven crystal classes: cubic, monoclinic, triclinic, orthorhombic, trigonal, tetragonal and hexagonal.

TABLE 11.1

Unit Cell Parameters for Polypropylene

Coordinate direction	Axial angles
x(a) 6.666 Å	α 90°
y(b) 20.87	β 98.2°
z(c) 6.488	γ 90°

From these dimensions, the volume of the unit cell can be computed and a density (d') based on a single monomer unit calculated.

$$d' = \frac{\text{mass of a monomer unit}}{\text{volume of the unit cell}}$$

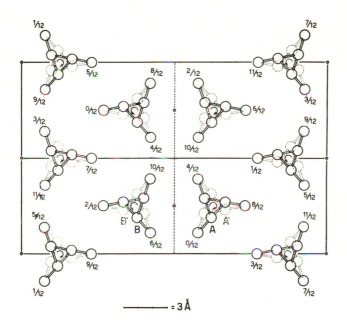

————— = 3 Å

FIGURE 11.21   A model of the packing of polypropylene chains in the unit cell. As can be seen from the center of each chain, the unit cell contains an equal number of right and left handed helices. The numbers in the figure represent relative heights as one traverses the depth of the unit cell. (Taken from G. Natta, P. Corradini and M. Cesari. "The Crystalline Structure of Isotactic Polypropylene," *Att. Acc. Naz. Linc.*, Ser. 8, **21**, p. 365 (1956))

The experimental density was obtained by careful measurement. For poly-propylene it was seen that the ratio

$$\frac{d_{\text{experimental}}}{d'} \cong 12.$$

Therefore there are 12 monomer units in a given cell for polypropylene which was used to calculate an ideal crystallographic density (d cryst) (Figure 11.21).

$$d_{\text{cryst}} = 12 \times d' = 0.9379 \text{ cm}^{-3} \text{ for isotactic polypropylene.}$$

Natta and Corradini then determined the mode of packing of isotactic polypropylene in order to assign the lattice to one of the 230 possible space groups. A space group is defined by the types and numbers of symmetry elements in the crystal lattice. Centers of symmetry, planes of symmetry, rotational axes, and glide planes are among the symmetry operations considered. Knowledge of the space group into which polypropylene falls allowed the scientists to interpret the packing parameters from the diffraction pattern and to propose a model showing how the helical isotactic polypropylene packs into a crystalline region of the fiber.

Nucleic Acid Structure

We can now return to our consideration of fibrous biopolymers. Fibers of nucleic acids received attention more than fifty years ago. It was necessary, however, for the techniques of isolation and chemical characterization to progress to the point where good fibers could be prepared and oriented.

The pioneering x-ray diffraction studies of DNA were performed by Wilkins, Franklin, and their collaborators. DNA fiber samples from various sources give similar x-ray patterns. The sodium salt of DNA exists in two modifications, A and B. Structure A is crystalline and occurs at 75% humidity while structure B is less ordered and is obtained at higher humidities. The transition from A to B is reversible.

We showed earlier that general rules were developed to deduce helical structures from x-ray diffraction patterns from fibers. By applying these rules, Watson and Crick realized that DNA forms helices. Using x-ray data of Wilkins and Franklin, base pairing requirements, and tinker-toy-type models for the chain, they proposed the double helix model for the B-form of sodium DNA (Figure 11.22). The structure consists of two right-handed helical poly-nucleotide chains intertwined to form a double helix (see Chapter III, p. 44).

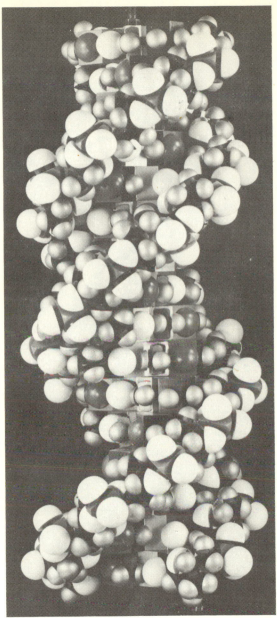

FIGURE 11.22   The double helix structure of DNA proposed by J. D. Watson and
F. Crick. The polynucleotide chains are hydrogen-bonded together by cross-chain
bridges between the pyrimidine and purine bases. (See p. 44)

The phosphate groups are on the outside of the helix and the bases are on the inside. The planes of the stacked base pairs are perpendicular to the helix axis. The axial distance between two base pairs on the same chain is 3.4 Å. There are ten residues per turn and therefore the helical repeat distance is 34 Å. The diameter of the double helix is about 20 Å.

The replication of DNA.

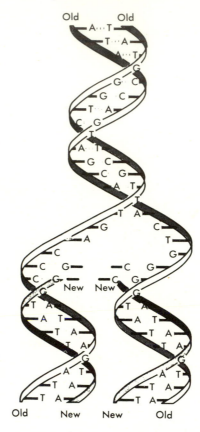

FIGURE 11.23   Representation of the mechanism of replication of DNA. New strands of DNA are formed from a pool of triphosphates and assembled by the enzyme DNA synthetase according to the old DNA double helix which serves as a template

The two strands of the helix are joined by hydrogen bonding between a purine base on one chain and a pyrimidine base on the other chain—guanine pairs with cytosine and adenine pairs with thymine (See Chapter III, P. 44).

The ratios of the base pairs (adenine to thymine, and guanine to cytosine) are almost always unity. It follows that the sequence in one chain automatically determines the sequence in the other chain. This suggests a mode of self-replication of DNA involving the separation of the chains, each chain acting as a template for the formation of its complementary chain (Figure 11.23).

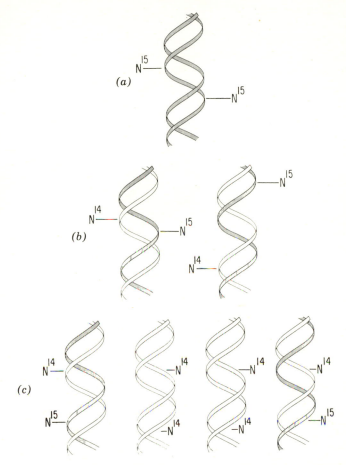

FIGURE 11.24   Schematic representation of the fate of DNA chains labeled with N^{15} as a function of replication with N^{14} building units. [Taken from a paper by M. Meselson and F. W. Stahl, published in *Proc. Natl. Acad. Sci.*, 1958, **44**, p. 671]

The structure of the A-form was also found to consist of a double helix. The pitch of the helix is 28 Å and the vertical distance between bases on the same chain is 2.55 Å. There are thus eleven residues per turn and a diameter of

18 Å. In this form the planes of the base pairs are inclined at an angle of 65° to the helix axis.

The mechanism of DNA replication proposed by Watson and Crick received support from the findings of Meselson and Stahl. Bacteria labeled with N^{15} were transferred to a medium containing N^{14}. In one generation half-labeled molecules of DNA had been formed at the expense of fully labeled molecules. After two generations, the bacteria contained equal amounts of half-labeled and unlabeled DNA. Meselson and Stahl were able to differentiate fully labeled, half-labeled and unlabeled DNA by means of extremely small differences in their densities using density gradient ultracentrifugation. These results demonstrate clearly that the heavy nitrogen of a DNA molecule is divided between two sub-units and after replication one parent sub-unit is present in each daughter molecule (Figure 11.24).

Much work has been carried out on the structure of RNA molecules. Since base pairing is not a general feature of these molecules, it is impossible to obtain highly crystalline fibers from most samples of ribosomal RNA. Diffraction studies by x-rays indicate that the type of RNA called "transfer RNA" has a structure similar to an A-form DNA. The double helix is formed by a single chain folding back on itself. Most structural data for RNA samples have been obtained by indirect evidence gained from solution studies of absorption spectra, light scattering, viscosity and ultracentrifugation experiments.

Mechanisms of Enzyme Action Using Crystal Structures

Lysozyme was the third protein and the first enzyme to have its three-dimensional structure elucidated by x-ray diffraction studies at high resolution. Many other enzymes have now had their structures similarly characterized including ribonuclease, carboxypeptidase, papain, chymotrypsin, and subtilisin. Lysozyme, however, remains the best example of how x-ray diffraction can be employed to explore the nature of enzyme action.

Mucopolysaccharides in bacterial cell walls are lysed or broken down by lysozyme. Phillips and his associates were able to diffuse a number of analogs of the natural substrate into crystals of the protein. These substances are low molecular weight compounds (oligomers) made up of the same building units as are found in the high molecular weight (polymer) natural substrate which is composed of alternating N-acetyl-glucosamine (NAG) and N-acetyl muramic acid (NAM) residues. It turned out that the NAG dimers and trimers form stable isomorphous complexes with lysozyme. Difference electron density maps (Fourier) were calculated between the native protein and a specific protein-inhibitor complex. Larger NAG oligomers are substrates and therefore could

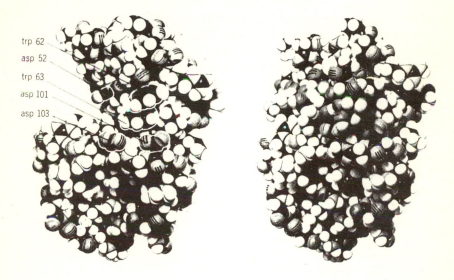

not be used to measure diffraction patterns. The natural substrate is much too large to be diffused into lysozyme crystals. From the above-mentioned protein-inhibitor complexes, the site of binding for substrate and substrate-related inhibitors was located.

Earlier we saw the high resolution structure proposed for lysozyme (see Figure 11.11). The substrate crevice is obvious in Figure 11.25. Difference Fouriers clearly showed that the inhibitors (particularly the $(NAG)_3$ trimer) are bound at one end of the crevice (Figure 11.25).

FIGURE 11.25   Space filling models of lysozyme with and without substrate fitted into the crevice containing the active site. [Taken from R. E. Dickerson and I. Geis. *The Structure and Action of Proteins*, published by Harper and Row, New York, Evanston, and London, 1969]

Model building based on these protein-inhibitor complexes led to a detailed proposal for the binding of substrates to the active site of the enzyme. It should be recalled that most of the nonpolar or hydrophobic amino acid residues are on the inside of the molecule; held that way by specific intramolecular hydrogen bonding involving specific side chain interaction. The majority of the hydrophobic side chains that occur on the outside of the protein line the crevice. Thus the crevice becomes a likely region for the active site.

A plausible mechanism was devised by Phillips after consideration of the possible interactions between substrate and side chains within the crevice. He reasoned that glutamic acid (Glu 35) attracts the glycosidic oxygen linking the NAG to NAM, weakens and then breaks the C—O bond. This leads to a carboxonium ion intermediate. A negatively charged side chain of aspartic acid (Asp 52) helps to stabilize the intermediate. A water molecule is involved to complete the reaction by donating a proton to Glu 35 and an hydroxyl to the carbonium ion intermediate. This process is schematically shown in Figure 11.26.

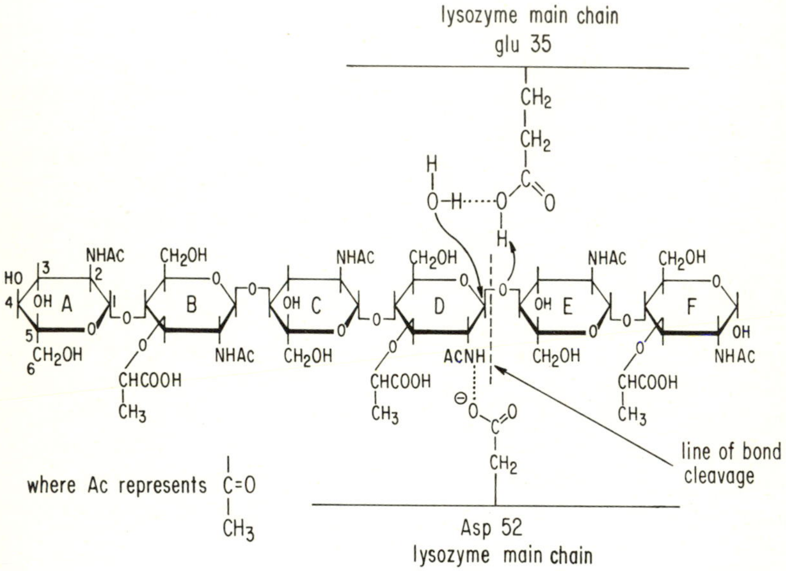

FIGURE 11.26 Schematic representation of the mechanism of polysaccharide substrate cleavage by lysozyme. The polysaccharide is composed of alternating residues of N-acetyl-glucosamine (NAG) and N-acetyl muramic acid (NAM). In this diagram, the cleavage occurs between the NAM (D-ring) and the NAG (E-ring). While the aspartic acid residue at position 52 is interacting with the amide and stabilizing the incipient carbonium ion, the glutamic acid residue at position 35 is able to protonate the glycocidic linkage between rings D and E. This allows a nucleophilic attack by a water molecule at position 1 of the D-ring, thus cleaving the chain. [Adapted from D. C. Phillips, *Sci. Am.*, Nov. 1966, p. 78]

Molecular Evolution

X-ray diffraction studies have created an unexpected entrée into the fascinating field of molecular evolution. The close relationship of the shape of myoglobin with the shape of the subunits of hemoglobin is not fortuitous. Three-dimensional structures for specific proteins from different species together with sequence studies provide a powerful insight into molecular evolution. All enzymes whose structures have been elucidated to date locate their active sites in crevices or clefts with remarkably similar shapes. Why? How did these globular spatial arrangements evolve?

As we saw in Chapter III, living organisms store genetic information in the DNA template. It has also been established that mutations can occur when DNA is irradiated (uv light, x-rays, etc.) or exposed to certain chemical agents. These mutations can lead to a cessation of vital biological functions; to altered sequences in the biosynthesis of proteins and other natural products; or even to the development of entirely new biological systems. It depends on the nature and number of the altered genes.

We can state with some certainty that most mutations do not improve the species. The vast majority of mutations lead nowhere and the altered organisms die out. Where beneficial mutations occur, they are few in number and conservative in that similar amino acids substitute for each other in a protein sequence. It is clear that such mutations provide a mechanism over evolutionary time spans for organisms to adapt to changing environments. Bacteria, for example, that show a resistance to an antibiotic do so by having developed mutant strains. Short-winged flies that survive the high winds of certain Pacific Ocean islands evolved by mutation. The older long-winged varieties were selected against by being blown out to sea and drowning. From bacteria to man, mutations were and are the molecular means for biological change.

Hemoglobin and cytochrome c represent proteins widely distributed among the phyla. X-ray diffraction and sequence studies for these proteins from many species allow us to trace the molecular evolution of both.

In mammals, hemoglobin is the oxygen carrier of red blood cells, while the closely related myoglobin is the oxygen carrier of muscle. The molecular structure and sequences of hemoglobins are now available from an insect, a marine annelid worm, and a sea lamprey (in addition to the human and horse varieties). Although their sequences differ from each other and from mammalian-derived molecules, the hemoglobins all possess the myoglobin fold (crevice). The obvious question is whether they evolved from a common precursor, an oxygen-carrying protein.

Biologists have shown that hemoglobins occur sporadically in the invertebrate phyla. Because of this, it is probable that oxygen-binding proteins developed independently many times. Each time nature utilized very similar and efficient heme-containing proteins with pockets or folds. Such crevices protect the heme and metal and create a proper hydrophobic binding site for oxygen and/or carbon dioxide. Alternative systems to transport oxygen have evolved in those invertebrates which do not have hemoglobin.

If the sequences of myoglobin (whale) and hemoglobin (horse and human) are compared, it seems apparent that both developed from a common ancestral globular protein. Figure 11.27 contains a suggested picture for these evolutionary differentiations.

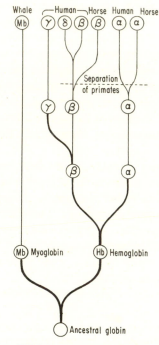

THE EVOLUTION OF THE GLOBINS

FIGURE 11.27   Because of the similarity of amino acid sequence and three-dimensional structure, hemoglobin and myoglobin are thought to have evolved from a common ancestral globin. [Taken with permission from *The Structure and Action of Proteins*, by R. E. Dickerson and I. Geis, published by Harper and Row, New York, Evanston, and London, 1969]

Through the world-wide collection of human blood samples, nearly one hundred different mutant hemoglobins have been uncovered. As a result, a picture of the molecular pathology is evolving. It appears that surface amino acids are readily mutated without generally affecting the biological function of hemoglobin. Mutations near the heme group, in the internal nonpolar regions of the protein and/or at the contacts among the subunits drastically affect the biological functions of hemoglobin.

Perutz and Lehman classified many mutations and correlated their findings with conformational changes and clinical consequences. For example, they reported that replacement of a specific internal phenylalanine which is in contact with the heme group by a valine causes the heme group to drop out of the protein. This pathological mutant is known as hemoglobin Torino and results in inclusion body anemia.* In another case, hemoglobin Boston arises when a histidine that interacts with the iron is replaced by a tyrosine. The hemoglobin of this mutant is difficult to reduce and causes cyanosis† and methemoglobinanemia.‡ It has been suggested from x-ray studies that the heme group is displaced somewhat in hemoglobin Boston.

Perutz and Lehman also considered mutations in which contacts between subunits are altered. Hemoglobin Kansas represents a mutation in which an asparagine is replaced by a threonine. In the oxy form, the protein dissociates to $\alpha\beta$ dimers (from the normal $\alpha_2\beta_2$ tetramer) leading to the pathological condition of cyanosis.

Perhaps the most widely known and studied hemoglobin mutation is hemoglobin S found in patients with the hereditary disease called sickle cell anemia. In this form, a valine replaced a specific glutamic acid on the exterior of the protein. The genetic trait is recessive. Therefore the clinical symptoms will only appear in homozygotes.§ The appearance of sickle cells is caused by the lowered solubility of the deoxyhemoglobin S. The non-polar side chain

---

* Anemia is a condition of the blood wherein red blood cells are defective in the oxygen and/or carbon dioxide transporting properties. One form of anemia results from production of iron granules or inclusion bodies. When iron "drops out" of a mutant hemoglobin, it aggregates. Oxygen transport is blocked and a severe anemia results.

† Cyanosis is a disease in which the skin and mucous membranes appear blue, especially in the extremities. It is caused by deficient oxygenation of the blood. Often the malady is directly proportional to the concentration of reduced hemoglobin. The disease can also come from other forms of hemoglobin in which oxygen transport is blocked or inhibited.

‡ Methemoglobin builds up from the oxidation of the heme group of hemoglobin. This compound cannot transport oxygen and leads to the disease, methemoglobinanemia. The most obvious clinical sign of this affliction is cyanosis.

§ A homozygote for a given trait has identical genes at the position controlling that trait on both chromosomes of a similarly shaped pair.

of the valine induces fibrous aggregation of the deoxyprotein which leads to the sickling of the red cells and early death.*

X-ray analysis of the oxygenated hemoglobin S shows it to be essentially identical in structure to normal oxygenated hemoglobin. The insolubility of deoxyhemoglobin S may not be caused by conformational changes but rather by nonpolar surface associations caused by the presence of the valine side chains.

Many other mutant hemoglobins have been described; none with oxygen transport properties superior to normal hemoglobin. At best the mutations appear harmless. In many cases they are pathological. We are not able to treat people with the genetic defects leading to impaired hemoglobins. The suggestion of transducing healthy genes remains but a long-term possibility.

Cytochrome c represents another protein whose molecular evolution has been studied. It is a relatively small protein (molecular weight about 12,000) containing an iron and a heme group (much like hemoglobin). Cytochrome c is an electron carrier involved in the conversion of energy (Chapter II, p. 39). This protein is found in all living organisms.

Margoliash and his associates have compared sequences of cytochrome c from 35 different species. Most changes of sequence have occurred on the surface of the protein and have been highly conservative. Some sequences are completely invariant over all species. It appears that mutations arising in these regions have always been lethal. Cytochrome c from one source when purified can be used in *in vitro* experiments with cytochrome system enzymes from a completely different source (e.g., yeast and horse derived materials). This shows that the essential features of cytochrome c function have remained nearly constant over the billion years of evolution.

X-ray diffraction results have begun to appear on cytochrome c derived from horse heart, bonita and tuna. The heme group resides in a cleft at right angles to the surface. In myoglobin and hemoglobin, the heme groups are parallel to the surfaces. The heme group in cytochrome c is firmly attached to the protein chain by means of two cysteine and one histidine residues. Histidine and methionine form other attachments to the heme group as ligands in the extraplanar positions. Such a "closed" structure cannot complex with small molecules. The function of this protein is to oxidize or reduce other molecules in an energy transport system.

Protein sequence comparisons for cytochrome c have been reported and a family tree for the evolution of the protein proposed. If sequence differences alone are considered, we see greater variability among the fungi than between

* Sickle cell anemia appears to have evolved in countries where the blood disease, malaria, is prevalent (see p. 98). The trait in its heterozygous form provides a selective advantage against the disease. In Americans, the trait is found almost exclusively amongst Blacks.

insects and vertebrates. It is not at all clear where to locate the apex of an anthropocentric scale of values based on cytochrome c. Is man at the top?

SUMMARY

Throughout this chapter our attention has centered on molecular structure. We developed the theme of x-ray diffraction analysis and demonstrated how the structures of small and large molecules were elucidated. Heavy emphasis was placed on molecules of biological interest such as proteins and nucleic acids. We proceeded to consider mechanisms of enzyme action based on models derived from analysis of three-dimensional x-ray diffraction patterns. By use of such models, together with protein sequence studies, it was possible to propose pathways for molecular evolution and briefly discuss some aspects of mutations and molecular disease.

SOURCE MATERIALS AND SUGGESTED READING

Dickerson, R. E., and I. Geis, 1969. *The Structure and Action of Proteins*, Harper and Row, New York, Evanston, and London.

Holmes, K. C., and D. M. Blow, 1965. *The Use of X-ray Diffraction in the Study of Protein and Nucleic Acid Structure*, Interscience Publishers, New York.

Kendrew, J. C., 1963. "Myoglobin and the Structure of Proteins," *Science*, **139**, part 2, pp. 1259–1266.

Natta, G., and P. Corradini, 1959. "Conformation of Linear Chains and Their Mode of Packing in the Crystal State," *J. Poly. Sci.*, **39**, pp. 29–46.

Neurath, H., ed., 1964. *The Proteins*, Vol. 2, second edition, Academic Press, New York and London.

Perutz, M. F., 1962. *Proteins and Nucleic Acids: Structure and Function*, Elsevier Publishing Co., Amsterdam and New York.

Perutz, M. F., 1963. "X-ray Analysis of Hemoglobin," *Science*, **140**, pp. 863–869.

Phillips, D. C., 1966. "The Three-dimensional Structure of an Enzyme Molecule," *Sci. Am.*, **215**, no. 5, pp. 78–90.

Taylor, C. A., and H. Lipson, 1964. *Optical Transforms*, Cornell University Press, Ithaca and New York.

Watson, J. D., 1968. *The Double Helix*, Atheneum Press, New York.

THE LANGUAGE OF ORGANIC CHEMISTRY
Structure and Function of Molecules

I. The Heart of the Atom

A. THE NUCLEUS AND THE PLANETARY ELECTRONS

Spectroscopists have provided much of the information on which our present understanding of atomic structure is based. We know that atoms are composed of a nucleus containing positively charged protons and uncharged neutrons. The planetary electrons revolve about the nucleus in specific orbits. The electrons are said to be quantized since definite energy differences exist among the orbits. The energy of the electrons increases with the distance of the orbit from the nucleus.

B. ORBITALS

Three scientists drastically altered the thinking about atomic structure through their work in the first two decades of this century. Niels Bohr of Denmark suggested the circular orbit of the planetary electrons. His theory worked well with hydrogen but exhibited serious defects for all other atoms because the motion of the electrons within each orbit was too rigidly defined. Werner Heisenberg of Germany realized that electrons move with wavelike patterns. Their positions could never be known exactly since extremely short-wavelength probes are required to place the electrons in space and time. These high energy photons cause excitation of an electron upon collision and thus change the electron's position. This limitation in the accuracy of locating any object exactly is known as the **Heisenberg uncertainty principle.** Erwin Schroedinger, also of Germany, formulated wave equations to describe the motion of the planetary electrons. In this manner, the severe limitations of the Bohr highly-defined, circular orbits were removed. Electrons are viewed as being distributed in orbitals or groups of orbitals. Strictly speaking, the term "orbital" is a mathematical representation for the description of the motion of an electron as it revolves about the nucleus. In a descriptive sense, we consider that orbitals preexist and that they define the locations where the planetary electrons are most likely to be found. The orbitals are grouped into shells and subshells which are distinguished from each other by the character of the motion (quantum numbers) of the electrons as they move around the nucleus.

The motion of an electron about a nucleus can be characterized by several quantum numbers. The principal quantum number, n, is a rough measure of the size of the electron cloud or the energy level of the orbit. It can assume a whole-number value 1, 2, 3, 4, ... but may not be zero. Spectroscopists have called these energy levels, shells, and denoted them by letters $K, L, M, N, \ldots$ in order of increasing energy.

TABLE A-1

Interrelationship of Atomic Quantum Numbers

Quantum number	Shell					
	K	L		M		
n	1	2		3		
l	0	0	1	0	1	2
m	0	0	$+1, 0, -1$	0	$+1, 0, -1$	$+2, +1, 0, -1, -2$
Symbol for electrons	$1s$	$2s$	$2p$	$3s$	$3p$	$3d$

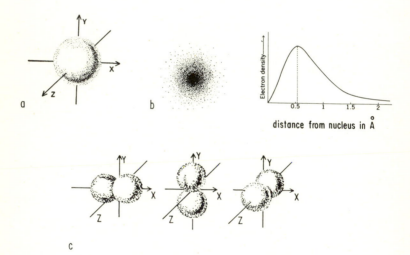

FIGURE A-1   Representations of the electron density distribution for electrons in the s and p orbitals. (a) Spherical distribution of an s electron in space. (b) The s electron density and its dependence upon the distance from the nucleus. (c) The three $2p$ electron distributions in space

The secondary quantum number, l, is a measure of the shape of the electron cloud, i.e., spherical, dumbbell-shaped, etc. The values assumed can be 0, 1, 2, 3, ..., $(n-1)$. The magnetic quantum number, m, depicts the orientation of the electron cloud in space. The values for this aspect of electronic distribution involve whole integers from $+l$ to $-l$. Lastly, there is a spin quantum number, s, which indicates the spin of the electron. Values of $+\frac{1}{2}$ or $-\frac{1}{2}$ are allowed.

Electrons with the same n belong to the same shell. Electrons with the same n and l are in identical subshells while electrons with the same n, l and m are in the same orbital.

From Table A-1 it is possible to see the relationship of the various quantum numbers in relation to each other. Figure A-1 contains representations of the distributions of electronic densities for the various electrons shown. Electrons in s orbitals are spherically symmetrical about the origin (Figure A-1a). The electron cloud and a plot of the electronic density of an electron as a function of the distance from the nucleus of an s-electron are shown in Figure A-1b. The perspective representations of the three p orbitals, which have nodes at the origin but are equally distributed on the X, Y, and Z axes, are shown in Figure A-1c. Since electrons within the same orbital (i.e., with the same n, l, and m) can have spins of either $+\frac{1}{2}$ or $-\frac{1}{2}$, only two electrons of opposite spin can be placed in each orbital. Such electrons are said to be paired. The concept for

TABLE A-2

Electronic Configurations for Atoms in the First Three Rows of the Periodic Chart

Element	K shell	L shell				M shell			
	$1s$	$2s$	$2p_x$	$2p_y$	$2p_z$	$3s$	$3p_x$	$3p_y$	$3p_z$
H	↑								
He	↓↑								
Li	↓↑	↑							
Be	↓↑	↓↑							
B	↓↑	↓↑	↑						
C	↓↑	↓↑	↑	↑					
N	↓↑	↓↑	↑	↑	↑				
O	↓↑	↓↑	↓↑	↑	↑				
F	↓↑	↓↑	↓↑	↓↑	↑				
Ne	↓↑	↓↑	↓↑	↓↑	↓↑				
Na	↓↑	↓↑	↓↑	↓↑	↓↑	↑			
Mg	↓↑	↓↑	↓↑	↓↑	↓↑	↓↑			
Al	↓↑	↓↑	↓↑	↓↑	↓↑	↓↑	↑		
Si	↓↑	↓↑	↓↑	↓↑	↓↑	↓↑	↑	↑	
P	↓↑	↓↑	↓↑	↓↑	↓↑	↓↑	↑	↑	↑
S	↓↑	↓↑	↓↑	↓↑	↓↑	↓↑	↓↑	↑	↑
Cl	↓↑	↓↑	↓↑	↓↑	↓↑	↓↑	↓↑	↓↑	↑
A	↓↑	↓↑	↓↑	↓↑	↓↑	↓↑	↓↑	↓↑	↓↑

Doublet	Octet	Octet

this electronic structural limitation has come to be known as the **Pauli exclusion principle** after the scientist who first postulated the idea. Another scientist, Hund, formulated a rule which stated that orbitals of equal energy, such as $2p_x$, $2p_y$, and $2p_z$, must each fill with one electron before pairing is allowed. Table A-2 contains the electronic pictures of the atoms in the first three rows of the periodic chart. It is clear that both the Pauli exclusion principle and **Hund's rule** are obeyed.

II. Chemical Bonding

A. IONIC BONDS

The stability of ionically bonded molecules such as sodium chloride arises from simple electrostatic interactions between the positively charged sodium and the negatively charged chlorine.

$$Na^\oplus \; :\!\ddot{\underset{..}{Cl}}\!:^\ominus$$

Ionic bonds form when elements on the left side of the periodic table (i.e., the electropositive alkali and alkaline earth metals) react with those on the right side of the periodic table (i.e., the nonmetallic oxygen and halogen elements). Individual molecules do not exist for ionic compounds in the solid state because aggregates are constructed by packing positively and negatively charged ions in regular arrays.

A great American chemist, G. N. Lewis, introduced the idea of the electron pair bond. He deduced that atoms become particularly stable when they possess electronic configurations of the noble gases. The $1s$ electrons of the K shell are complete with the paired electrons of the noble gas, helium (cf. Table A-2). Thus, lithium and sodium achieve the noble gas configurations of helium and neon by donating their $2s$ and $3s$ electrons respectively to elements such as the halogens. By accepting these electrons, the halogens also achieve the noble gas electronic configurations. Fluorine resembles neon while chlorine becomes like argon (cf. Table A-2).

B. COVALENT BONDS

1. *σ-bonds (single)*

G. N. Lewis also developed the concept of the sharing of electrons to achieve stable electronic configurations. From this idea, we have our present views of bonding for most organic molecules. It is useful to regard the molecular bonding orbital as the overlap of two atomic orbitals. Thus, two hydrogen atoms can form a hydrogen molecule by pairing their $1s$ electrons with opposite spin. Figure A-2 shows a schematic picture of this process.

The bonds which are formed by increased electronic density between the two nuclei are called sigma (σ) bonds (cf. Figure A-2b). It is, of course, also possible to form σ-bonds by joining s and p orbitals or two p orbitals. Before discussing such bonding, we will digress to consider the phenomenon of **orbital hybridization.**

If the electronic configuration of carbon, as shown in Table A-2, is used to form bonds, two ionic or covalent bonds would be expected. A more favorable electronic arrangement can be devised whereby four covalent bonds can be formed. One of the two $2s$ electrons is unpaired from the other and promoted to the empty $2p_z$ orbital. This requires energy. A new tetrahedral orbital

a   b

FIGURE A-2   Pictorial representation of the covalent bond joining two hydrogen atoms into a hydrogen molecule. (a) Schematic drawing of overlap. (b) Electron density

system is created by raising the energy of the $2s$ orbital and lowering the energy of the three $2p$ orbitals. In this manner, four isoenergetic orbitals are established with energies between the original s and p orbitals:

energy

$2p_x$ $2p_y$ $2p_z$

$2s$

Original s and p orbital states

sp^3 hybridized orbitals

In accordance with Hund's rule, each hybridized orbital contains only one electron. The net small requirement for the input of energy to form the sp^3 hybridized orbitals is more than offset by the formation of four covalent bonds. When four σ-bonds are formed with atoms such as hydrogen, 397 kcal/mole are liberated. Were carbon to form two bonds from the original unhybridized p orbitals, a maximum of 198 kcal/mole would be expected. Hybridization is thus favored since more energy is liberated.

The sp^3 orbitals are symmetrically oriented about the origin in a tetrahedral array as is shown in Figure A-3.

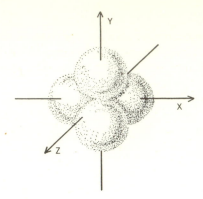

FIGURE A-3   Representation of the orientation of the sp^3 hybrid orbitals in space

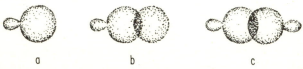

FIGURE A-4   Representation of a single sp^3 orbital (a) and its overlap with an s orbital (b) and another sp^3 hybridized orbital (c)

FIGURE A-5   Representative σ-bonding for compounds of beryllium (a) and boron (b)

Returning to the structure of σ-bonds, we can see that hybridized orbitals can share electrons in much the same manner as the hydrogen atoms shown in Figure A-2. A single sp^3 orbital is oriented in space as shown in Figure A-4a. Most of the electron density exists on one side of the nucleus although a small amount is actually found on the other side. Bonding is achieved when overlapping occurs so that the electrons can be shared.

Of course, other elements can also make σ-bonds. Beryllium hybridizes an s and a p orbital and thus forms linear molecules. Boron hybridizes one s and two p orbitals to form planar compounds with bond angles of 120° as seen in Figure A-5.

2. π-bonds (multiple)

Unsaturated compounds involve another type of molecular bonding in which the maximum electron densities occur above and below the plane joining the nuclei. These linkages are called π-bonds. They are best illustrated by molecules such as ethylene and acetylene. In both cases, σ-bonds are made from hybridized orbitals to join the atoms of the molecules. For ethylene, four C—H and one C—C σ-bonds are formed from sp^2 hybridized orbitals. This leaves an unshared electron in the remaining p orbital on each carbon as is shown in Figure A-6a. All the bond angles in ethylene are equal to 120°. The

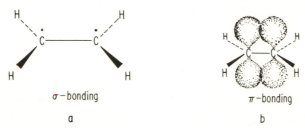

FIGURE A-6   Representation of the σ- and π-bonding for ethylene

FIGURE A-7   Representations of the π-bonding in ethylene (a) acetylene (b)

atoms lie in one plane with the unshared electrons either above or below this plane. The unshared electron densities actually overlap. As a result, bonding can occur if the spins of the electrons are opposite to each other. A delocalized molecular orbital is created above and below the plane of the carbon and hydrogen atoms which is called a π-bond (see Figure A-7a).

Similar considerations apply in the case of acetylene where a linear bonding exists because of the use of sp hybridization. The remaining unshared electrons can overlap above and below, in front and behind the linear array of σ-bonds in acetylene. This leads to a cylindrical form for the π-bonding (cf. Figure A-7b).

There are, of course, numerous additional classes of molecules containing π-bonds. We will consider many of them below.

III. Bonding of the Carbon Atom

A. CARBON–HYDROGEN AND CARBON–CARBON BONDING

The carbon atom is the fundamental building unit for organic molecules. To obtain a filled valence shell of eight electrons, the carbon atom must share electrons with four hydrogen atoms.

$$
\cdot \overset{\displaystyle \cdot}{\underset{\displaystyle \cdot}{C}} \cdot \quad + \quad 4H \cdot \quad \longrightarrow \quad H : \overset{\displaystyle H}{\underset{\displaystyle H}{\overset{..}{C}}} : H
$$

The hybridization and sharing processes fill the carbon atom's valence shell, and also the hydrogen atoms' valence shells. This constitutes an extremely stable electronic configuration and a stable molecule, CH_4, called methane.

As the symbol-dot representation indicates, the electrons are shared in pairs. Each pair of shared electrons comprises a covalent bond (see p. 282). For convenience, each shared pair, or covalent bond, is usually represented as a line connecting the bonded atoms.

$$
H : \overset{\displaystyle H}{\underset{\displaystyle H}{\overset{..}{C}}} : H \quad = \quad H - \overset{\displaystyle H}{\underset{\displaystyle H}{C}} - H \quad = \quad CH_4 \quad = \quad \text{methane}
$$

The four lines surrounding the carbon atom represent the tetravalent nature of the carbon atom. Similarly, the single line between each hydrogen atom and the carbon atom represents the monovalent nature of each hydrogen atom.

None of these simple representations, however, indicates the real, three-dimensional geometry of the methane molecule. Methane is tetrahedral because of hybridization. Its bond angles (HCH) are all 109.5° (cf. Figure A-8). The tetrahedral geometry allows each hydrogen atom to remain as far away as possible from the nearest neighboring hydrogen atom. This minimizes the repulsive forces between neighboring hydrogen atoms.

FIGURE A-8   The tetrahedral structure for methane

The carbon–hydrogen covalent bonds in methane are strong bonds. To pull one gram molecular weight (1 mole) of methane (16 gm) apart into its constituent carbon and hydrogen atoms would require 404 kilocalories of energy. But in real life, how big is a mole of methane and how much energy is 404 kilocalories? Methane is natural gas, and a mole of any gas at ordinary temperature and pressure occupies about 1 cubic foot. One kilocalorie of energy is enough energy to heat a little more than a quart of water 1 degree centigrade. Therefore, the energy released in the formation of a cubic foot of methane from its atoms would heat about a gallon of liquid water from ice temperature to boiling.

Since there are four carbon–hydrogen bonds in methane, each has an average bond strength of 101 kcal per mole. This is considered to be a very strong covalent bond.

1.   *Alkanes*

If one hydrogen atom is removed from a molecule of methane, the remainder of the molecule is called a methyl radical. A radical is anything which contains an unpaired, unshared electron.

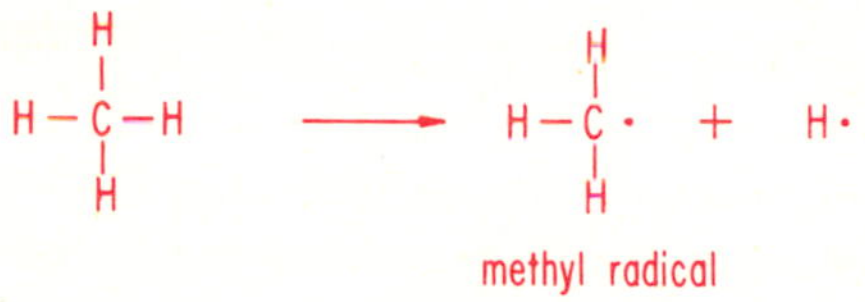

methyl radical

In this sense, a hydrogen atom is a radical, too, but single atoms are usually referred to simply as atoms rather than as radicals.

If two methyl radicals come together, they couple to form a carbon–carbon bond, and also a new molecule called ethane.

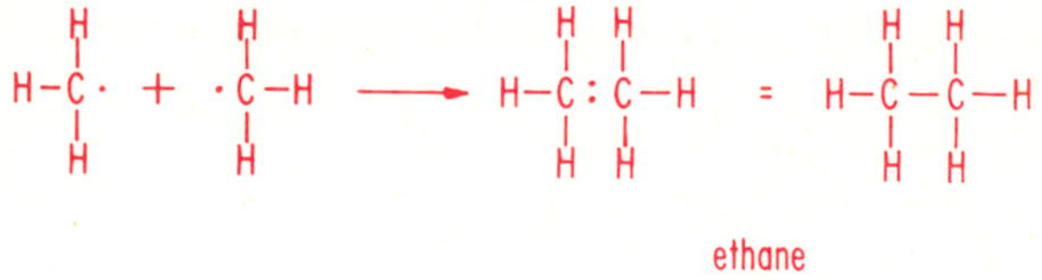

ethane

The carbon–carbon bond in ethane has a strength of 83 kcal per mole. This is also considered a strong bond.

If the radical-coupling process above were repeated on ethane, the product would be a four-carbon molecule called butane.

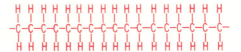

ethyl radical

butane

Continued repetition of this process would eventually produce an extremely long chain of CH_2 units with a methyl unit at each end.

As far-fetched as this proposition may sound, the macromolecule represented above is the now common material known as polyethylene. In building the chain from a few carbon atoms to more than four, the material changes from a gas to a volatile liquid. More carbon atoms in the chain further change the material to an oil, then to a wax at about 18 carbon atoms. The thousands of carbon atoms in polyethylene result in a widely-used plastic. Such long arrays of covalent bonds produce a material strength which is impossible to find in short-chain molecules. Of all the elements in the periodic chart, only carbon atoms form such strong bonds with each other and with other elements. This is why the number of possible organic molecules is practically unlimited, and why more carbon compounds exist than all compounds of all other elements. In Table A-3, we present a listing of some of the unbranched alkanes. Their molecular formulas follow the general relationship of C_nH_{2n+2}. Common names are used for the first four alkanes, i.e., methane, ethane, propane, and butane. Beyond this point, Roman numerical designations indicate the number of carbons in the chain. The n indicates that the molecule is unbranched.

Many alkanes can be formed in which the carbon chains are branched. At least four carbons are necessary for alkanes to exhibit such **isomerism**. (Chemical isomers have the same empirical formula, but different structures.) Table A-4 contains some representative examples of branched alkanes.

TABLE A-3
Unbranched Alkanes

Name	Formula	Bp	Mp	State at 25°C and 1 atmosphere
Methane	CH_4	−162°C	−183°C	Gas
Ethane	C_2H_6	−89	−183	Gas
Propane	C_3H_8	−42	−187	Gas
n-Butane	C_4H_{10}	0	−138	Gas
n-Pentane	C_5H_{12}	36	−130	Liquid
n-Hexane	C_6H_{14}	69	−95	Liquid
n-Heptane	C_7H_{16}	98	−91	Liquid
n-Octane	C_8H_{18}	126	−57	Liquid
n-Nonane	C_9H_{20}	151	−54	Liquid
n-Decane	$C_{10}H_{22}$	174	−30	Liquid
n-Undecane	$C_{11}H_{24}$	196	−26	Liquid
n-Dodecane	$C_{12}H_{26}$	216	−10	Liquid
n-Tridecane	$C_{13}H_{28}$	235	−6	Liquid
n-Tetradecane	$C_{14}H_{30}$	254	6	Liquid
n-Pentadecane	$C_{15}H_{32}$	271	10	Liquid
n-Hexadecane	$C_{16}H_{34}$	287	18	Liquid
n-Heptadecane	$C_{17}H_{36}$	302	22	Liquid
n-Octadecane	$C_{18}H_{38}$	316	28	Solid
n-Nonadecane	$C_{19}H_{40}$	330	32	Solid
n-Eicosane	$C_{20}H_{42}$	343	36	Solid
n-Heneicosane	$C_{21}H_{44}$	356	40	Solid
n-Docosane	$C_{22}H_{46}$	369	44	Solid
n-Tricosane	$C_{23}H_{48}$	380	48	Solid
n-Tetracosane	$C_{24}H_{50}$	391	51	Solid
n-Pentacosane	$C_{25}H_{52}$	402	53	Solid

These branched alkanes are of course isomeric with the linear alkanes of the same number of carbons shown in Table A-3. The number of isomers possible for any given alkane depends on the number of carbons in the molecule. There are 2 butanes, 3 pentanes, 5 hexanes, 9 heptanes, 18 octanes, 35 nonanes and 75 decanes. For the alkanes with 26 carbons, 366,319 isomers are possible.

Cyclic alkanes are also important. These compounds follow the general formula C_nH_{2n}. Representative structures are shown below.

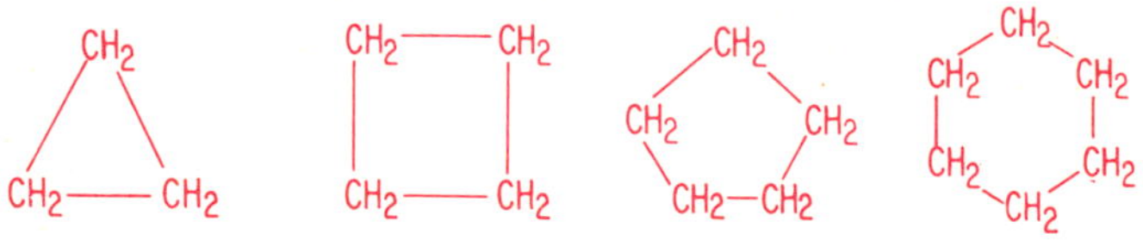

cyclopropane   cyclobutane   cyclopentane   cyclohexane

TABLE A-4

Some Typical Branched Alkanes

Name	Structure	Parent alkane
Isobutane	CH_3 $\mid$ CH_3CHCH_3	Butane
Isopentane	CH_3 $\mid$ $CH_3CHCH_2CH_3$	Pentane
Neopentane	CH_3 $\mid$ CH_3CCH_3 $\mid$ CH_3	Pentane
2-Methylpentane	CH_3 $\mid$ $CH_3CHCH_2CH_2CH_3$	Hexane
3-Methylpentane	CH_3 $\mid$ $CH_3CH_2CHCH_2CH_3$	Hexane
2,2-Dimethylbutane	CH_3 $\mid$ $CH_3CCH_2CH_3$ $\mid$ CH_3	Hexane
2,3-Dimethylbutane	CH_3 CH_3 $\mid$ $\mid$ $CH_3CH—CHCH_3$	Hexane

2. *Alkenes*

The alkene series of hydrocarbons differs from the alkanes in that each member of the family contains a double bond as in ethylene. They therefore have the general formula C_nH_{2n}. If two hydrogen atoms were somehow pulled out of methane by a high energy process, a transitory, highly reactive product, methylene diradical would result.

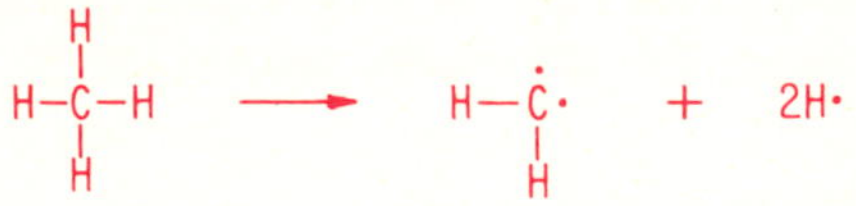

$$H-\overset{\overset{\textstyle H}{|}}{\underset{\underset{\textstyle H}{|}}{C}}-H \longrightarrow H-\overset{\overset{\textstyle \cdot}{}}{\underset{\underset{\textstyle H}{|}}{C}}\cdot \quad + \quad 2H\cdot$$

When two methylene diradicals couple, they form a carbon–carbon double bond and the molecule ethylene.

methylene diradicals   ethylene

As noted above, the double bond of ethylene is constructed from a σ- and a π-bond. π-bonds are weaker than σ-bonds as can be seen by the fact that the **bond strength** of a carbon–carbon single bond is 83 kcal/mole while the total strength for the total carbon–carbon double bond is only 142 kcal/mole. Even though all double bonds are made up of a σ- and a π-bond, they are almost always represented by two linear bonds joining the atoms.

Table A-5 contains a listing of some of the alkenes with double bonds linking carbon atoms 1 and 2 of the chain.

TABLE A-5

Unbranched 1-Alkenes[a]

Name	Formula	Bp
Ethene or ethylene	$H_2C{=}CH_2$	−102.4°C
Propene or propylene	$H_2C{=}CHCH_3$	−47.7
1-Butene or α-butylene	$H_2C{=}CHCH_2CH_3$	−6.5
1-Pentene or α-pentylene	$H_2C{=}CHCH_2CH_2CH_3$	30.1
1-Hexene or α-hexylene	$H_2C{=}CHCH_2CH_2CH_2CH_3$	63.5
1-Heptene or α-heptylene	$H_2C{=}CHCH_2CH_2CH_2CH_2CH_3$	93.1
1-Octene or α-octylene	$H_2C{=}CHCH_2CH_2CH_2CH_2CH_2CH_3$	122.5
1-Nonene or α-nonylene	$H_2C{=}CHCH_2CH_2CH_2CH_2CH_2CH_2CH_3$	146
1-Decene or α-decylene	$H_2C{=}CHCH_2CH_2CH_2CH_2CH_2CH_2CH_2CH_3$	171

[a] The number in front of the name refers to the carbon atom where the double bond is located. The letter α in the other naming system similarly designates the location of the double bond between the first and second carbons.

The alkenes listed in Table A-5 show that the boiling points increase with the increasing number of carbon atoms in the chain. Of course, it is possible to have branching. Internal double bonds can lead to geometric isomerization (see page 305).

a. *Dienes and cyclic compounds* Compounds containing two carbon–carbon double bonds are called dienes; three carbon–carbon double bonds, trienes, etc. Cyclic compounds with double bonds are also possible.

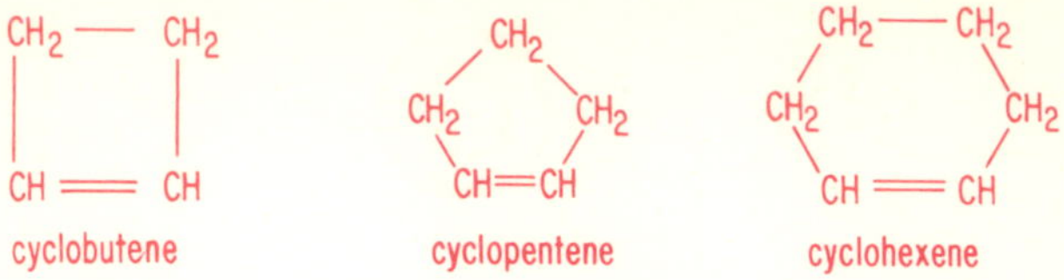

cyclobutene   cyclopentene   cyclohexene

b. *Benzene and resonance* The molecule, benzene, is actually cyclo-hexatriene.

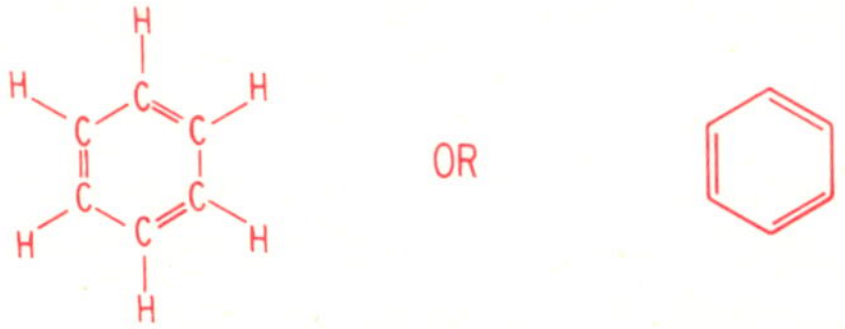

OR

This compound is far more stable than would be predicted from its formula. With this molecule, we encounter π-electron delocalization beyond the two carbon atoms making up the double bond. This kind of delocalization of electrons is called **resonance**. The special stability of resonance arises from the fact that dispersed electron density releases energy. The structure for benzene above may be written in the equivalent form.

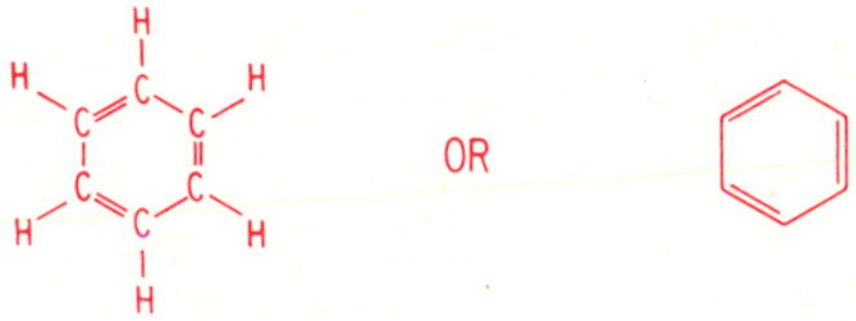

OR

The molecule does not exist in either of these forms but rather as a structure intermediate between them. All the carbon–carbon bonds in benzene are identical, as are the carbon–hydrogen bonds.

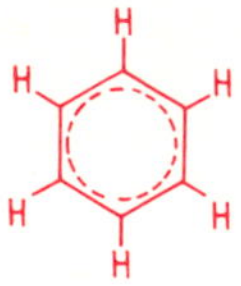

A three-dimensional representation of the benzene structure is shown in Figure A-9.

Alkenes can react with hydrogen in the presence of a catalyst to give alkanes.

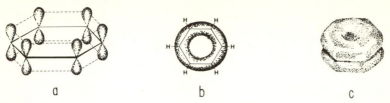

FIGURE A-9. Pictorial representations of the delocalization of the π-electrons in benzene: (a) π-orbitals with indicated interactions; (b) top view of delocalized orbital; (c) perspective view of delocalized orbitals above and below the plane of the σ-bonds

This **hydrogenation** reaction occurs because of the weakness of the π-bond and because of the spatial accessibility of the π-electrons. For example, ethylene adds hydrogen (hydrogenation) to form ethane.

With benzene, the amount of resonance stabilization can be determined by comparing the **heat of hydrogenation** of benzene to yield cyclohexane with the heat of hydrogenation of an ordinary "localized" double bond, such as the one in ethylene. When an ordinary double bond is hydrogenated, it releases almost 30 kcal of energy per mole. Three double bonds then would release almost 90 kcal. When benzene is hydrogenated, it releases only 50 kcal. This means that the three delocalized double bonds in benzene contain almost 40 kcal less energy than three ordinary localized double bonds. This 40 kcal is the stabilization energy of benzene.

If benzene were put together from its constituent carbon atoms and hydrogen atoms, the 40 kcal would be released along with the rest of the energy because of bond formation. Therefore, the carbon–carbon bonds in benzene are 40 kcal stronger collectively than they would be if benzene contained three localized double bonds.

3. *Alkynes*

As has been shown above, carbon atoms can also form triple bonds with each other. A triple bond contains one σ-bond and two π-bonds. This class of compounds is known as alkynes and follows the general formula C_nH_{2n-2}.

As before, each π-bond is really composed of two half-bond portions. With two π-bonds, there are four half-bond portions completely surrounding the σ-bond. They are constructed from one orbital above and below, and another behind and in front. The overall effect resembles a single hollow cylinder of electrons wrapped around the σ-bond (see Figure A-7b). The bond strength of the carbon–carbon triple bond is 186 kcal per mole, indicating that the second π-bond is a weak bond also. Acetylene is somewhat less reactive than ethylene in the usual addition reactions.

B. CARBON–OXYGEN BONDING

Carbon atoms also form single and multiple covalent bonds with other kinds of atoms. The two most common kinds are oxygen and nitrogen atoms.

An oxygen atom has six electrons in its valence shell. It needs two more electrons to complete the shell. An oxygen atom therefore has a valence of two and can satisfy its covalent bonding requirement by forming two single bonds with two other atoms or by forming one double bond with one other atom. Thus, oxygen atoms can form a series of methane derivatives.

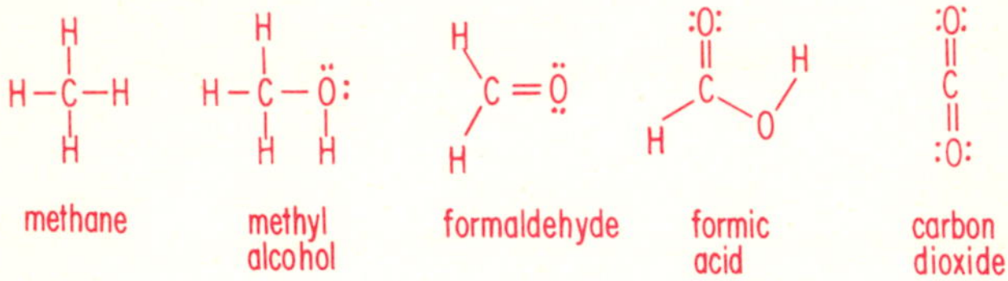

| methane | methyl alcohol | formaldehyde | formic acid | carbon dioxide |

The other possibilities, namely methylene dialcohol, methinyl trialcohol,

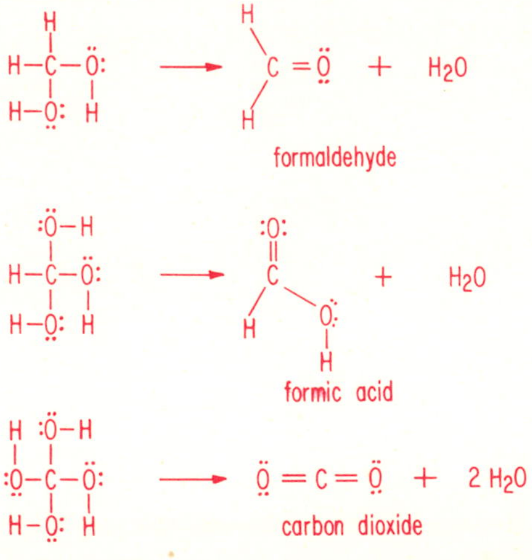

formaldehyde

formic acid

carbon dioxide

and carbon tetra-alcohol, are unstable hydrates of formaldehyde, formic acid, and carbon dioxide, respectively.

As might be expected, the introduction of an oxygen atom into the methane molecule changes the character and behavior of the molecule tremendously. Methyl alcohol bears little resemblance to methane. Methyl alcohol, also called wood alcohol, is a liquid at room temperature and dissolves in water. Methane is a gas, even when chilled to −160°C, and does not dissolve in water. This difference in properties arises from the fact that an oxygen atom is more electronegative than a carbon atom or a hydrogen atom. A difference in electronegativity between bonded atoms causes bond polarization. The more electronegative atom pulls the bonding electrons closer to itself. This produces a slight negative charge, $\delta(-)$, on the more electronegative atom and a slight positive charge, $\delta(+)$, on the less electronegative atom.

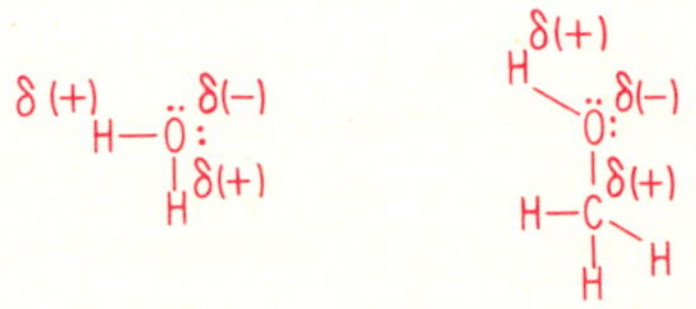

Polar bonds in water and in methyl alcohol

1. *Hydrogen bonding*

Since opposite charges attract each other, the oxygen atom in one water molecule attracts a hydrogen atom in another water molecule. This inter-molecular association, known as hydrogen bonding, sets up a rather tight three-dimensional network of molecules, which accounts for the high boiling points of molecules like water and methyl alcohol.

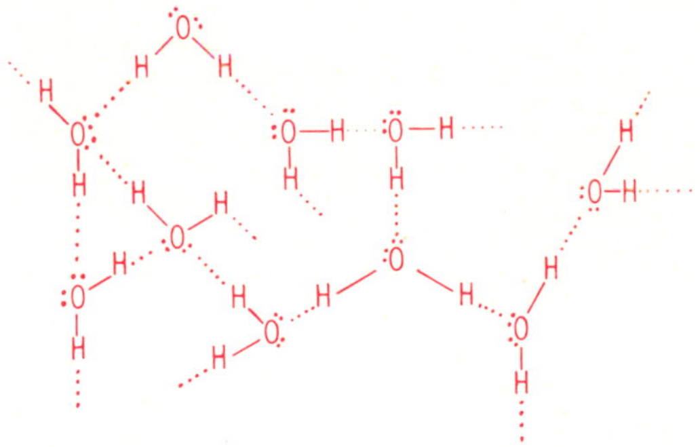

Hydrogen bonding between water molecules

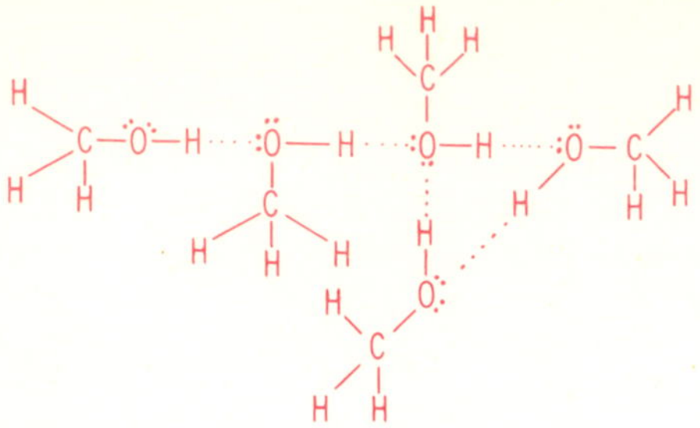

Hydrogen bonding between methyl alcohol molecules

Water and methyl alcohol can also associate with each other, which makes them mutually soluble.

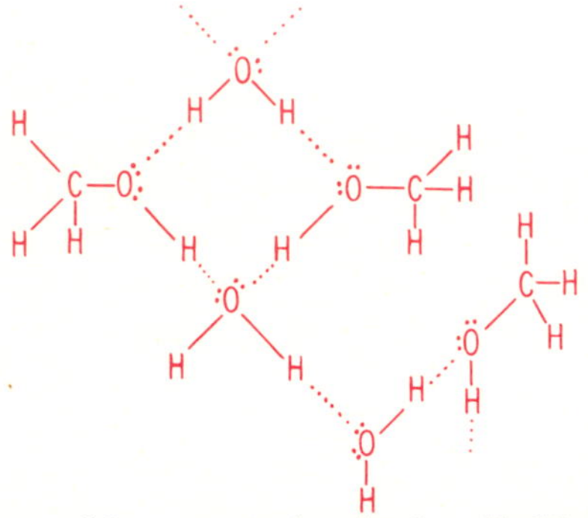

The ability to associate with water molecules is called **hydrophilicity**. Methane does not have this ability, and is therefore called **hydrophobic**. Also, since hydrogen and carbon atoms have similar electronegativities, carbon–hydrogen bonds have little polarity, and intermolecular association is weak. This accounts for the low boiling point of methane, compared with water and methyl alcohol.

Hydrogen bonding also allows hydrogen-ion exchange between molecules. In the networks shown on the previous page, the hydrogens do not belong to any specific oxygen. Since a hydrogen ion is simply a proton, it is extremely mobile and can transfer from one molecule to another via a hydrogen-bond with little difficulty.

Proton exchange through hydrogen bonds

2. *Carbonyl groups*

The situation is somewhat different with formaldehyde. The formaldehyde molecule does not contain an hydroxyl group (O—H group), as do water and methyl alcohol, and so it cannot hydrogen bond with another formaldehyde molecule. However, it does contain a polar carbon–oxygen double bond. This provides the molecule with weak intermolecular attraction, and a receptor site for a hydrogen bond.

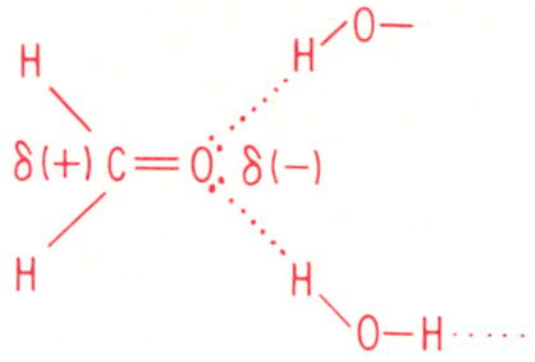

The carbon–oxygen double bond is called a carbonyl group. Its geometry is the same as that of a carbon–carbon double bond, that is, the bond angles in formaldehyde are 120°, and all the atoms lie in the same plane.

The next molecule in the series, formic acid, contains an hydroxyl group and a carbonyl group. This combination on the same carbon atom is called a carboxyl group, and the molecule is a carboxylic acid. The carboxyl group engages in strong bimolecular hydrogen bonding.

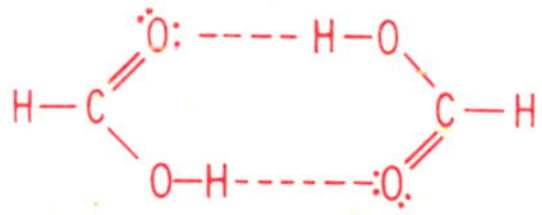

It is also extremely hydrophilic, since it contains both the hydrophilic groups seen in methyl alcohol and in formaldehyde.

This special combination of oxygen atoms on the same carbon atom gives the molecule a special power—the power to release a proton (H^+) much more easily than water does. A molecule which is a stronger proton donor than water is called an **acid**, hence the name formic acid.

The principal reason for the acidity of the carboxyl group is that release of a proton leaves a very stable negative ion, the carboxylate anion.

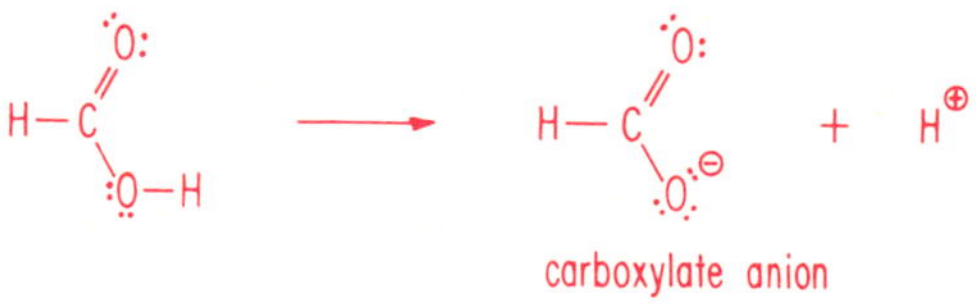

carboxylate anion

The carboxylate anion is unusually stable because both oxygen atoms bear an equal share of the negative charge simultaneously. For this reason, a more accurate representation of the carboxylate anion would show each oxygen atom with half the charge, and each oxygen atom bonded to the carbon atom by equal bonds intermediate between a single and a double bond.

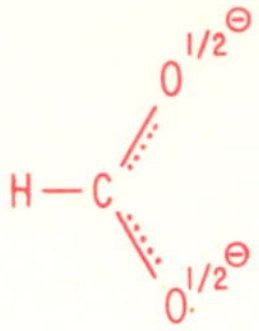

Carboxylate anion as it really exists

3. *Resonance*

The kind of delocalization of electrons mentioned above is called resonance (see p. 292). The special stability of resonance arises from the fact that dispersal of charge releases energy. This may be seen more clearly from the converse fact that to concentrate charge at a point requires an input of energy.

4. *Carbon dioxide*

The fourth and final oxygen-containing derivative of methane is carbon dioxide. This molecule contains four carbon–oxygen bonds (two σ-bonds and two π-bonds), and therefore represents the maximum degree of oxidation possible for a carbon atom. When an organic molecule undergoes complete combustion, every carbon atom in the molecule is converted into a molecule of carbon dioxide. This makes it easy to tell how much carbon is present in a sample of organic material—simply burn the sample and weigh the carbon dioxide produced. The weight of the carbon atoms in the sample equals the weight of carbon dioxide times the ratio C/CO_2, which equals 12/44.

Carbon dioxide is like formaldehyde in that it contains carbonyl groups but no hydroxyl groups. Therefore, it can accept hydrogen bonds, but it cannot donate them. Even so, like formaldehyde, it is a hydrophilic molecule. When carbon dioxide, a gas at room temperature, dissolves in water, it forms an unstable hydrate usually referred to as carbonic acid.

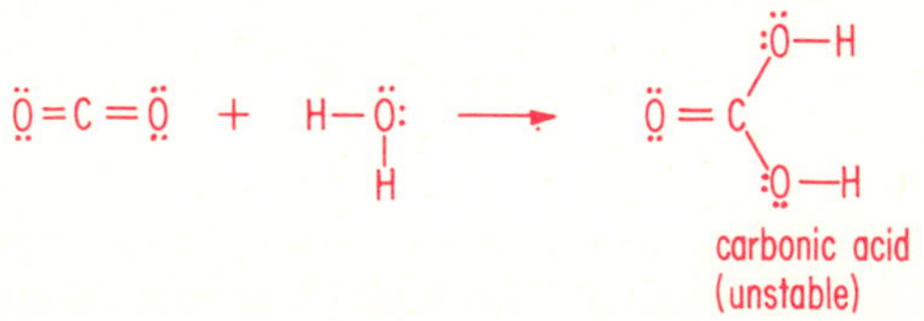

carbonic acid
(unstable)

As in the case of the hypothetical molecules seen earlier—methylene dialcohol, methinyl trialcohol, and carbon tetra-alcohol—carbonic acid is unstable because it contains more than one hydroxyl group on the same carbon atom. However, the sodium salts of carbonic acid, namely sodium bicarbonate and sodium carbonate, are stable and can be isolated.

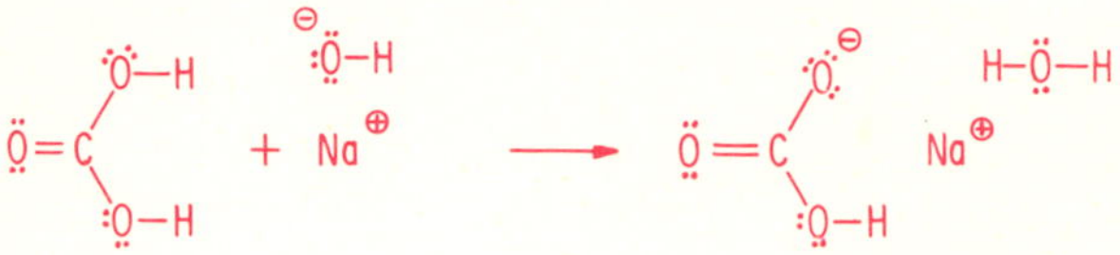

sodium bicarbonate

sodium carbonate

Like the carboxylate anion, the carbonate anion, CO_3^{--}, is stabilized by resonance, usually represented by double-headed arrows between *non-delocalized* structures which *do not actually exist*.

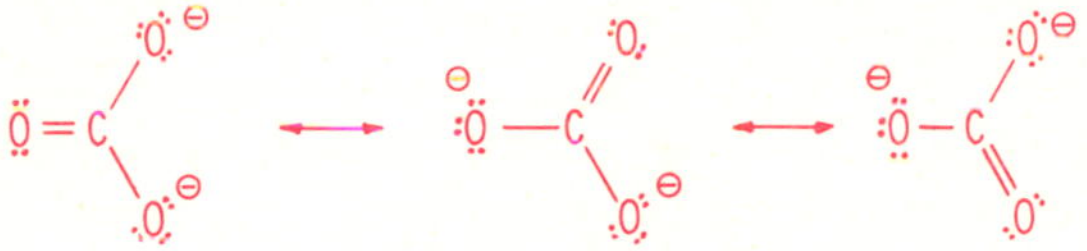

Representation of resonance in the carbonate anion

In reality, the anion contains three equal carbon–oxygen bonds intermediate between single and double bonds, and each oxygen atom bears an equal share of the ionic charge. The anion is planar and the bond angles equal 120°.

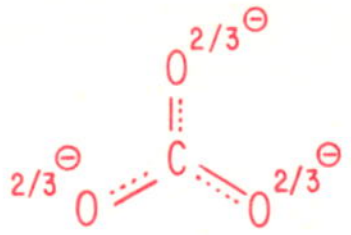

Carbon dioxide itself is a linear molecule, since the carbon atom is involved in two π-bonds, as in acetylene. The individual carbon–oxygen bonds are polarized in the usual way, but the linearity of the molecule causes the individual bond polarities to oppose each other, and so the molecule as a whole is nonpolar.

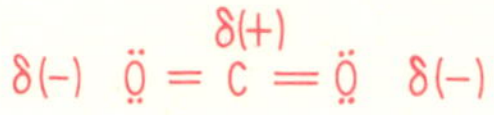

$$\delta(-) \quad \ddot{O} = C = \ddot{O} \quad \delta(-)$$

with $\delta(+)$ over the central C.

Bond polarization in carbon dioxide

C. CARBON–NITROGEN BONDING

The introduction of nitrogen atoms into the methane molecule produces a new set of molecules with properties strikingly different from those of the oxygen-atom derivatives. A nitrogen atom contains five electrons in its valence shell and so needs to share three more electrons to fill the shell. This gives a nitrogen atom a valence of three, allowing it to form single, double, and triple bonds.

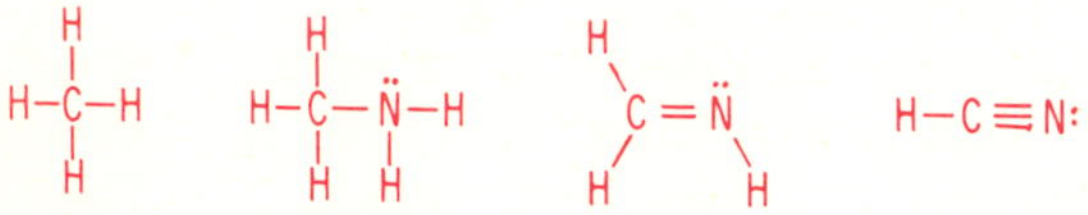

methane   methylamine   methylene imine   hydrogen cyanide

A nitrogen atom is more electronegative than a carbon atom or a hydrogen atom, but it is less electronegative than an oxygen atom, since electronegativity increases from left to right in the periodic table of the elements.

Li Be B C N O F →

electronegativity increases

This means that carbon–nitrogen bonds and hydrogen–nitrogen bonds are polar bonds. In general, a nitrogen atom makes an organic molecule hydrophilic, in much the same way that an oxygen atom does.

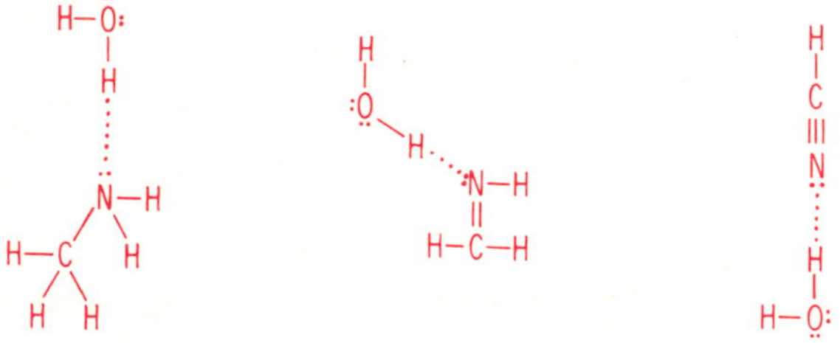

However, the weaker electronegativity of a nitrogen atom compared with an oxygen atom makes nitrogen-atom derivatives different from oxygen-atom derivatives in many ways. Primarily, these differences arise from the fact that an oxygen atom holds its electrons more tightly than a nitrogen atom does. This is due to the greater positive charge on the oxygen nucleus. For this reason, an unshared pair of electrons on a nitrogen atom is more available for sharing than an unshared pair on an oxygen atom.

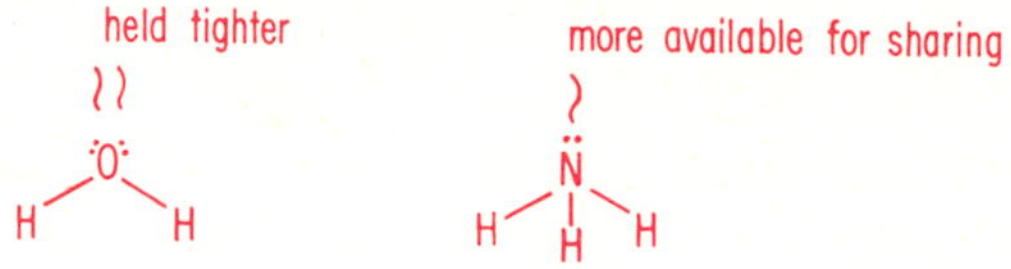

An oxygen atom is more electronegative and holds its
electrons more tightly than a nitrogen atom

The practical result of this difference is that, in general, a nitrogen atom is more **basic** and more **nucleophilic** than an oxygen atom. Basicity and nucleophilicity are closely related. **Basicity** is the willingness to share a pair of electrons, and **nucleophilicity** is the ability to attack a positive center with a pair of electrons.

For example, addition of a strong acid such as HCl to a solution of ammonia and water produces more ammonium ions than hydronium ions. The stronger base (ammonia) captures more of the protons than does the weaker base (water).

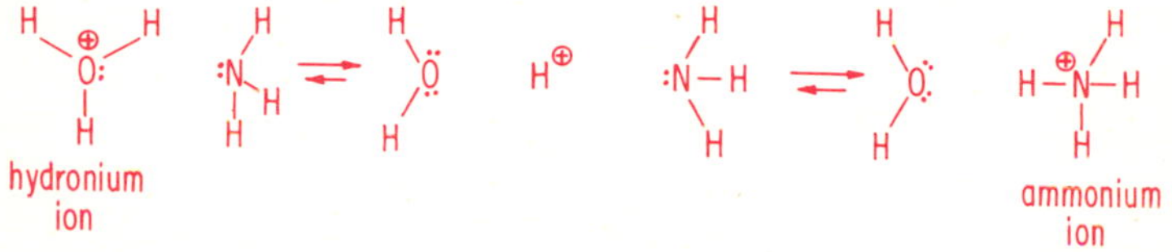

Equilibrium favors the ammonium ion over the hydronium ion

The geometries of the nitrogen-atom derivatives of methane are similar to the geometries of the corresponding oxygen-atom derivatives. The bond angles in methylamine are all close to the tetrahedral angle of 109°. Methylene imine is a planar molecule with bond angles of 120°. Hydrogen cyanide, with its two π-bonds on the carbon atom, is a linear molecule like acetylene and carbon dioxide.

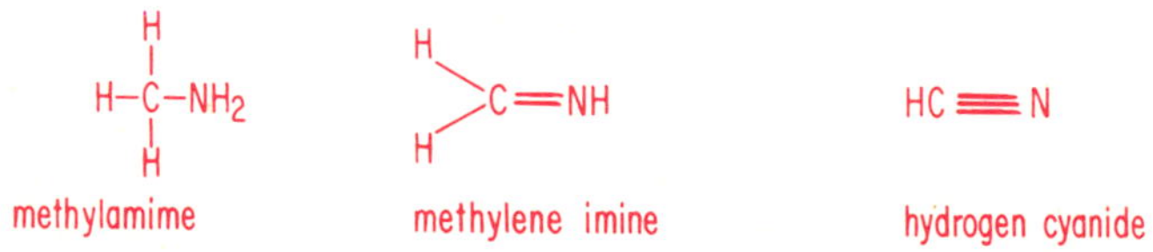

methylamime   methylene imine   hydrogen cyanide

The bond energies for the types of molecules and bonding we have considered are listed in Table A-6. This summary clearly shows that the nature of a given bond influences its reactivity.

TABLE A-6

A Summary of Bond Energies for
Typical Organic Linkages

	kcal/mole
C—C	83
C=C	142
C≡C	186
C—H	99
C—O	81
C=O	127
C—N	62
C=N	121
C≡N	191

IV. Historical Background of Chemical Structure

The stereochemical theories of brilliant chemists of the 19th century greatly facilitated our understanding of chemical phenomena. The field of stereochemistry deals with the relative spatial placement of atoms within molecules. Since the three-dimensional placement of atoms affects the physical properties of a molecule, an awareness and understanding of stereochemistry is important to the understanding of molecular behavior. Four men were responsible for major breakthroughs in this field: Louis Pasteur of France, Friedrich August Kekulé of Prussia, Jacobus van't Hoff of Holland, and Joseph Le Bel of France.

A. CHEMICAL ISOMERS

During the early part of the 19th century, chemists puzzled as to why compounds with the same empirical formula could have such different properties. For example, both dimethyl ether and ethanol have an empirical formula of C_2H_6O and a molecular weight of 46. Their properties differ markedly, however. Dimethyl ether is a gas, with a boiling point of $-24°C$; ethanol is a liquid with a boiling point of $79°C$.

In a classic paper published in 1858, Kekulé offered an explanation as to why such different properties can exist for chemical isomers. Taking into account the observation by an English chemist, Sir Edward Frankland, that each atom can combine with only so many other atoms, Kekulé postulated that the valence of carbon must be four. He further suggested that carbon atoms could be connected to each other to form chains. This idea first occurred to Kekulé in a dream while he was riding on the top deck of a horse-drawn bus in London. He later described this dream as follows

Atoms danced before my eyes. In the past, whenever these small bodies had appeared to me, they had always been in motion. Now, however, I saw how two smaller atoms often united to form a pair; how a larger one embraced two smaller ones; how still larger ones kept hold of three or even four of the smaller while the whole kept whirling in a giddy dance. I saw how the larger ones formed a chain, dragging the smaller ones after them.... The cry of the conductor, "Clapham Road," awakened me from my dream....

Thus, our concept of chemical bonding was born. Kekulé represented the bonds between atoms by lines and postulated that chemical isomers resulted from different bonding patterns. For example, the structural formulae for the chemical isomers mentioned above, ethanol and dimethyl ether, can be represented by the following structural formulae.

ethanol   dimethyl ether

To explain the structural formula of some molecules, double and triple bonds were postulated by Kekulé. Carbon dioxide, for example, must have two double bonds. The nitrogen atom in hydrogen cyanide must be connected to the carbon atom by a triple bond.

carbon dioxide   hydrogen cyanide

Kekulé's structural formulae showed molecules in two dimensions. In 1874 two chemists, van't Hoff and Le Bel, independently postulated that the carbon atom should be viewed as a three-dimensional entity. They envisioned that it could best be represented by a tetrahedron, the points of which were the bonds of carbon. The single, double, and triple bonds of carbon–carbon bonding could therefore be shown as a connection between points, between edges, and between faces of carbon tetrahedra.

$$-\overset{|}{\underset{|}{C}} - \overset{|}{\underset{|}{C}} - \qquad \overset{|}{\underset{|}{C}} = \overset{|}{\underset{|}{C}} \qquad -C \equiv C-$$

Another perplexing problem to nineteenth-century chemists was the apparent existence of two classes of organic molecules: the fatty (**aliphatic**) and the **aromatic** molecules. Those of the first group were known to be relatively stable and included substances like soaps, alcohols, and lubricants. The aromatics, on the other hand, were generally volatile, fragrant compounds. These molecules always had at least six carbon atoms and a higher proportion of carbon with respect to the other atoms than the aliphatic group.

In 1864 Kekulé offered an explanation for the aromatic compounds. As with his work on chemical bonding, he first conceived of his idea in a dream. He said,

There I sat, trying to work on my textbook but it did not go very well; my mind was elsewhere. I turned the chair toward the fireplace and dozed off. Again the atoms danced before my eyes. This time, the smaller groups remained modestly in the background. My inner eye, sharpened by similar visions I had had before, now distinguished bigger forms of manifold configurations. Long rows, more densely joined; everything in movement, contorting and turning like snakes. And behold, what was that? One of the snakes took hold of its own tail and whirled derisively before my eyes. I woke up as though I had been struck by lightning; again I spent the rest of the night working out the consequences of the hypothesis."

Thus, Kekulé postulated that the structural unit of the aromatics was the benzene ring, formed by the joining of six carbon atoms together. Since the empirical formula for benzene was known to be C_6H_6, Kekulé suggested that the structural formula could be represented by either of two ways (see p. 292).

These representations are frequently abbreviated by the following chemical shorthand.

The carbon atoms of the benzene molecule are conventionally numbered as shown below.

Since all the carbon atoms of a benzene ring should be similar, Kekulé postulated that only one type of substitution product for a given chemical group (i.e., —OH, —NH$_2$, —CH$_3$) would exist. On the other hand, three types of disubstitution products should occur, depending on the relative positions of the substituted groups. The substituted groups could be on adjacent carbon atoms of the benzene ring (*ortho* position), alternate carbons (*meta* position), or opposite carbons (*para* position). In the examples below, alternate names for a given molecule are listed. Note the great difference in melting points of the disubstituted products—all of which have the same empirical formula.

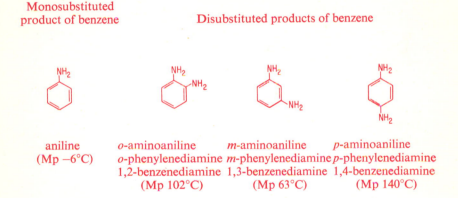

Monosubstituted product of benzene   Disubstituted products of benzene

aniline	*o*-aminoaniline	*m*-aminoaniline	*p*-aminoaniline
(Mp −6°C)	*o*-phenylenediamine	*m*-phenylenediamine	*p*-phenylenediamine
	1,2-benzenediamine	1,3-benzenediamine	1,4-benzenediamine
	(Mp 102°C)	(Mp 63°C)	(Mp 140°C)

B. GEOMETRICAL ISOMERS

Stereoisomers are compounds with identical empirical and structural formulae. They differ from each other in the arrangement of atoms in three-dimensional space. One type of stereoisomerism can occur if a molecule contains double-bonded carbon atoms, and two different atoms or groups are bound to the carbon atoms. This is called geometrical isomerism, or *cis–trans* isomerism. Note that, as with chemical isomers, geometrical isomers have different physical properties—such as different boiling points. Examples are shown below.

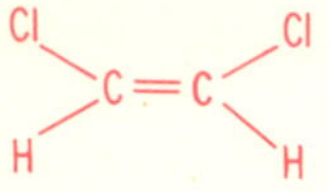

cis-1,2-dichloroethylene
Bp 60°C

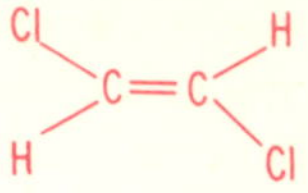

trans-1,2-dichloroethylene
Bp 48°C

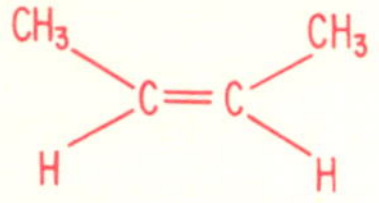

cis-2-butene
Bp 1°C

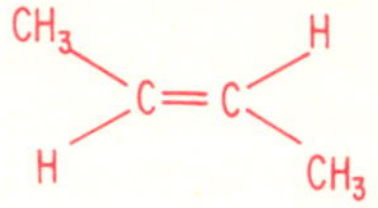

trans-2-butene
Bp 2.5°C

The terms *cis* and *trans* are Latin prepositions meaning "on this side" and "across". The number before butene refers to the placement of the double bond.

C. OPTICAL ISOMERS

Another type of stereoisomerism is defined by the ability of certain molecules to rotate plane-polarized light. These molecules are called optical isomers. The study of molecular optical activity was initiated by the discovery in 1815 by a French physicist, Biot, that solutions of some naturally occurring compounds are optically active. Following this discovery, chemists routinely investigated the optical activity of new compounds. Laboratory-synthesized substances were found to be optically inactive; but many naturally occurring compounds were optically active. Some solutions rotated plane polarized light to the right and were called dextrorotatory (*d*-form) and some of the solutions rotated plane-polarized light to the left and were called levorotatory (*l*-form).

Louis Pasteur elucidated this phenomenon of chemical optical activity with his work on tartaric and racemic acids. His investigations were an important milestone in our understanding of chemical structure and properties. It was even more remarkable in that Pasteur's work predated Kekulé's, van't Hoff's, and Le Bel's ideas on chemical bonding.

Pasteur was aware that both racemic and tartaric acid had the same empirical formula, $C_4H_6O_6$. Tartaric acid was known to be optically active; racemic acid was known to be optically inactive, and was much less soluble than tartaric acid. In 1848 Pasteur investigated the crystalline forms of all 19 salts of tartaric acid. He noticed something which had been overlooked in previous studies: the crystals of all the 19 salts were hemihedral; that is, they were only half symmetrical. Dissymmetric crystals have two hemihedral forms: a right-facing and a left-facing. Different hemihedral crystals for dissymmetric molecules are mirror images of each other, just as the right and left hand are mirror images of each other.

Pasteur also noted that all the crystals of all the 19 salts of tartaric acid were hemihedral in the same direction. This was not surprising, since the solutions of all the salts rotated polarized light in the same direction (to the right).

Since racemic acid was not optically active, Pasteur assumed that crystals of a salt of racemic acid, sodium ammonium racemate, would be symmetrical. To his surprise, however, when he allowed crystallization to occur below 28°C, he found hemihedral crystals. But, some of the crystals were hemihedral to the right; some were hemihedral to the left. Pasteur was able to separate by hand the right- and left-facing crystals. When he converted the right-handed sodium ammonium racemate to the acid, he found that he had tartaric acid. Thus, he concluded that racemic acid is a mixture of dextrorotatory tartaric and levorotatory tartaric acid. Optical isomers such as the *d*- and *l*-form of tartaric acid are called **enantiomorphs, enantiomers**, or **optical antipodes**. They exhibit the same chemical and physical properties, except for the direction of rotation of plane-polarized light.

As mentioned above, the concept of structural formulae had not yet been developed. Nonetheless Pasteur correctly concluded that there must be a stereochemical difference between the molecules of *d*- and *l*-tartaric acids. He noted, "But it cannot be a subject of doubt that there exists an asymmetric arrangement having a nonsuperposable image. It is not less certain that the atoms of the *levo* acid possess precisely the inverse asymmetric arrangement" (from the *dextro* acid).

It has subsequently been shown that tartaric acid can be represented by the following structural formulae.

$$
\begin{array}{cccc}
\text{COOH} & \text{COOH} & \text{COOH} & \text{COOH} \\
| & | & | & | \\
\text{H—C—OH} & \text{HO—C—H} & \text{H—C—OH} & \text{HO—C—H} \\
| & | & | & | \\
\text{H—C—OH} & \text{HO—C—H} & \text{HO—C—H} & \text{H—C—OH} \\
| & | & | & | \\
\text{COOH} & \text{COOH} & \text{COOH} & \text{COOH}
\end{array}
$$

| mesotartaric acid (not optically active, due to plane of symmetry between top and bottom half) | dextrorotatory enantiomorph *d*-tartaric acid (naturally occurring form) | levorotatory enantiomorph *l*-tartaric acid |

Racemic acid is a mixture of the *d*- and *l*-forms of tartaric acid.

1.   *Fischer convention (D-L nomenclature)*

In 1891 Emil Fischer introduced a system for naming carbohydrates with asymmetric carbon atoms based on the configuration of saccharic acid. He arbitrarily assigned the position of the hydroxyl group of the fifth carbon to be on the right of the projection formula of saccharic acid.

Because dextrorotatory glucose could be converted into dextrorotatory saccharic acid, Fischer reasoned that dextrorotatory glucose must have the same steric configuration. He named this glucose, D-glucose. The placement of the other groups on the other asymmetric carbon atoms of glucose was assigned relative to the C-5 carbon. The mirror image of D-glucose was designated as L-glucose.

$$
\begin{array}{ccc}
\overset{1}{C}HO & CHO & \\
H-\overset{2}{\underset{*}{C}}-OH & HO-\overset{*}{C}-H & \\
HO-\overset{3}{\underset{*}{C}}-H & H-\overset{*}{C}-OH & \\
H-\overset{4}{\underset{*}{C}}-OH & HO-\overset{*}{C}-H & {}^{*}\text{asymmetric} \\
H-\overset{5}{\underset{*}{C}}-OH & HO-\overset{*}{C}-H & \text{carbon} \\
\overset{6}{C}H_2OH & CH_2OH & \\
\text{D-glucose} & \text{L-glucose} &
\end{array}
$$

Not all sugars which are classified in the D-family—because of the configuration of groups around the next-to-last carbon—are dextrorotatory. Some are levorotatory. Thus, both D and L and $+$ and $-$ are used to characterize a given sugar. D($-$)-ribose, for example, is levororatory, but belongs to the D-family because of the steric configuration around the next-to-last carbon, C-4.

$$
\begin{array}{c}
CHO \\
H-\overset{*}{C}-OH \\
H-\overset{*}{C}-OH \\
H-\overset{*}{C}-OH \\
CH_2OH \\
\text{D($-$)-ribose}
\end{array}
$$

Since there are four asymmetric carbon atoms in glucose, a molecule with only one asymmetric center, glyceraldehyde, was eventually chosen to be the standard reference used in designating D or L configurations.

$$
\begin{array}{cc}
\text{CHO} & \text{CHO} \\
| & | \\
\text{H}-\overset{*}{\text{C}}-\text{OH} & \text{HO}-\overset{*}{\text{C}}-\text{H} \\
| & | \\
\text{CH}_2\text{OH} & \text{CH}_2\text{OH}
\end{array}
$$

D(+)-glyceraldehyde   L(−)-glyceraldehyde

Compounds related to (+)-glyceraldehyde are said to belong to the D-family; those related to (−)-glyceraldehyde, to the L-family. Using the rules enumerated below it is possible to extend the Fischer convention (D-L nomenclature) to include molecules which are not carbohydrates. For example, the amino acid alanine can be classified in the D- or L-family, depending upon the three-dimensional configuration of the alanine in question.

Rules for use of the Fischer projection formulae

The convention to insure unambiguous structural assignments with the Fischer projection formulae involves the following rules:

1) Vertical bonds are considered to be *behind* the plane of the paper; horizontal bonds in *front* of the plane of the paper.

2) The molecule is arranged so that the hydrogen and hetero-atom of the asymmetric center are in the horizontal plane.

3) The vertical groups are arranged so that the most oxidized group is at the top.

4) The interchange of the position of any two groups at the asymmetric center converts that asymmetric carbon to the opposite configuration.

5) The interchange of any three groups about the asymmetric center maintains the configuration.

6) Because of the three-dimensional orientation, it is not possible to lift the molecule out of the plane of the paper.

Consider the example of alanine.

$$
\begin{array}{ccc}
 & \text{COOH} & \text{COOH} \\
 & | & | \\
\equiv \quad \text{H}_2\text{N}-\text{C}-\text{H} \quad \equiv & \text{H}-\text{C}-\text{CH}_3 \\
 & | & | \\
 & \text{CH}_3 & \text{NH}_2
\end{array}
$$

model of
L(+) − alanine

Fischer projection
of L(+) − alanine

model of
D(−) − alanine

Fischer projection
of D(−) − alanine

2. Cahn–Ingold–Prelog convention (R-S nomenclature)

Since it is impossible to use the D-L nomenclature without specifying the orientation of the projection formula, and since some compounds have more than one asymmetric carbon, another system for designating the three-dimensional configuration of molecules was developed in 1956 by R. S. Cahn D. K. Ingold, and V. Prelog. Their system is called R-S nomenclature, where R stands for *rectus* (right) and S for *sinister* (left). (Note that R and S are also Cahn's first and middle initials).

The steric configuration of groups around each asymmetric carbon is designated according to the following rules:

1) Note the atomic number of each of the atoms directly attached to the asymmetric carbon in question.

2) Rank these attached atoms in order of highest to lowest atomic number.

3) If there are two atoms of the same atomic number attached to the asymmetric carbon (two other carbon atoms, for example), then consider the atomic number of the substituents on the attached atoms. The atom with substituents of higher atomic number is ranked over that with substituents of lower atomic number. The priority order of frequently encountered groups on asymmetrical carbons is as follows: I, Br, Cl, SH, OH, NO_2, NH_2, COOR, COOH, CHO, CR_2OH, CHOHR, CH_2OH, C_6H_5, CH_2R, CH_3, H. Double and triple bonded atoms are counted twice or three times respectively. For example,

$$C{=}N; \text{ same as } C\!\!\diagup^{N}_{\diagdown N}; \qquad C{\equiv}N; \text{ same as } C\!\!\diagup^{N}_{\diagdown N}\!\!-N.$$

4) Look at the asymmetric carbon so that the atom of lowest atomic number (H in most cases) is in the rear. In the two-dimensional projection, the bottom position is equivalent to the rear position of the three-dimensional model.

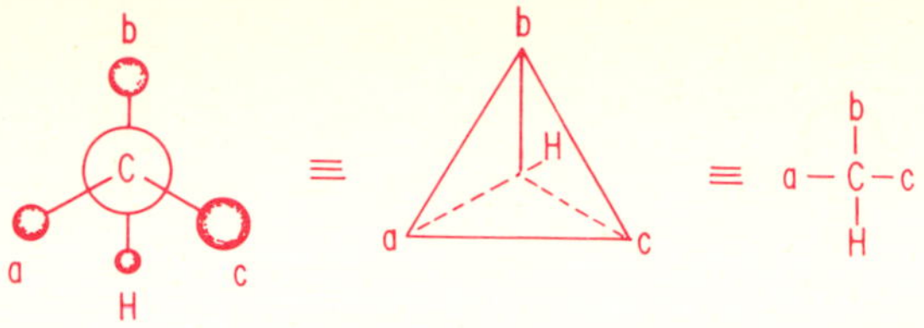

Note that two pairs of any Fischer two-dimensional projection formula may be exchanged or the position of three groups changed in order, without changing the actual three-dimensional structure. For example, the positions of H, OH, and CH_2OH in the Fischer projection formula of D(+)-glyceraldehyde may be viewed differently.

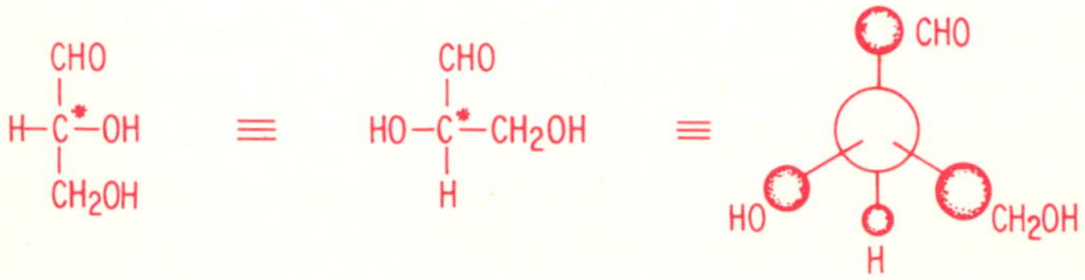

(1)	(2)	
Fischer projection of D(+)-glyceraldehyde	alternate projection of D(+)-glyceraldehyde	model of D(+)-glyceraldehyde viewed to give projection (2)

If the model is rotated 120° to the right, it would give projection (1).

5) Consider the three front-facing atoms attached to the asymmetric carbon. (Recall that the atom of lowest atomic number is in the rear.) Determine whether a clockwise (right-handed) or counter-clockwise (left-handed) direction will enable you to progress from the atom with the highest atomic number to that with the middle-sized atomic number to that with the smallest atomic number. If the direction is clockwise, the steric notation for that particular asymmetric carbon is R; if the direction is counterclockwise, the notation is S.

For example, the groups of glyceraldehyde attached to the asymmetric carbon, ordered according to the above rules, are OH, CHO, CH_2OH, and H. To determine whether or not the asymmetric carbon is R or S, orient the molecules so that the H is at the bottom in a two-dimensional formula or at the rear in a three-dimensional representation. (See rule 4.)

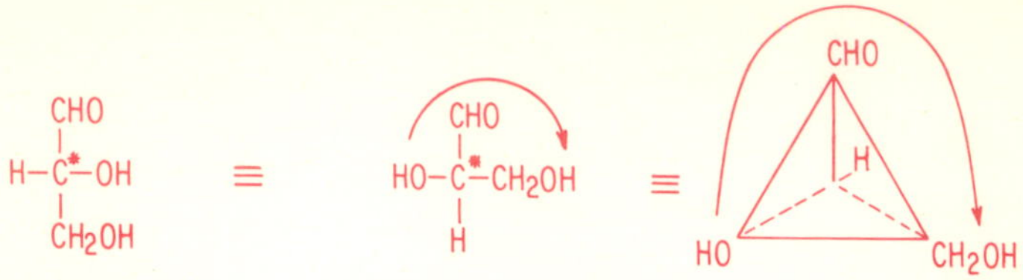

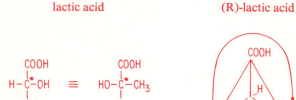

Fischer projection of glyceraldehyde   **alternate projection of glyceraldehyde with H at the bottom**   **three-dimensional representation of alternate projection**

It is necessary to go in a clockwise direction to get from the highest ranked group, OH, to the lowest ranked group, CH_2OH (not counting the group in the rear). Thus the asymmetric carbon of glyceraldehyde is designated R, and the molecule is called (R)-glyceraldehyde in this convention.

Other examples are shown below.

lactic acid   **(R)-lactic acid**

$$H-\overset{*}{\underset{CH_3}{\overset{COOH}{C}}}-OH \equiv HO-\overset{*}{\underset{H}{\overset{COOH}{C}}}-CH_3$$

Ranking of groups:
OH, COOH, CH_3, H

bromobutane   **2(S)-bromobutane**
                                             (The 2 notes that the Br is on C-2)

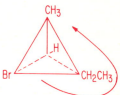

$$H-\overset{*}{\underset{\underset{CH_3}{CH_2}}{\overset{CH_3}{C}}}-Br \equiv Br-\overset{*}{\underset{H}{\overset{CH_3}{C}}}-CH_2CH_3$$

Ranking of groups:
Br, CH_2CH_3, CH_3, H

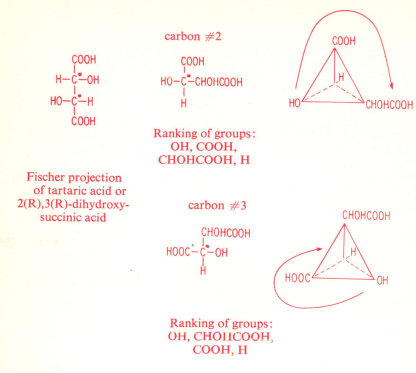

carbon #2

Ranking of groups:
**OH, COOH,
CHOHCOOH, H**

Fischer projection
of tartaric acid or
2(R),3(R)-dihydroxy-
succinic acid

carbon #3

Ranking of groups:
**OH, CHOHCOOH,
COOH, H**

V. Reactions

A. OXIDATION AND REDUCTION

From the field of inorganic chemistry there evolved a straightforward definition of oxidation and reduction. The transfer of electrons in a reaction is often characterized as oxidation and reduction. **Oxidation** refers to a *loss of electrons* and **reduction** denotes a *gain of electrons*. In the reaction between metallic iron and cupric ion, the iron is oxidized while the copper is reduced.

$$Fe + Cu^{\oplus\oplus} \longrightarrow Fe^{\oplus\oplus} + Cu$$

The steps are most easily seen by the half reaction or reaction couples. Iron loses two electrons and is thus oxidized.

$$Fe - 2e^{\ominus} \longrightarrow Fe^{\oplus\oplus}$$

The cupric ion gains the two electrons and is thereby reduced.

$$Cu^{\oplus\oplus} + 2e^{\ominus} \longrightarrow Cu$$

When an **oxidizing agent** reacts with any compound, *it is itself reduced*. There can be no oxidation without reduction. All such reactions are not only balanced stoichiometrically but also by electronic charge. The overall reaction of stannous chloride with potassium permanganate is shown below.

$$5Sn^{++} + 16H^{\oplus} + 2MnO_4^{\ominus} \longrightarrow 5Sn^{(4+)} + 2Mn^{++} + 8H_2O$$

The half reactions are broken down in the following manner.

$$5Sn^{++} \longrightarrow 5Sn^{(4+)} + 10e^{\ominus} \quad \text{oxidation}$$
$$2MnO_4^{\ominus} + 16H^{\oplus} + 10e^{\ominus} \longrightarrow 2Mn^{++} + 8H_2O \quad \text{reduction}$$

In organic chemistry, the terms oxidation and reduction are not so easily defined. It is necessary to use electron density within a given molecule to decide on its oxidation state. For practical purposes there are no differences in the electronic distributions about the carbon atoms of methane, ethane or the other alkanes. For ethylene, the π-bond is so arranged in space that less of it belongs to each of the carbons than is found in the σ-bonding of ethane (see p. 285). The π-bonds in acetylene provide an even lower electron density about the two carbons than in ethylene or ethane. Thus, we can construct an order of oxidation states.

acetylene > ethylene > ethane

On page 294 derivatives of methane containing carbon–oxygen bonds are listed in order of increasing oxidation state of the carbon. Carbon dioxide represents a compound with carbon at its highest oxidation state.

Combustion of a hydrocarbon taken to completion leads to carbon dioxide and water.

$$CH_3-CH_2-CH_2-CH_2-CH_3 + 8O_2 \longrightarrow 5CO_2 + 6H_2O$$

In contrast to inorganic oxidation–reduction reactions, it is somewhat more difficult to write half reactions for a combination of organic compounds; but the fact that the alkane is oxidized while the oxygen is reduced can readily be seen.

Proper conditions and reagents are often used to devise controlled oxidations. For example, ethanol can be converted to acetaldehyde or acetic acid by use of dichromate as an oxidizing agent.

$$CH_3CH_2OH \xrightarrow{Cr_2O_7^{\ominus}} CH_3CH=O \xrightarrow{Cr_2O_7^{\ominus}} CH_3COOH$$
$$\text{ethanol} \qquad\qquad \text{acetaldehyde} \qquad\qquad \text{acetic acid}$$

Since the acetaldehyde is easily oxidized to acetic acid, it can be isolated only if it is continuously removed by distillation as soon as it is formed. If isopropanol is the starting alcohol, acetone is the only oxidation product.

$$CH_3-\underset{\underset{OH}{|}}{CH}-CH_3 \xrightarrow{Cr_2O_7{}^\ominus} CH_3-\underset{\overset{O}{||}}{C}-CH_3$$

The reaction stops at this point because further oxidation would require breakage of carbon–carbon bonds, a high energy requiring process.

The rapid oxidation of aldehydes noted above is the basis of tests to identify such structures. Many sugars contain aldehydes. They can easily be identified by oxidation using the cupric salt of tartaric acid.

$$
\begin{array}{c}
CHO \\
| \\
H-C-OH \\
| \\
HO-C-H \\
| \\
H-C-OH \\
| \\
H-C-OH \\
| \\
CH_2OH \\
\text{D-glucose}
\end{array}
+ 2Cu\,(\text{tartrate})^\ominus + 5OH^\ominus \rightarrow Cu_2O\downarrow +
\begin{array}{c}
COOH \\
| \\
H-C-OH \\
| \\
HO-C-H \\
| \\
H-C-OH \\
| \\
H-C-OH \\
| \\
CH_2OH \\
\text{D-gluconic acid}
\end{array}
+ 4\,\text{tartrate}^\ominus + 3H_2O
$$

Cuprous oxide precipitates, indicating oxidation of the aldehyde group; the resulting acid decomposes under these conditions.

One of the most useful reactants of organic chemistry is the **Grignard reagent**. Alkyl halides are allowed to react with metallic magnesium under dry conditions in an ether solvent to form an alkyl magnesium halide, the Grignard reagent.

$$R-Br + Mg \xrightarrow[\text{dry}]{\text{ether}} RMgBr$$

This compound can react with carbon dioxide to prepare carboxylic acids which effectively reduce the oxidation state of the carbon in carbon dioxide. The magnesium metal is converted to the magnesium doubly-positive ion as the oxidation part of the overall reaction.

$$RMgBr + CO_2 \xrightarrow[\text{dry}]{\text{ether}} R-CO_2MgBr$$
$$\downarrow H_3O^\oplus$$
$$R-CO_2H + MgBr(OH)$$

If aqueous acid is allowed to react with the Grignard reagent directly, hydrocarbons result.

$$RMgBr + H_2O \xrightarrow{H^{\oplus}} R\text{-}H + MgBr(OH)$$

There are reactions in the armamentarium of organic chemistry by which carboxylic acids are reduced to carbonyl containing compounds, alcohols, and even hydrocarbons. Aldehydes and ketones can also be reduced. Metallic reagents such as lithium aluminum hydride and diborane can cause selective reductions.

$$CH_3\text{-}(CH_2)_n\text{-}CH{=}CH_2 + B_2H_6 \xrightarrow[\text{as solvent}]{CH_3O(CH_2CH_2O)_2CH_3} [CH_3\text{-}(CH_2)_n\text{-}CH_2\text{-}CH_2]_3B$$

$$[CH_3\text{-}(CH_2)_n\text{-}CH_2\text{-}CH_2]_3B \xrightarrow{H_2O_2,\ H^{\oplus}} \underset{\text{alcohol}}{CH_3\text{-}(CH_2)_n\text{-}CH_2\text{-}CH_2\text{-}OH}$$

$$[CH_3\text{-}(CH_2)_n\text{-}CH_2\text{-}CH_2]_3B \xrightarrow{H^{\oplus}} \underset{\text{alkane}}{CH_3\text{-}(CH_2)_n\text{-}CH_2\text{-}CH_3}$$

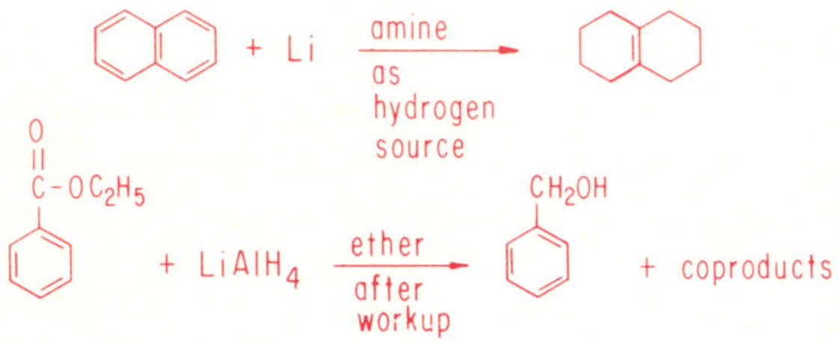

Oxidation and reduction reactions can also involve carbon–nitrogen bonded compounds. A series of nitrogen-containing derivatives of methane are shown on p. 300. Of the compounds listed, hydrogen cyanide represents carbon in its highest oxidation state. Carbodiimide is more closely related to carbon dioxide.

$$RN{=}C{=}NR$$

Such compounds are known, but are extremely reactive.

Imine- or nitrile-containing compounds can be reduced by reagents quite similar to those mentioned for oxygen compounds. The product of reduction for both nitrogen compounds is usually amines.

$$R-C\equiv N \quad \xrightarrow[C_2H_5OH]{Na} \quad R-CH_2NH_2$$

CH=NR

Schiff base

$$\xrightarrow[Pt]{H_2}$$

CH$_2$—NHR

benzylalkylamine

We cannot end this section without some comment on reduction by addition of hydrogen. This was shown earlier for ethylene (p. 293) and here for the imine (Schiff base). The noble metals such as platinum or palladium can adsorb hydrogen gas and catalyze its addition to multiple bonds.

B. DISPLACEMENT AND ADDITION REACTIONS

Reaction rates or kinetics provide information about the sequence of steps or mechanism of reactions. A chemical transformation may involve many individual steps, only one of which is slow compared with the others. This slow conversion is the rate-determining step which limits the rate of reaction. The vast majority of all chemical reactions are either **unimolecular** or **bimolecular**, which means that one or two reactants participate in the rate-determining step. Experimentally, the rates of reactions in solution generally depend on the concentrations of one or more of the reactants or products. This is usually expressed as a rate law which holds that the rate is the product of the concentrations of the reactants and a rate constant k.

$$Rate = k\,[A]^X\,[B]^Y$$

The exponents x and y indicate the order of the reaction with respect to each reactant. It should be noted that the order of a reaction represents the experimentally determined dependence of the rate of the reaction on the concentration of each reactant. The molecularity, on the other hand, denotes the number of reactants undergoing the reaction. This quantity is deduced from the mechanism.

The displacement reaction between an alcoholate and an alkyl bromide to form an ether is termed an S_N2 reaction. This abbreviation stands for *substitution, nucleophilic, bimolecular*. The reaction and rate law are as follows.

$$RO^{\ominus} + R'-Br \longrightarrow R-OR' + Br^{\ominus}$$

$$Rate = k\,[RO^{\ominus}]\,[R'Br]$$

This reaction is not only bimolecular, but second order (i.e., first order in each reactant). A **nucleophile** is described as a reactant with an electron pair that attacks an atom with lower electron density.

S_N2 reactions have been characterized as single step, concerted bond-making and bond-breaking reactions. If the displaced group is attached to an asymmetric center, inversion of that center occurs.

$$Y^{\ominus} \overset{R'}{\underset{R''}{C}} - X \longrightarrow Y \cdots \overset{R'}{\underset{R''\ R''}{C}} \cdots X \longrightarrow Y - C \overset{R'}{\underset{R'''}{\diagdown}} + X^{\ominus}$$

transition state

A transition state can be envisaged in which the attacking group, Y, and the leaving group, X, are both partially bonded to a planar carbon containing the three substituent groups. From the reaction above, it is readily seen that the three alkyl groups in the products are arranged in space in a mirror image of the same groups as in the reactant. An elegant example in support of the S_N2 mechanism is the reaction of optically active 2-octyliodide and radioactive iodide ion.

$$CH_3-(CH_2)_5-\underset{\overset{|}{I}}{\overset{I}{C}}H-CH_3 \; + \; \overset{*}{I}{}^{\ominus} \longrightarrow CH_3-(CH_2)_5-\underset{\overset{|}{I^*}}{C}H-CH_3 \; + \; I^{\ominus}$$
(radioactive)

The starting iodide is the D-isomer. Reaction with the radioactive iodide produces the L-isomer and racemization. Within experimental error, the rate of racemization is the same as the rate of formation of radioactive alkyl iodide.

It is possible that a displacement reaction occurs in two separate steps. Initially the molecule ionizes producing a carbonium ion which then reacts with the nucleophile.

$$R \overset{\frown}{-} X \; \xrightarrow{\text{slow}} \; R^{\oplus} + X^{\ominus}$$
$$R^{\oplus} + Y^{\ominus} \; \xrightarrow{\text{fast}} \; R-Y$$
$$\text{Rate} = k\,[RX]$$

Since the R—X is the only molecule in the rate-determining step, the reaction is unimolecular. By measuring the rate of reaction at several concentrations, it was established that the rate is first order with respect to R—X. This type of reaction is termed S_N1, which stands for *substitution, nucleophilic, unimolecular*. From the reaction mechanism, it is readily apparent that the reaction rate is independent of the kind and amount of nucleophile used.

Alkyl halides, which can somehow stabilize carbonium ions, react via S_N1 reactions. Tertiary halides and some secondary halides are typical examples. (A **primary halide** has the general formula of RCH_2X; a **secondary halide**,

$$\overset{R'}{\underset{R}{\diagdown}}CHX; \text{ and \textbf{tertiary halide}, } R'' - \overset{R'}{\underset{R}{C}} - X.)$$

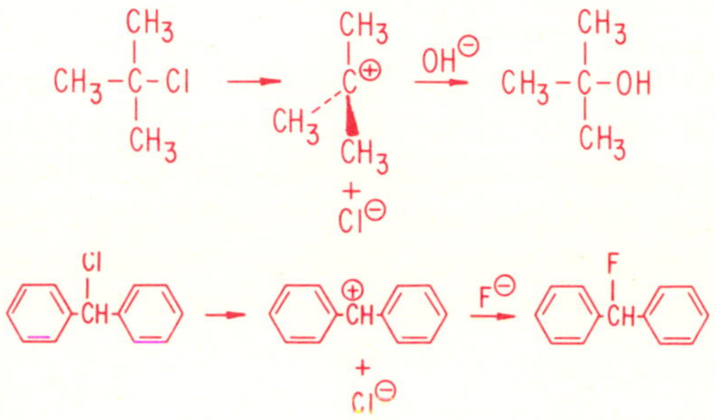

Since a symmetrical intermediate is involved in the S_N1 pathway, total racemization will always be encountered if the replaced group was attached to an asymmetric center.

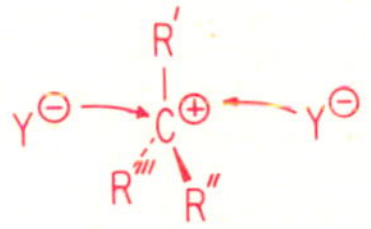

The nucleophile, $Y^{\ominus}$, has equal opportunity to attack the planar carbonium ion from either side.

A molecule containing a carbon–carbon double bond is called **unsaturated**, because of its ability to add other atoms. These addition reactions are generally **exothermic**, i.e., they release energy. This means that the products contain less energy—*are more stable*—than the reactants, principally because σ-bonds are stronger than π-bonds. However, the hydrogenation and polymerization reactions require catalysts because the uncatalyzed reaction is too slow at ordinary temperatures.

Hydrogenation $H_2C = CH_2 + H_2 \longrightarrow H_3C - CH_3$
                     ethylene   ethane

Hydrochlorination $H_2C = CH_2 + HCl \longrightarrow H_3C - CH_2 - Cl$
                                                          ethyl chloride

Polymerization $n\ H_2C = CH_2 \longrightarrow \left(CH_2 - CH_2 \right)_n$
                                                          polyethylene

Carbonyl compounds are also unsaturated and can be attacked by nucleophiles. Formaldehyde in water exists almost completely as the diol (or hydrate).

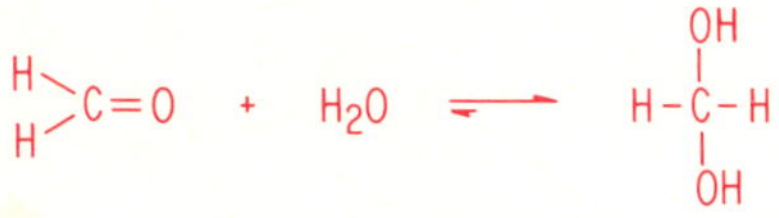

$$\begin{array}{c}H\\H\end{array}\!\!\!\!>C=O \ + \ H_2O \ \rightleftharpoons \ \begin{array}{c}OH\\|\\H-C-H\\|\\OH\end{array}$$

Since the carbon of the carbonyl group is less electron-poor with alkyl substituents than with hydrogens, acetaldehyde is only about 50% hydrated in water while acetone remains unhydrated.

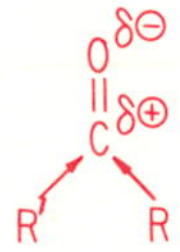

$$\begin{array}{c}O^{\delta\ominus}\\||\\C^{\delta\oplus}\\R \quad R\end{array}$$

The formation of **Schiff bases** represents an addition reaction in which an amine serves as the nucleophile.

$$\begin{array}{c}R\\R'\end{array}\!\!\!\!>C=O \ + \ H_2NR'' \ \rightleftharpoons \ \begin{array}{c}R\\R'\end{array}\!\!\!\!>C=N\!\!\begin{array}{c}\\R''\end{array} + \ H_2O$$

Related reactions include addition of hydroxylamine to form oximes, or hydrazine to form hydrazides.

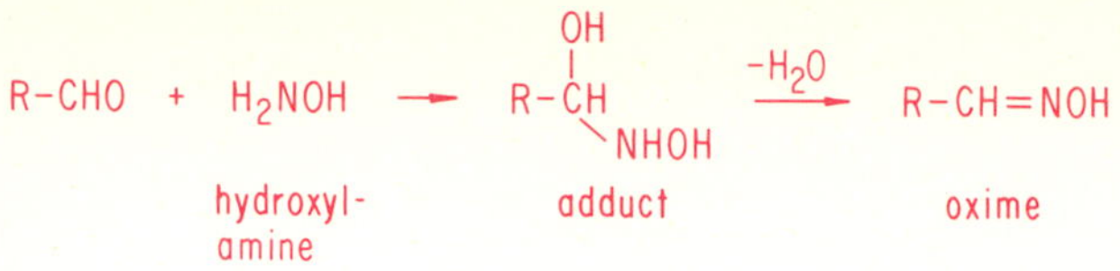

$$R-CHO + H_2NOH \longrightarrow R-\underset{\underset{NHOH}{|}}{\overset{\overset{OH}{|}}{CH}} \xrightarrow{-H_2O} R-CH=NOH$$

hydroxyl-amine   adduct   oxime

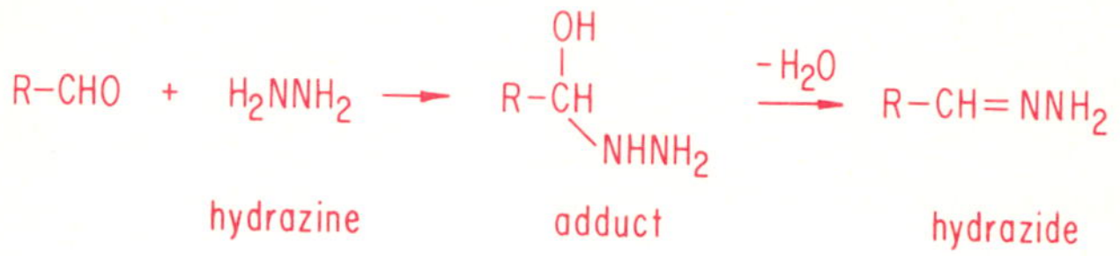

$$R-CHO + H_2NNH_2 \longrightarrow R-\underset{\underset{NHNH_2}{|}}{\overset{\overset{OH}{|}}{CH}} \xrightarrow{-H_2O} R-CH=NNH_2$$

hydrazine   adduct   hydrazide

A typical target molecule for nucleophilic attack (addition) is a carboxylic acid chloride such as acetyl chloride.

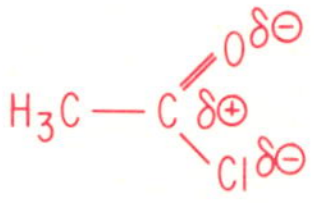

$$H_3C - C \overset{\delta\ominus}{\underset{Cl^{\delta\ominus}}{\overset{O}{\underset{\delta\oplus}{\diagup}}}}$$

acetyl chloride

The carbon atom in the carbonyl group is polarized positively because of the electron-withdrawing effect of the more electronegative oxygen and chlorine atoms bonded to it. Thus, the carbon atom is particularly susceptible to nucleophilic attack. Attack by ammonia produces acetamide.

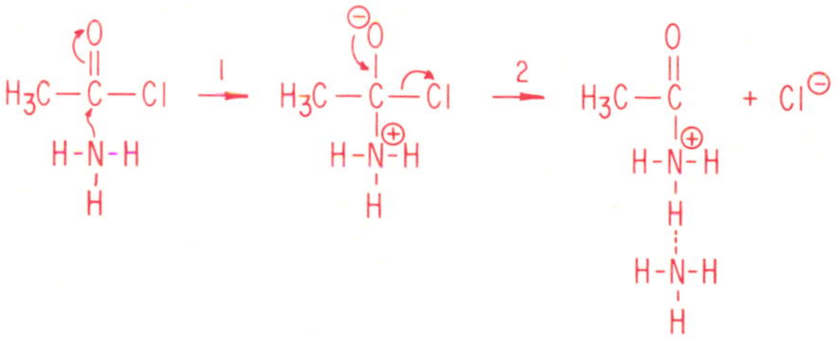

(1) Ammonia attacks and the weak π-bond breaks.
(2) The π-bond reforms and pushes out the chloride ion.
(3) The protonated product transfers a proton to a nearby solvent molecule.

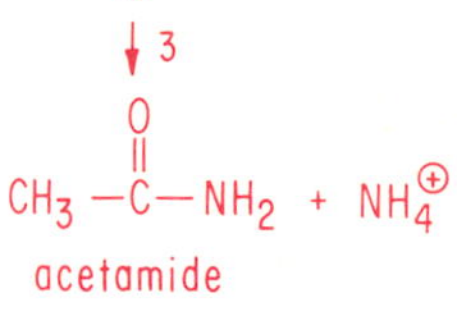

$$CH_3 - \overset{\overset{O}{\|}}{C} - NH_2 + NH_4^{\oplus}$$

acetamide

Attack by water produces acetic acid.

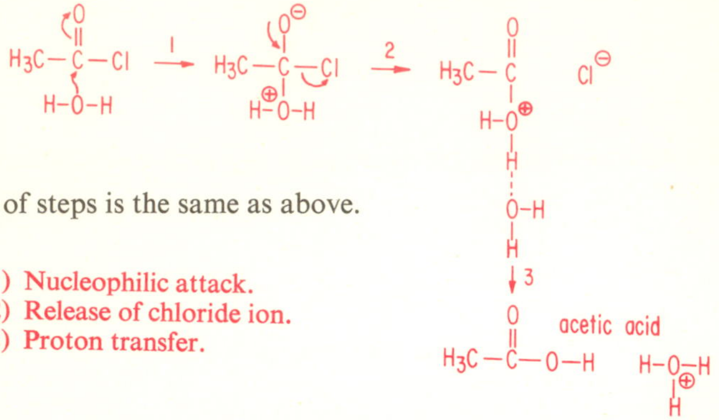

Sequence of steps is the same as above.

    (1) Nucleophilic attack.
    (2) Release of chloride ion.
    (3) Proton transfer.

Since the electron pair of ammonia is less tightly held than that of water (i.e., ammonia is the stronger nucleophile), acetamide forms faster than acetic acid.

The differences in basicity and nucleophilicity seen between ammonia and water are also seen in their methane derivatives—methylamine and methyl alcohol. Methylamine is basic and nucleophilic like ammonia, and methyl alcohol is less basic and less nucleophilic like water.

Also, methylene imine resembles formaldehyde in that its double-bonded carbon is positively polarized and susceptible to nucleophilic attack. Hydrogen cyanide is like a carboxylic acid in that it is slightly acidic. However, the anion produced by release of a proton from hydrogen cyanide is not resonance-stabilized, and so hydrogen cyanide is a considerably weaker acid than an average carboxylic acid.

Esters are formed by the reaction of acids with alcohols. When an ester is treated with acid, the ester can be cleaved to its component acid and alcohol. This process is known as **hydrolysis** and is reversible.

$$R-\overset{O}{\overset{\|}{C}}-OR_l + H_2O \underset{\longleftarrow}{\overset{H^+}{\longrightarrow}} R-\overset{O}{\overset{\|}{C}}-OH + R_lOH$$

If the cleavage is run under basic conditions, the reaction is called **saponification.** The basic reaction is not reversible because the product acid reacts further with base to form a carboxylate ion which repels nucleophilic attack by virtue of its anionic charge.

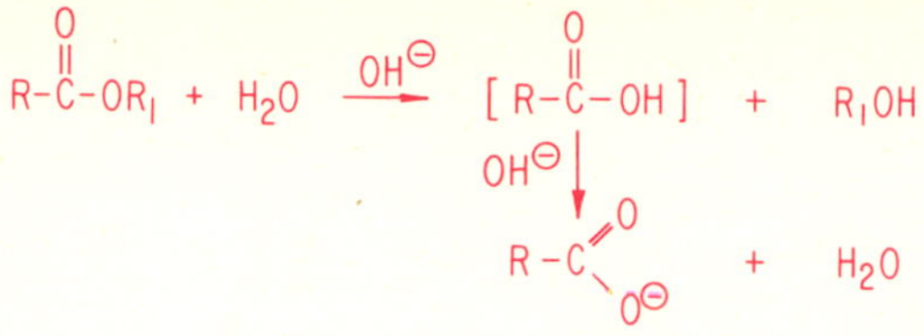

$$R-\overset{\overset{\displaystyle O}{\|}}{C}-OR_1 + H_2O \xrightarrow{\ OH^{\ominus}\ } [\ R-\overset{\overset{\displaystyle O}{\|}}{C}-OH\ ] + R_1OH$$

$$OH^{\ominus} \downarrow$$

$$R-C\overset{\diagup O}{\diagdown O^{\ominus}} + H_2O$$

For typical esters, the cleavage reactions generally involve acyl bond breaking.

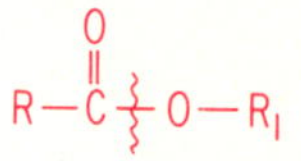

$$R-\overset{\overset{\displaystyle O}{\|}}{C}\ \}\ O-R_1$$

acyl bond cleavage

This can be shown by labeling experiments with O^{18}. If an ester is hydrolyzed using water enriched with O^{18}, the resulting alcohol does not contain any of the label.

$$R-\overset{\overset{\displaystyle O}{\|}}{C}-OR_1 + H_2O^{18} \longrightarrow R-\overset{\overset{\displaystyle O}{\|}}{C}-O^{18}H + R_1OH$$

Similarly, saponification of an O^{18}-containing ester of the following type provides evidence for acyl bond cleavage.

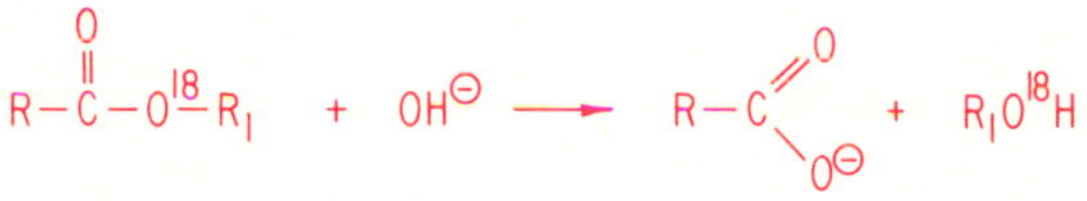

$$R-\overset{\overset{\displaystyle O}{\|}}{C}-O^{18}R_1 + OH^{\ominus} \longrightarrow R-C\overset{\diagup O}{\diagdown O^{\ominus}} + R_1O^{18}H$$

There are special cases in which alkyl bond cleavage of esters is encountered.

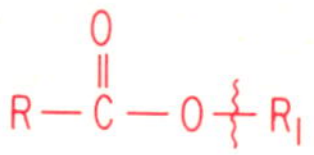

$$R-\overset{\overset{\displaystyle O}{\|}}{C}-O\ \}\ R_1$$

alkyl bond cleavage

Such cleavages are found if the ester used can ionize to yield a stable carbonium ion.

$$\begin{array}{c} R \\ | \\ R-C-O-\overset{\overset{\displaystyle O}{||}}{C}-R_1 \\ | \\ R \end{array} \rightleftharpoons \begin{array}{c} R \\ | \\ R-\overset{\oplus}{C} \\ | \\ R \end{array} + \overset{\ominus}{O}-\overset{\overset{\displaystyle O}{||}}{C}-R_1$$

$$\Big\downarrow H_2O^{18}$$

$$\begin{array}{c} R \\ | \\ R-C-O^{18}H \\ | \\ R \end{array} + R_1C\overset{\displaystyle O}{\underset{}{\diagup}}{-}OH$$

Acid-catalyzed esterification, hydrolysis and saponification proceed by an addition-elimination pathway. The mechanism is based on O^{18} labeling studies.

$$R-\overset{\overset{\displaystyle O^{18}}{||}}{C}-OR_1 \quad \overset{\ominus}{O}H \quad \underset{\text{addition}}{\overset{\text{nucleophilic}}{\rightleftharpoons}} \quad R-\overset{\overset{\displaystyle \overset{\ominus}{O}^{18}}{|}}{\underset{\underset{\displaystyle OH}{|}}{C}}-OR_1 \rightarrow \text{products}$$

Scrambling of the label occurs because of facile proton transfer.

$$R-\overset{\overset{\displaystyle \overset{\ominus}{O}^{18}}{|}}{\underset{\underset{\displaystyle OH}{|}}{C}}-OR_1 \rightleftharpoons R-\overset{\overset{\displaystyle O^{18}H}{|}}{\underset{\underset{\displaystyle O^{\ominus}}{|}}{C}}-OR_1$$

Since the nucleophilic addition step is reversible, leakage of O^{18} to the solvent occurs.

$$R-\overset{\overset{\displaystyle O^{18}H}{|}}{\underset{\underset{\displaystyle O^{\ominus}}{|}}{C}}-OR_1 \rightleftharpoons R-\overset{\overset{\displaystyle O}{||}}{C}-OR_1 + \overset{\ominus}{O}^{18}H$$

Only an addition–elimination mechanism can account for the leakage of the label. Under acidic conditions, the same results are observed.

$$R-\overset{\overset{\displaystyle O}{||}}{C}-OR_1 + H^{\oplus} \rightleftharpoons R-\overset{\overset{\displaystyle \overset{\oplus}{O}^{18}H}{||}}{C}-OR_1 \overset{H_2O}{\rightleftharpoons} R-\overset{\overset{\displaystyle O^{18}H}{|}}{\underset{\underset{\displaystyle H^{\overset{\oplus}{O}}H}{|}}{C}}-OR_1 \overset{-H^{\oplus}}{\rightleftharpoons}$$

$$R-\overset{\overset{\displaystyle O^{18}H}{|}}{\underset{\underset{\displaystyle OH}{|}}{C}}-OR_1 \rightleftharpoons \text{products}$$

symmetrical adduct

Reversal of the reactions leading to the symmetrical adduct must create a leakage of the O^{18} label to the solvent. In these acidic and basic experiments, the O^{18} label appears in the solvent prior to the hydrolysis or saponification of the ester. This could only happen if water or an hydroxide ion is added to the carbonyl group.

Mechanistically, both reactions proceed to completion by cleavage of the alkoxyl group from the adducts.

$$
\begin{array}{l}
R-\overset{\overset{\overset{\ominus}{O}}{\underset{}{\|}}}{\underset{\underset{OH}{|}}{C}}\text{—} \overset{}{O}R_1 \longrightarrow [RCOOH] + \overset{\ominus}{O}R_1 \\[2ex]
\text{adduct in base} \qquad\quad RCOO^{\ominus} + HOR_1
\end{array}
$$

$$
\begin{array}{l}
R-\overset{\overset{\overset{+}{O}-H}{\underset{}{\|}}}{\underset{\underset{OH}{|}}{C}}\text{—}OR_1 \longrightarrow R-COOH + HOR_1 \\[2ex]
\text{adduct in acid}
\end{array}
$$

We have already considered amide formation (pp. 320 and 321). Hydrolysis and saponification of amides proceed in completely analogous reactions to those of the esters discussed above.

Previous sections of our survey on chemical reactions have been based on heterolytic cleavage of bonds.

$$
A-B \;\rightleftharpoons\; \begin{array}{l} A^{\oplus} + B^{\ominus} \text{ or} \\ A^{\ominus} + B^{\oplus} \end{array}
$$

Heterolysis of a bond is defined as scission in which the electrons remain paired and associated with one of the atoms. Ions are the consequence of heterolysis.

Bonds may also cleave **homolytically**.

$$
A-B \;\rightleftharpoons\; A\cdot + B\cdot
$$

In this reaction the electron pair is broken and free radicals result. Often thermal and photochemical cleavages of bonds are homolytic. Once formed, radicals can create chain reactions.

The example of a radical chain reaction on p. 326 shows several characteristics of radicals. The bromine atom can displace a methyl radical from methane.

Initiation of a
chain reaction

$$Br-Br \xrightarrow{\text{heat}} Br\cdot + Br\cdot$$

Propagation of a
chain reaction

$$Br\cdot + CH_4 \longrightarrow CH_3\cdot + HBr$$

$$CH_3\cdot + Br_2 \longrightarrow CH_3Br + Br\cdot$$

Termination of a
chain reaction

$$CH_3\cdot + CH_3\cdot \longrightarrow CH_3CH_3$$

$$CH_3\cdot + Br\cdot \longrightarrow CH_3Br$$

$$Br\cdot + Br\cdot \longrightarrow Br_2$$

The methyl radical can displace a bromine atom from a bromine molecule. Radicals can easily recombine. Thus, if two radicals collide they form a bond exothermically. Chain reactions proceed until the concentration of molecules such as the methane of the example above is greatly reduced. When this occurs, recombination is dominant and the chain ceases.

Radicals can also add to unsaturated molecules. Such reactions are the basis for many polymerizations. Radicals provide initiating species which can then add to the double bond of an alkene, generating a new radical intermediate. They then can add repetitively to the alkene monomers. A steady-state, low concentration of radicals is established because of the rapid addition of radicals to alkene molecules. Termination becomes important when the alkene concentration is greatly reduced. The termination reaction involves the joining of two chains by recombination of the free radicals at the end of each chain.

Initiation: $RO-OR \longrightarrow RO\cdot + RO\cdot$
peroxide

Propagation: $RO\cdot + CH_2=\overset{X}{C}H \longrightarrow RO-CH_2-\overset{X}{C}H\cdot$

$RO-\overset{X}{C}H-\overset{}{C}H\cdot + CH_2=\overset{X}{C}H \longrightarrow RO-(CH_2-\overset{X}{C}H)_2\cdot$

$RO(CH_2-\overset{X}{C}H)_{n-1}\cdot + CH_2=\overset{X}{C}H \longrightarrow RO(CH_2-\overset{X}{C}H)_n\cdot$

Termination: $RO-(CH_2-\overset{X}{C}H)_n\cdot + \cdot(\overset{X}{C}H-CH_2)_m-OR \longrightarrow$

$RO(CH_2-\overset{X}{C}H)_n-(\overset{X}{C}H-CH_2)_m OR$

C. ENOLIZATIONS AND TAUTOMERISMS

Our survey of chemical structure and reactions must include a consideration of a class of compounds in which two structurally distinct isomers are in rapid equilibrium with each other. This phenomenon is termed **enolization** or **tautomerism** (*tauto* from the Greek, meaning same). The shifts we discuss generally involve the movement of a proton from one atom to another three atoms away in the same molecule. A schematic representation is shown below.

$$H{-}X{-}Y{-}Z \rightleftharpoons X{-}Y{-}Z{-}H$$

Carbonyl compounds containing a hydrogen on the carbon adjacent to the carbonyl group are well known to undergo tautomerism.

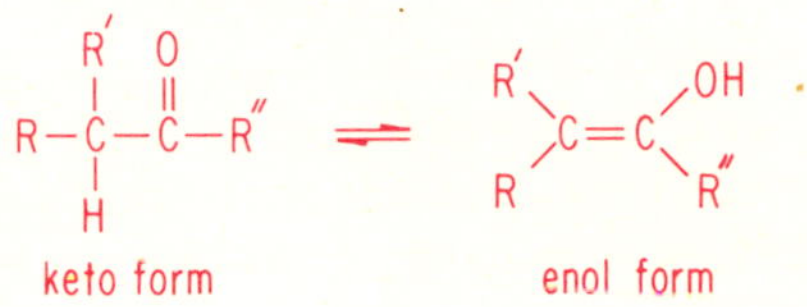

keto form   enol form

This type of **tautomerism** is termed enolization. (Enol is a contraction of the ends of the words alk*ene* and alcoh*ol*.) The above equilibrium generally lies far toward the keto form. With certain features of structure, the equilibrium can be shifted toward the enol structure. For example, a group such as acyl or nitrile attached to the carbon containing the hydrogen favors enolization.

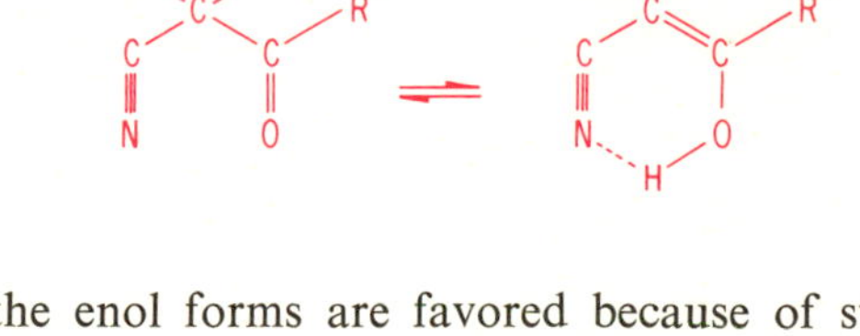

In both cases, the enol forms are favored because of stabilization by an intramolecular hydrogen bond of the chelate type (from the Greek *chele*, meaning claw or pincers). Table A-7 contains some representative carbonyl structures and their enol contents.

TABLE A-7

Enol Content of Some Carbonyl Compounds

Compound	% Enol content
$CH_3-\overset{\overset{O}{\|\|}}{C}-CH_3$	1.5×10^{-4}
$CH_3-\overset{\overset{O}{\|\|}}{C}-CH_2-\overset{\overset{O}{\|\|}}{C}-CH_3$	77
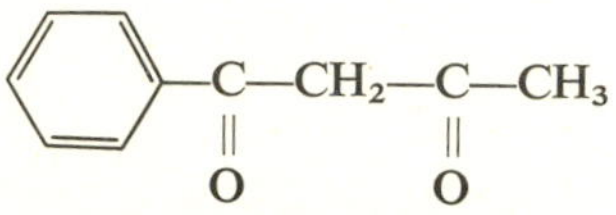 $C_6H_5-\overset{\overset{O}{\|\|}}{C}-CH_2-\overset{\overset{O}{\|\|}}{C}-CH_3$	90
$CH_3-\overset{\overset{O}{\|\|}}{C}-CH_2-\overset{\overset{O}{\|\|}}{C}-OC_2H_5$	8
$C_2H_5O-\overset{\overset{O}{\|\|}}{C}-CH_2-\overset{\overset{O}{\|\|}}{C}-OC_2H_5$	7.7×10^{-3}
$N\equiv C-CH_2-\overset{\overset{O}{\|\|}}{C}-OC_2H_5$	0.25

In the presence of a strong base, the tautomeric hydrogen can be abstracted, producing an enolate anion.

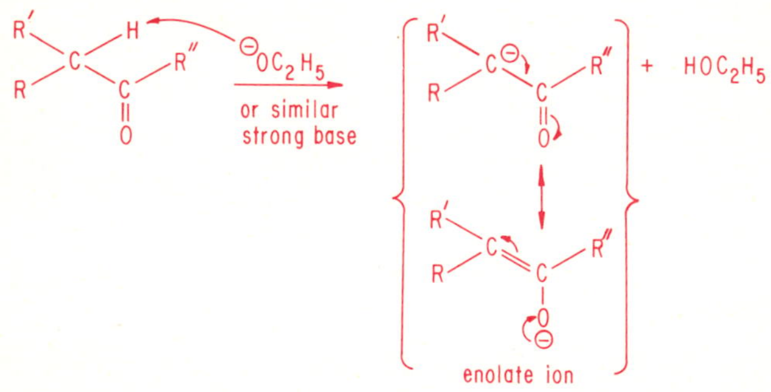

The atoms of the enolate ion remain in exactly the same place in both formulae above. Only the electrons have moved. Such delocalizations are examples of resonance (see p. 292). The true structure is better represented as:

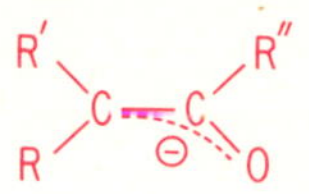

Numerous additional systems in organic chemistry exhibit analogs of the keto-enol tautomerism. The planarity of the peptide linkages of proteins creates many of their three-dimensional structural features. Pauling realized that tautomerism is one of the most important factors in maintaining the peptide group in planar form.

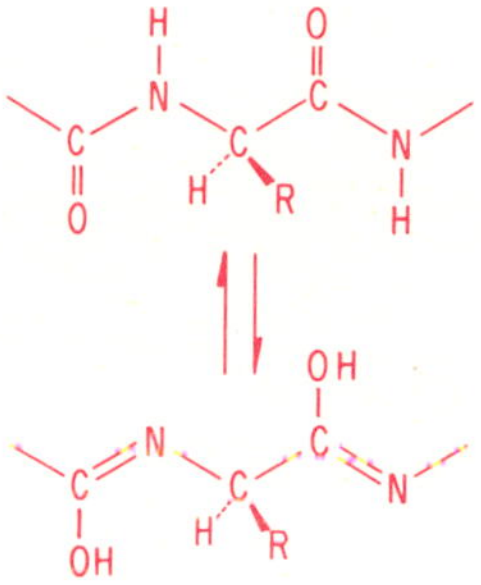

keto form

imino-ol form

Nucleotide structures involve tautomerism in the joining of the pyrimidine bases to the ribose grouping of nucleosides. For example, consider the pyrimidine uracil.

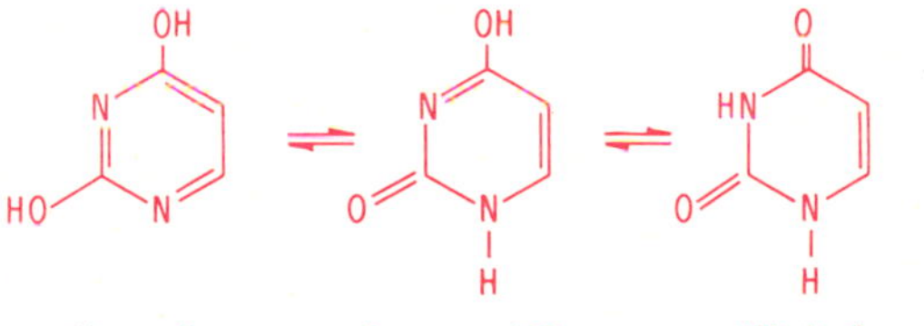

dienol form   keto-enol form   diketo form

Various tautomers of uracil

Three different structures can be drawn. It is actually the keto-enol structure which forms the nucleoside (uridine) that goes into biological systems.

Another important case in which tautomerism plays an important role is the hydroxypyridine–pyridone system.

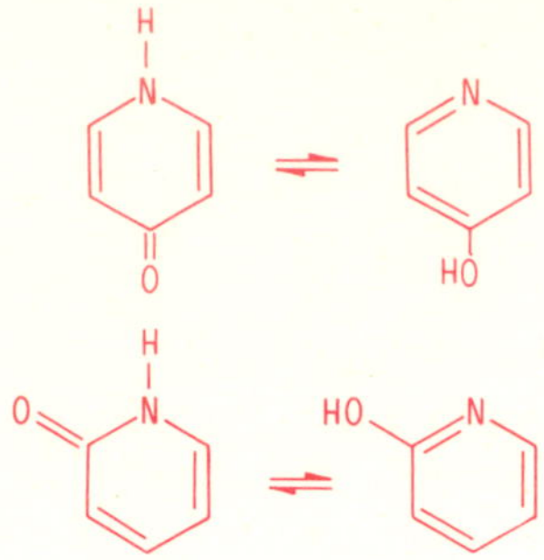

Such tautomerisms are involved in many biological functions of nicotinamide derivatives.

An imine–enamine tautomerism is often encountered. Imines are frequently called Schiff bases and are synthesized by allowing carbonyl compounds to react with amines. The tautomeric shifts occur if a hydrogen is present on the carbon adjacent to the imino grouping.

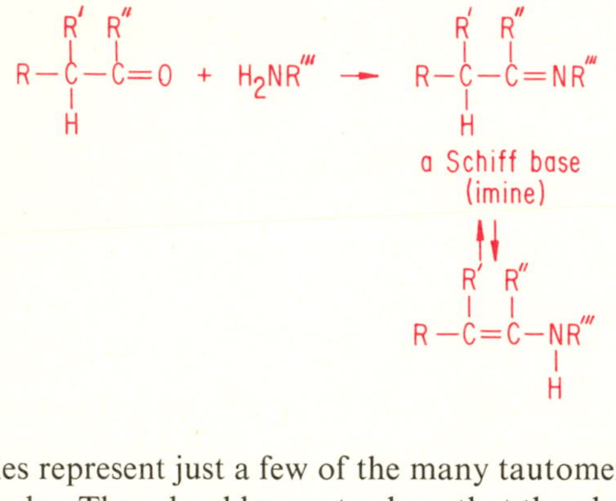

These examples represent just a few of the many tautomerisms observed with organic molecules. They should serve to show that the phenomenon is general and significant.

D. DIELS–ALDER ADDITIONS

The thermal cycloaddition between a diene and a dienophile has long been known as the Diels–Alder reaction. It is the basis for many important syntheses in organic chemistry because of its stereospecificity and its ability to join two carbon groupings together. Generally, a six-membered ring forms through 1,4-addition of an unsaturated compound (dienophile) to a conjugated ring.

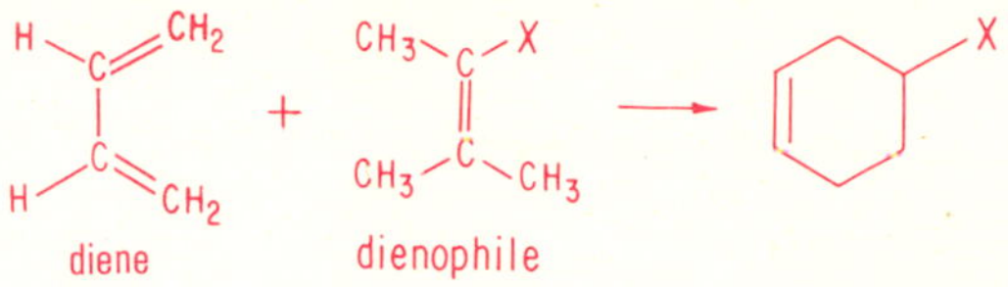

diene   dienophile

The ability of a dienophile to add to a diene is enhanced by electron withdrawing (electronegative) groups while the diene is activated by electron donating substituents. For example, the Diels–Alder reaction for isoprene and maleic anhydride occurs rapidly and smoothly without very much heat.

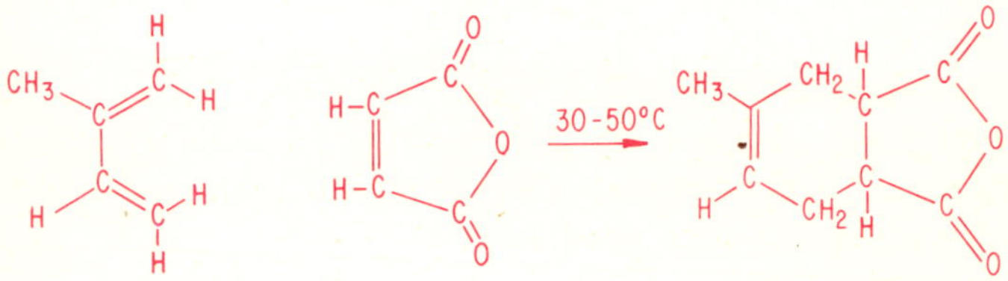

Bridged bicyclic products can be obtained by this Diels–Alder process.

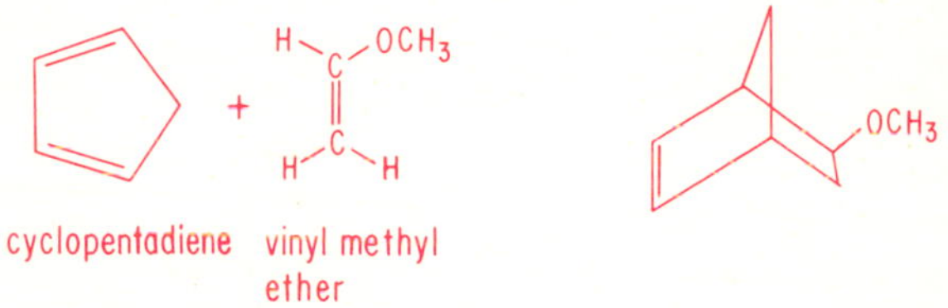

cyclopentadiene   vinyl methyl
                  ether

It is now generally agreed that the mechanism of the Diels–Alder reaction requires concerted bond formation and rearrangement of the double bond.

These reactions are examples of cycloadditions with no detectable intermediate forms. There is simply a rearrangement of three π-bonds to one π-bond and two new σ-bonds.

E. AROMATIC SUBSTITUTION

In the discussion of displacement and addition reactions we noted that alkenes react by adding groups to the double bond (pp. 319–320). Because of resonance stabilization, however, aromatic compounds undergo substitution, rather than addition. The net effect is that a hydrogen is replaced by another chemical group. Common examples of these groups are $-NO_2$, $-Br$, $-Cl$, $-CH_3$, $-CH_2CH_3$, $-SO_3H$ and $-C(CH_3)_3$.

Aromatic substitution is actually a two-step process. An electrophilic group is first attacked by the π-cloud of an aromatic ring which then rearranges to a σ-bonded adduct, forming an sp^3 hybrid at that carbon. This upsets the resonance structure of the aromatic, and the ring becomes positively charged. Using bromination as an example of electrophilic aromatic substitution, we note that the bromine must initially interact with a catalyst (typically a Lewis acid) such as ferric bromide. This generates a positive bromonium ion which is electrophilic.

$$Br_2 + FeBr_3 \rightleftharpoons Br^\oplus (FeBr_4)^\ominus$$

Aromatic systems contain characteristic π-electron clouds which are able to complex with reactive electrophilic agents.

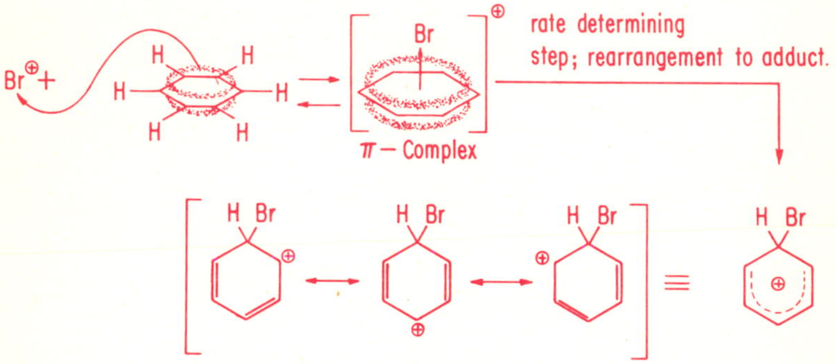

Stabilized resonance representations of the adduct

The π-cloud of benzene is not sufficiently reactive to form more than a rapid reversible steady-state concentration of the π-complex. It is the rearrangement to the stabilized σ-bonded intermediate adduct which allows the reaction to proceed. In the second part of this reaction, a proton is lost and the catalyst is regenerated while the substitution product establishes an aromatic structure.

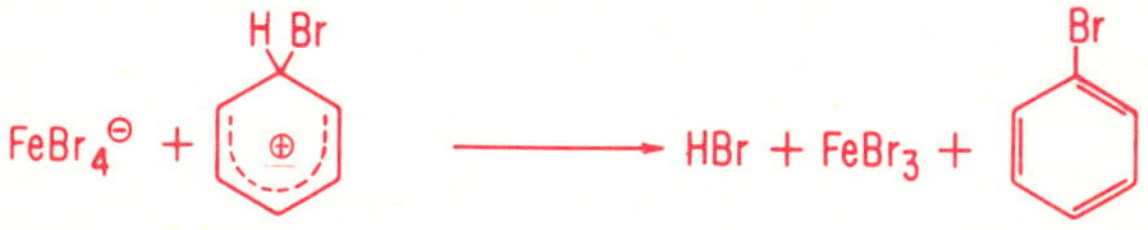

The above example is typical for the **halogenation** of aromatic compounds. Other classes of substitution reactions include **nitration, sulfonation, alkylation,** and **acylation.** The facility of these substitutions depends on the nature of the electrophile and the availability of aromatic π-electrons to attack them. Aromatic compounds with enhanced π-electron character exhibit rapid rates for electrophilic substitution.

   Nitration is similar to halogenation and other aromatic substitution reactions in that it is a two-step process, and the first step is the rate-determining one. Also, like other aromatic substitution reactions, it requires a catalyst. Sulfuric acid is commonly used to generate the nitronium ion, NO_2^+, from nitric acid. In concentrated sulfuric acid (i.e., 92%) nitric acid exists almost completely as the nitronium ion. As a result the rate expressions for such nitrations are of the second order.

$$Rate = k(aromatic)\,(NO_2^+)$$

Under these conditions the mechanism for nitration is as follows:

$$HNO_3 + 2H_2SO_4 \rightleftharpoons NO_2^{\oplus} + H_3O^{\oplus} + 2HSO_4^{\ominus}$$

π — Complex

$$H^{\oplus} + HSO_4^{\ominus} \rightleftharpoons H_2SO_4$$

   When nitronium ions are slowly and incompletely produced and the aromatic reacts at least as readily as benzene, the nitronium ions will be consumed as formed. Such situations obtain when nitrations using nitric acid are run in acetic acid or nitromethane as solvents. The rate depends on the concentration of the aromatic and on the efficiency of generating nitronium ions. When the aromatic compound does not react as fast as benzene (i.e. ethyl benzoate, halobenzenes, etc.,) the rate of nitration in a large excess of weak acid and nitric acid depends only on the concentration of the aromatic compound (pseudo first-order reaction).

$$Rate = k(aromatic)$$

Once an aromatic ring is nitrated, the π-electron system becomes less available to attack a nitronium ion. Toluene itself is easily nitrated between 0° and 25°C. The product, in toluene, is much more difficult to nitrate than toluene itself under the same conditions. With sufficient heat, 2,4-dinitrotoluene and even 2,4,6-trinitrotoluene (TNT) can be formed.

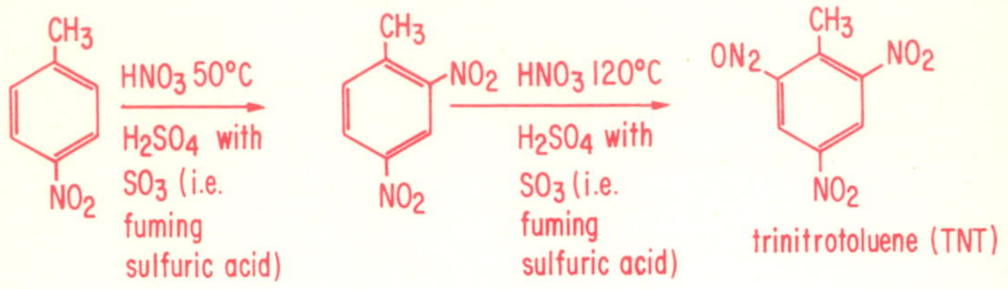

Sulfonation provides us with an example of substitution with a neutral electrophilic agent, SO_3. The electrophilic sulfur atom is clearly seen from the following resonance hybrids of sulfur trioxide.

Fuming sulfuric acid (H_2SO_4—SO_3) is usually employed for sulfonation of benzene and other aromatics:

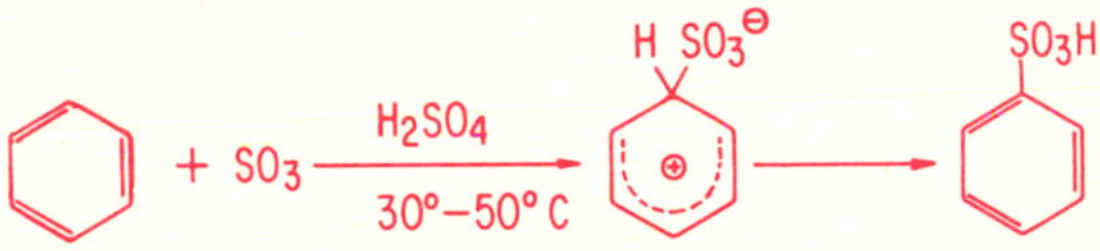

If concentrated rather than fuming sulfuric acid is used, elevated temperatures are required for sulfonations. Sulfonations using sulfur trioxide in nitrobenzene has been demonstrated to be third order.

$$\text{Rate} = k(\text{ArH}) (\text{SO}_3)^2$$

A mechanism based on such a kinetic expression must involve a dimeric SO_3 complex with the aromatic in the rate determining step.

In 1887 the French chemist, Charles Friedel, and his American student, James Crafts, discovered an important general synthesis of organic chemistry, which is now called Friedel–Crafts *alkylation*. They combined an alkyl halide and an aromatic compound using a small amount of metal halide as a catalyst. Since then, chemists have discovered that alkylation will occur whenever a carbonium ion is available to be attacked by an aromatic ring. Thus, alcohols and alkenes can be used instead of alkyl halides. Examples of Friedel–Crafts alkylation reactions are shown on the following pages.

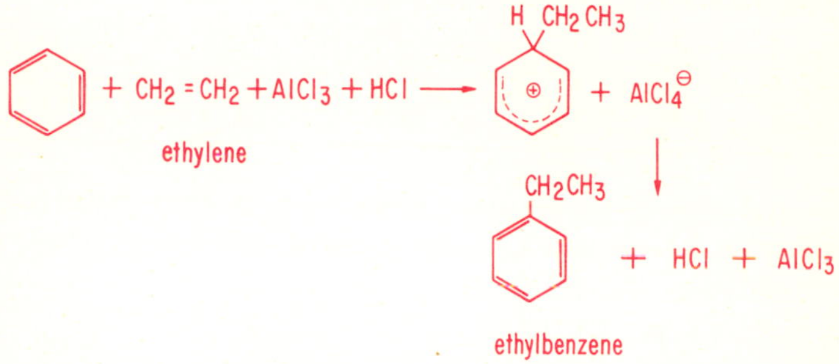

The catalyst for the Friedel–Crafts alkylation shown above, aluminum chloride, can be used in small non-stoichiometric amounts since it is regenerated in the reaction. Aso, the products of alkylation of benzene generally react more rapidly than benzene itself. Thus care must be exercised to remove monoalkylbenzene rapidly when formed to prevent further alkylation.

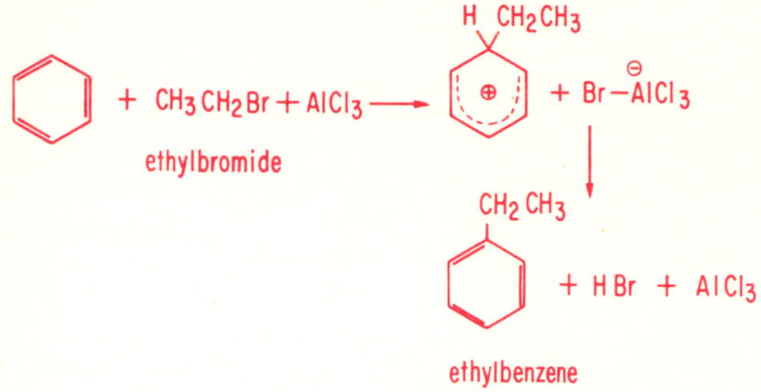

Loss of a molecule of hydrogen from ethylbenzene (formed by either of the above routes) by a catalyst and heating to very high temperature leads to styrene, a compound used in the synthesis of plastics and in synthetic rubber (see p. 71).

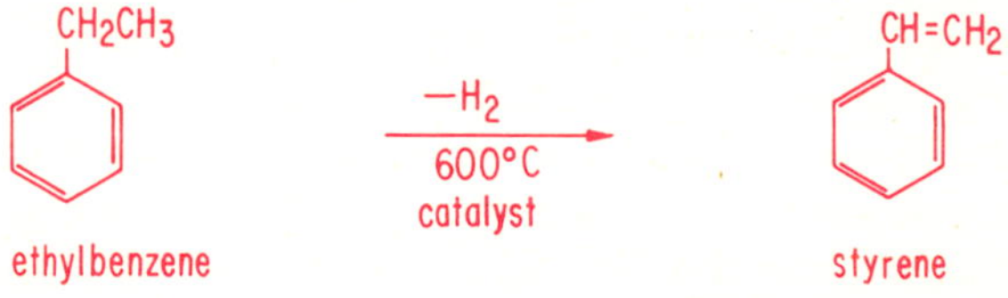

When an alkylhalide such as *n*-propyl halide is used, a primary carbonium ion can be conceptually postulated which rapidly rearranges to a much more stable secondary carbonium ion (reaction 3).

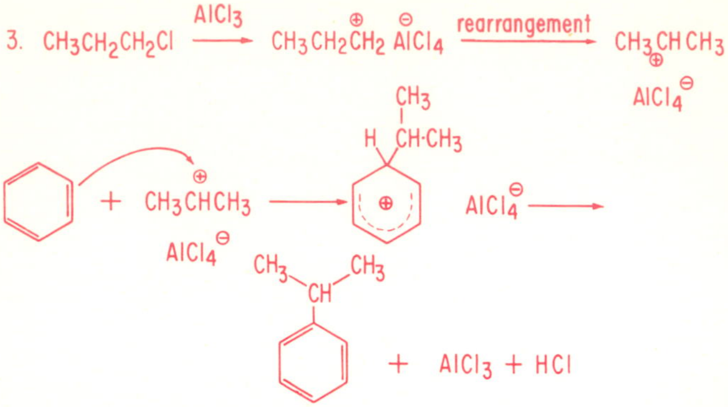

The catalysts most frequently employed in alkylation are Lewis acids such as aluminum chloride ($AlCl_3$) and boron trifluoride (BF_3). If an alkyl halide is involved, it is generally a chloro- or a bromo-derivative. In spite of most precautions, polysubstituted products usually occur and extensive purification is required to obtain a specific product from the several that are synthesized. The number of products is further increased by the tendency of the *n*-alkyl halides to form several possible branched substituent groups. (Note reaction 3 above.) This branching is temperature-dependent and occurs readily at higher temperatures.

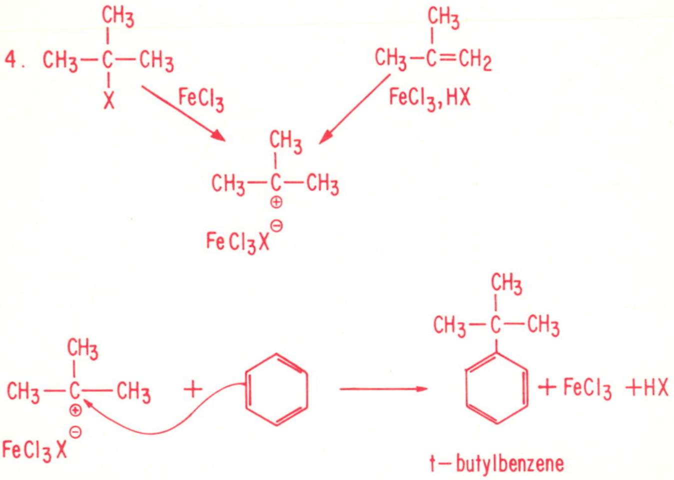

Aliphatic alcohols, alkyl halides and branched alkenes which can generate tertiary carbonium ions provide facile precursors for generating electrophiles (reaction 4). Aromatic compounds can attack such electrophiles to form alkyl aromatics containing a quaternary substituted carbon atom.

Another Friedel–Crafts reaction which is more specific than alkylation is Friedel–Crafts **acylation.** In this class of reactions a ketone is synthesized from

1)

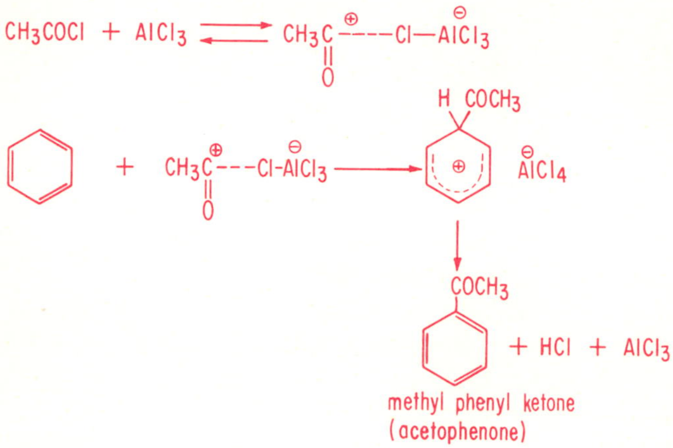

$CH_3COCl + AlCl_3 \rightleftharpoons CH_3\overset{\oplus}{C}\underset{\underset{O}{\|}}{}----Cl—\overset{\ominus}{AlCl_3}$

methyl phenyl ketone
(acetophenone)

2)

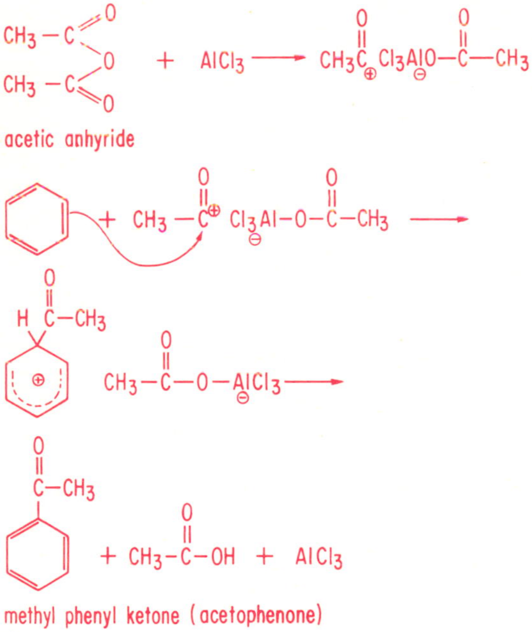

acetic anhyride

methyl phenyl ketone (acetophenone)

an acid chloride, RCOCl, or anhydride, $(RCO)_2O$, and an aromatic hydrocarbon. The most frequently used catalyst is $AlCl_3$, which acts to generate the electrophilic groups.

This class of reactions differs from alkylation in that a pure product is formed, rather than several different ones; also, a larger amount of catalyst is required for acylation, since much of it becomes complexed with the ketone synthesized.

If an acid chloride is used as the reagent, slightly over one molar equivalent of catalyst is needed; if an anhydride is the reagent, slightly over two molar equivalents of catalyst are used.

The presence of groups on an aromatic ring influences the ease of substitution by other groups. This influence arises from the electronic effect of the attached groups on the π-electrons of the aromatic ring. If the atom of a group attached to the aromatic is positively charged or electron-attracting, then the π-electrons of the ring are attracted to that atom and are less likely to attack an electrophile. Examples of substituent groups which cause such deactivation are —NO_2, —SO_3H, —Cl, esters, and ketones. (The deactivating effect of ketones explains why pure, monosubstituted products result from Friedel–Crafts acylation.)

In contrast, substitution is enhanced by electron-releasing atoms of groups attached to an aromatic ring. These groups make the ring more electron rich and thus the ring can more readily attack an electrophile. Examples of such activating groups are —NH_2, —OH, —OCH_3, and alkyl groups. (We explained polysubstitution of aromatics by the activating effect of the first-substituted alkyl group.) Substituent groups not only affect the ease with which other groups are added to an aromatic, but also their placement on the ring. Recall that in an earlier section (p. 305) we defined the terms *ortho*, *meta*, and *para*, as referring to adjacent, alternate or opposite positions, respectively.

Because of resonance between the original substituent group and the ring itself, either the *meta* carbons or the *ortho–para* carbons become electron rich. Consider the example of resonance hydrids for methoxybenzene:

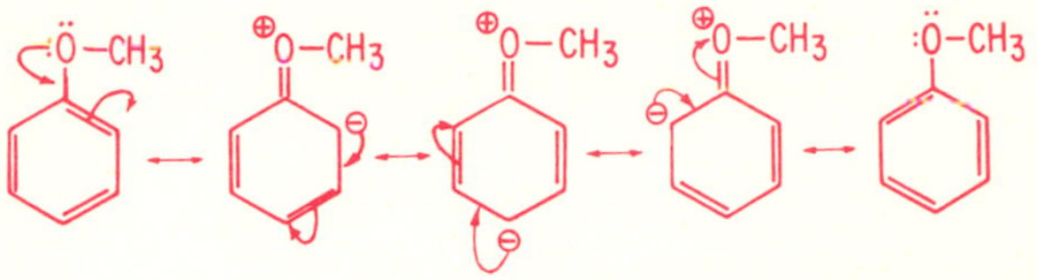

We would predict that the methoxyl group would activate the ring for substitution and new electrophilic groups would be attached primarily at the *ortho* and/or *para* positions. Thus —OCH_3 is called an *ortho–para* director. **Most ortho–para directors are activators** (see Table A-8). Halide groups are the major exception; they are *ortho–para* directors but are deactivators because of their large electronegativity.

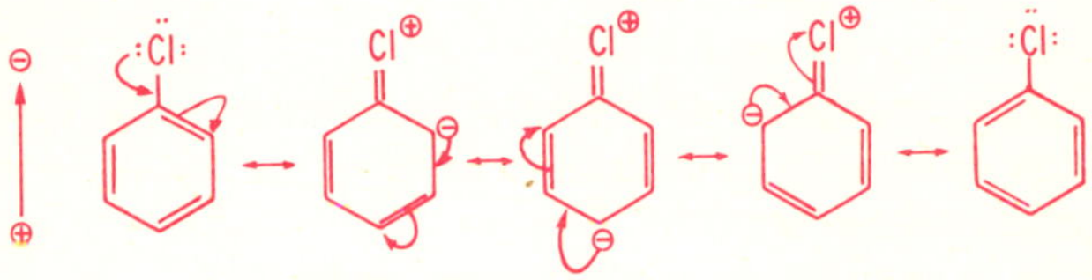

Resonance hybrids for chlorobenzene

All meta directors are deactivators. Although deactivators decrease the ease with which an aromatic ring attacks an electrophile, substitution is still possible. If substitution occurs, it will take place primarily at the *meta* positions. For example, the nitro group of nitrobenzene is a deactivator and a *meta*-director. It causes an electron shift so that the *ortho* and *para* carbons are less electron rich than the *meta* positions. Thus, an electrophilic group would be less likely to be attacked by the *ortho* or *para* positions than by the *meta* carbons. Under such circumstances, the specificity of directing is less than encountered with *ortho–para* directors.

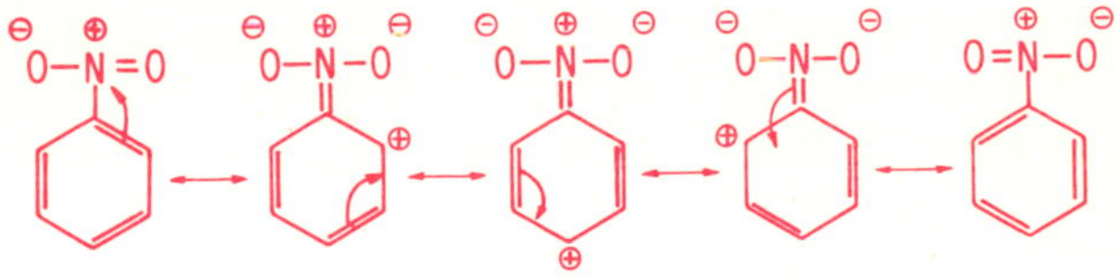

Resonance hybrids for nitrobenzene

As shown earlier in this section, the orientation of substitution is primarily determined by resonance effects. Steric hindrance is also important, however,

TABLE A-8
Effect of Aromatic Ring Substituents[a] on Reactivity and Placement of Electrophiles

Activators *ortho–para* directors	Deactivators, *ortho–para* directors	Deactivators, *meta*-directors
$-O^-$	$-F$	$-N^+(CH_3)_3$
$-N(CH_3)_2$	$-Cl$	$-NO_2$
$-NH_2$	$-Br$	$-CN$
$-OH$	$-I$	$-SO_3H$
$-OCH_3$		$-CHO$(aldehyde)
$-$alkyl		$-COCH_3$(ketone)
$-$aromatic		$-COOH$(acid)
		$-COOCH_3$(ester)

[a] In order of decreasing activity.

and explains why many *ortho–para* directors favor substitution at the *para* position. For example, the large tertiary butyl group orients electrophiles almost exclusively at the *para* position. There is not much room for a group to squeeze in at the *ortho* positions.

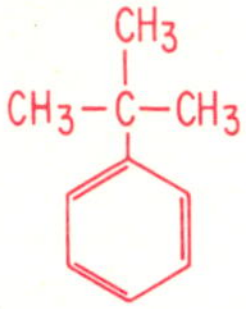

Naturally this appendix cannot be complete. It is meant to assist students who need easy reference to many of the terms and concepts used in the book. We urge the student to undertake additional reading in order to enjoy the fascinating story of the structure and dynamics of organic molecules.

BACKGROUND MATERIAL AND SUGGESTED READING

Cason, J., 1966. *Principles of Modern Organic Chemistry*, Prentice-Hall, Inc., Englewood Cliffs, New Jersey.

Cram, D. J., and G. S. Hammond, 1964. *Organic Chemistry*, McGraw-Hill Book Co., Inc., New York.

Fieser, L. F., and M. Fieser, 1957. *Introduction to Organic Chemistry*, D. C. Heath and Co., Boston.

Morrison, R. T., and R. N. Boyd, 1959. *Organic Chemistry*, Allyn and Bacon, Inc., Boston.

Noller, C. R., 1965. *Chemistry of Organic Compounds*, W. B. Saunders Co., Philadelphia.

Patai, S., 1962. *Glossary of Organic Chemistry*, John Wiley and Sons, New York.

Roberts, J. D., and M. C. Caserio, 1964. *Basic Principles of Organic Chemistry*, W. A. Benjamin, Inc., New York.

INDEX